Elementary
Differential Geometry

Elementary Differential Geometry

Barrett O'Neill

DEPARTMENT OF MATHEMATICS
UNIVERSITY OF CALIFORNIA
LOS ANGELES, CALIFORNIA

ACADEMIC PRESS   New York   San Francisco   London

A Subsidiary of Harcourt Brace Jovanovich, Publishers

ACADEMIC PRESS, INC.
111 Fifth Avenue, New York, New York 10003

United Kingdom Edition published by
ACADEMIC PRESS, INC. (LONDON) LTD.
24/28 Oval Road, London NW1

LIBRARY OF CONGRESS CATALOG CARD NUMBER: 66-14468

PRINTED IN THE UNITED STATES OF AMERICA

Preface

This book is an elementary account of the geometry of curves and surfaces. It is written for students who have completed standard first courses in calculus and linear algebra, and its aim is to introduce some of the main ideas of differential geometry.

The traditional undergraduate course in differential geometry has changed very little in the last few decades. By contrast, geometry has been advancing very rapidly at the research level, and there is general agreement that the undergraduate course needs to be brought up to date. I have tried to think through the classical material, to prune and augment it, and to write down the results in a reasonably clean and modern mathematical style. However, I have used a new idea only if it really pays its way by simplifying and clarifying the exposition.

Chapter I establishes the language of the book—a language compounded of familiar parts of calculus and linear algebra. Chapter II describes the method of "moving frames," which is introduced, as in elementary calculus, to study curves in space. In Chapter III we investigate the rigid motions of space, in terms of which congruence of curves (or surfaces) in space is defined in the same fashion as congruence of triangles in the plane.

Chapter IV requires special comment. The main weakness of classical differential geometry was its lack of any adequate definition of *surface*. In this chapter we decide just what a surface is, and show that each surface has a differential and integral calculus of its own, strictly comparable with the familiar calculus of the plane. This exposition provides an introduction to the notion of *differentiable manifold*, which has become indispensable to those branches of mathematics and its applications based on the calculus.

The next two chapters are devoted to the geometry of surfaces in 3-space. Chapter V stresses intuitive and computational aspects to give geometrical meaning to the theory presented in Chapter VI. In the final chapter, although our methods are unchanged, there is a radical shift of viewpoint. Roughly speaking, we study the geometry of a surface *as seen by its inhabi-*

tants, with no assumption that the surface is to be found in ordinary three-dimensional space.

No branch of mathematics makes a more direct appeal to the intuition than geometry. I have sought to emphasize this by a large number of illustrations, which form an integral part of the text. A set of exercises appears at the end of each section; these range from routine tests of comprehension to more seriously challenging problems.

In teaching from preliminary versions of this book, I have usually covered the background material in Chapter I rather rapidly, and have not devoted any classroom time to Chapter III (hence also Section 8 of Chapter VI). A course in the geometry of curves and surfaces in space might consist of: Chapter II, Chapter IV (omit Sections 6 and 8), Chapter V, and Chapter VI (omit Sections 6 and 7). This is essentially the content of the traditional undergraduate course in differential geometry, with clarification of the concepts of surface and mapping of surfaces.

The omitted sections in the list above are used only in Chapter VII. This final chapter, an extensive account of two-dimensional Riemannian geometry, is in a sense the goal of the book. Rather than shift the discourse to higher dimensions, I have preferred to retain dimension 2, so that this more sophisticated view of geometry will develop directly from the special case of surfaces in 3-space. Chapter VII is long, and on a first reading Theorem 5.9 and Sections 6 and 7 may well be omitted. Serious use of differential equations theory has been largely avoided in the early chapters; however, some acquaintance with the fundamentals of the subject will be helpful in Chapter VII.

Los Angeles, California   B. O'N.

Contents

Chapter IV. **Calculus on a Surface**

Chapter V. **Shape Operators**

Chapter VI. **Geometry of Surfaces in E^3**

Chapter VII. **Riemannian Geometry**

Elementary
Differential Geometry

Introduction

This book presupposes a reasonable knowledge of elementary calculus and linear algebra. It is a working knowledge of the fundamentals that is actually required. The reader will, for example, frequently be called upon to *use* the chain rule for differentiation, but its proof need not concern us.

Calculus deals mostly with real-valued functions of one or more variables, linear algebra with functions (linear transformations) from one vector space to another. We shall need functions of these and other types, so we give here general definitions which cover all types.

A *set* S is a collection of objects which are called the *elements* of S. A set A is a *subset* of S provided each element of A is also an element of S.

A *function* f from a set D to a set R is a rule that assigns to each element x of D a unique element $f(x)$ of R. The element $f(x)$ is called the *value* of f at x. The set D is called the *domain* of f; the set R is often called the *range* of f. If we wish to emphasize the domain and range of a function f, the notation $f: D \rightarrow R$ is used. Note that the function is denoted by a single letter, say f, while $f(x)$ is merely a value of f.

Many different terms are used for functions—mappings, transformations, correspondences, operators, and so on. A function can be described in various ways, the simplest case being an explicit formula such as

$$f(x) = 3x^2 + 1,$$

which we may also write as $x \rightarrow 3x^2 + 1$.

If both f_1 and f_2 are functions from D to R, then $f_1 = f_2$ means that $f_1(x) = f_2(x)$ for all x in D. This is not a definition, but a logical consequence of the definition of function.

Let $f: D \rightarrow R$ and $g: E \rightarrow S$ be functions. In general, the *image* of f is the subset of R consisting of all elements of the form $f(x)$; it is usually denoted by $f(D)$. Now if this image also happens to be a subset of the domain E of g, it is possible to combine these two functions to obtain the *composite function* $g(f): D \rightarrow S$. By definition, $g(f)$ is the function whose value on each element x of D is the element $g(f(x))$ of S.

If $f: D \to R$ is a function and A is a subset of D, then the *restriction* of f to A is the function $f \mid A: A \to R$ defined by the same rule as f, but applied only to elements of A. This seems a rather minor change, but the function $f \mid A$ may have properties quite different from f itself.

Here are two vital properties which a function may possess. A function $f: D \to R$ is *one-to-one*, provided that if x and y are any elements of D such that $x \neq y$, then $f(x) \neq f(y)$. A function $f: D \to R$ is *onto* (or *carries D onto R*) provided that for every element y of R there is at least one element x of D such that $f(x) = y$. In short, the image of f is the entire set R. For example, consider the following functions, each of which has the real numbers as both domain and range:

(1) The function $x \to x^3$ is both one-to-one and onto.
(2) The exponential function $x \to e^x$ is one-to-one, but not onto.
(3) The function $x \to x^3 + x^2$ is onto, but not one-to-one.
(4) The sine function $x \to \sin x$ is neither one-to-one nor onto.

If a function $f: D \to R$ is both one-to-one and onto, then for each element y of R there is one and only one element x such that $f(x) = y$. By defining $f^{-1}(y) = x$ for all x and y so related, we obtain a function $f^{-1}: R \to D$ called the *inverse* of f. Note that the function f^{-1} is also one-to-one and onto, and that *its* inverse function is the original function f.

Here is a short list of the main notations used throughout the book, in order of their appearance in Chapter I:

p, q,...............	points	(Sec. 1)
$f, g,$...............	real-valued functions	(Sec. 1)
v, w,...............	tangent vectors	(Sec. 2)
$V, W,$...............	vector fields	(Sec. 2)
$\alpha, \beta,$...............	curves	(Sec. 4)
$\phi, \psi,$...............	differential forms	(Sec. 5)
$F, G,$...............	mappings	(Sec. 7)

In Chapter I we define these concepts for Euclidean 3-space. (Extension to arbitrary dimensions is virtually automatic.) In Chapter IV we show how these concepts can be adapted to a surface.

A few references are given to the brief bibliography at the end of the book; these are indicated by numbers in square brackets.

CHAPTER **1**

Calculus on Euclidean Space

As mentioned in the Preface, the purpose of this initial chapter is to establish the mathematical language used throughout the book. Much of what we do is simply a review of that part of elementary calculus dealing with differentiation of functions of three variables, and with curves in space. Our definitions have been formulated so that they will apply smoothly to the later study of surfaces.

1 Euclidean Space

Three-dimensional space is often used in mathematics without being formally defined. It is said to be the space of ordinary experience. Looking at the corner of a room, one can picture the familiar process by which rectangular coordinate axes are introduced and three numbers are measured to describe the position of each point. A precise definition which realizes this intuitive picture may be obtained by this device: instead of saying that three numbers *describe the position* of a point, we define them to *be* a point.

1.1 Definition *Euclidean 3-space* $\mathbf{E}^3$ is the set of all ordered triples of real numbers. Such a triple $\mathbf{p} = (p_1, p_2, p_3)$ is called a *point* of $\mathbf{E}^3$.

In linear algebra, it is shown that $\mathbf{E}^3$ is, in a natural way, a vector space over the real numbers. In fact, if $\mathbf{p} = (p_1, p_2, p_3)$ and $\mathbf{q} = (q_1, q_2, q_3)$ are points of $\mathbf{E}^3$, their *sum* is the point

$$\mathbf{p} + \mathbf{q} = (p_1 + q_1, p_2 + q_2, p_3 + q_3).$$

The *scalar product* of a point $\mathbf{p} = (p_1, p_2, p_3)$ by a number a is the point

$$a\mathbf{p} = (ap_1, ap_2, ap_3).$$

3

It is easy to check that these two operations satisfy the axioms for a vector space. The point $\mathbf{0} = (0,0,0)$ is called the *origin* of $\mathbf{E}^3$.

Differential calculus deals with another aspect of $\mathbf{E}^3$ starting with the notion of differentiable real-valued functions on $\mathbf{E}^3$. We recall some fundamentals.

1.2 Definition Let x, y, and z be the real-valued functions on $\mathbf{E}^3$ such that for each point $\mathbf{p} = (p_1, p_2, p_3)$

$$x(\mathbf{p}) = p_1, \qquad y(\mathbf{p}) = p_2, \qquad z(\mathbf{p}) = p_3.$$

These functions x, y, z are called the *natural coordinate functions* of $\mathbf{E}^3$. We shall also use index notation for these functions, writing

$$x_1 = x, \qquad x_2 = y, \qquad x_3 = z.$$

Thus the value of the function x_i on a point $\mathbf{p}$ is the number p_i, and so we have the identity $\mathbf{p} = (p_1, p_2, p_3) = (x_1(\mathbf{p}), x_2(\mathbf{p}), x_3(\mathbf{p}))$ for each point $\mathbf{p}$ of $\mathbf{E}^3$. Elementary calculus does not always make a sharp distinction between the *numbers* p_1, p_2, p_3 and the *functions* x_1, x_2, x_3. Indeed the analogous distinction on the real line may seem pedantic, but for higher-dimensional spaces such as $\mathbf{E}^3$, its absence leads to serious ambiguities. (Essentially the same distinction is being made when we denote a function on $\mathbf{E}^3$ by a single letter f, reserving $f(\mathbf{p})$ for its value at the point $\mathbf{p}$.)

We assume that the reader is familiar with partial differentiation and its basic properties, in particular the chain rule for differentiation of a composite function. We shall work mostly with first-order partial derivatives $\partial f/\partial x$, $\partial f/\partial y$, $\partial f/\partial z$ and second-order partial derivatives $\partial^2 f/\partial x^2$, $\partial^2 f/\partial x\, \partial y$, $\cdots$. In a few situations, third-, and even fourth-, order derivatives may occur, but to avoid worrying about exactly how many derivatives we can take in any given context, we establish the following definition.

1.3 Definition A real-valued function f on $\mathbf{E}^3$ is *differentiable* (or, *infinitely differentiable*, or *of class* C^∞) provided all partial derivatives of f, of all orders, exist and are continuous.

Differentiable real-valued functions f and g may be added and multiplied in a familiar way to yield functions that are again differentiable and real-valued. We simply add and multiply their values at each point—the formulas read

$$(f + g)(\mathbf{p}) = f(\mathbf{p}) + g(\mathbf{p}), \qquad (fg)(\mathbf{p}) = f(\mathbf{p})g(\mathbf{p}).$$

The phrase "differentiable real-valued function" is unpleasantly long. Hence we make the convention that *unless the context indicates otherwise,* "function" shall mean "real-valued function," and (unless the issue is explicitly raised) the functions we deal with will be assumed to be differentiable. We do not intend to overwork this convention; for the sake of emphasis the words "differentiable" and "real-valued" will still appear fairly frequently.

Differentiation is always a *local* operation: To compute the value of the function $\partial f/\partial x$ at a point $\mathbf{p}$ of $\mathbf{E}^3$, it is sufficient to know the values of f at all points $\mathbf{q}$ of $\mathbf{E}^3$ which are sufficiently near $\mathbf{p}$. Thus Definition 1.3 is unduly restrictive; the domain of f need not be the whole of $\mathbf{E}^3$, but need only be an *open set* of $\mathbf{E}^3$. By an *open set* $\mathcal{O}$ of $\mathbf{E}^3$ we mean a subset of $\mathbf{E}^3$ such that if a point $\mathbf{p}$ is in $\mathcal{O}$, then so is every other point of $\mathbf{E}^3$ that is sufficiently near $\mathbf{p}$. (A more precise definition is given in Chap. II.) For example, the set of all points $\mathbf{p} = (p_1, p_2, p_3)$ in $\mathbf{E}^3$ such that $p_1 > 0$ is an open set, and the function $yz \log x$ defined on this set is certainly differentiable, even though its domain is not the whole of $\mathbf{E}^3$. Generally speaking, the results in this chapter remain valid if $\mathbf{E}^3$ is replaced by an arbitrary open set $\mathcal{O}$ of $\mathbf{E}^3$.

We are dealing with *three-dimensional* Euclidean space for no better reason than that this is the dimension we use most often in later work. It would be just as easy to work with *Euclidean n-space* $\mathbf{E}^n$, for which the points are n-tuples $\mathbf{p} = (p_1, \cdots, p_n)$, and which has n natural coordinate functions $x_1, \cdots, x_n$. All the results in this chapter are valid for Euclidean spaces of arbitrary dimensions, although we shall rarely take advantage of this except in the case of the *Euclidean plane* $\mathbf{E}^2$. In particular, the results are valid for the *real line* $\mathbf{E}^1 = \mathbf{R}$. Many of the concepts introduced are specifically designed to deal with higher dimensions, however, and are thus apt to be overelaborate when reduced to dimension 1.

EXERCISES

1. Let $f = x^2 y$ and $g = y \sin z$ be functions on $\mathbf{E}^3$. Express the following functions in terms of x, y, z:

   (a) fg^2.   (b) $\dfrac{\partial f}{\partial x} g + \dfrac{\partial g}{\partial y} f$.

   (c) $\dfrac{\partial^2 (fg)}{\partial y\, \partial z}$.   (d) $\dfrac{\partial}{\partial y} (\sin f)$.

2. Find the value of the function $f = x^2 y - y^2 z$ at each point:

   (a) $(1, 1, 1)$.   (b) $(3, -1, \tfrac{1}{2})$.

   (c) $(a, 1, 1 - a)$.   (d) (t, t^2, t^3).

3. Express $\partial f / \partial x$ in terms of x, y, and z if

  (a) $f = x \sin (xy) + y \cos (xz)$.

  (b) $f = \sin g$, $g = e^h$, $h = x^2 + y^2 + z^2$.

4. If g_1, g_2, g_3 and h are real-valued functions on $\mathbf{E}^3$, then

$$f = h(g_1, g_2, g_3)$$

is the function such that

$$f(\mathbf{p}) = h(g_1(\mathbf{p}), g_2(\mathbf{p}), g_3(\mathbf{p})) \qquad \text{for all } \mathbf{p}.\dagger$$

Express $\partial f / \partial x$ in terms of x, y, and z, if $h = x^2 - yz$ and

  (a) $f = h(x + y, y^2, x + z)$.

  (b) $f = h(e^x, e^{x+y}, e^x)$.

  (c) $f = h(x, -x, x)$.

2   Tangent Vectors

Intuitively, a vector in $\mathbf{E}^3$ is an oriented line segment, or "arrow." Vectors are used widely in physics and engineering to describe forces, velocities, angular momenta, and many other concepts. To obtain a definition that is both practical and precise, we shall describe an "arrow" in $\mathbf{E}^3$ by giving its starting point $\mathbf{p}$ and the change, or vector $\mathbf{v}$, necessary to reach its end point $\mathbf{p} + \mathbf{v}$. Strictly speaking, $\mathbf{v}$ is just a point of $\mathbf{E}^3$.

  2.1 Definition‡   A *tangent vector* $\mathbf{v}_p$ to $\mathbf{E}^3$ consists of two points of $\mathbf{E}^3$: its *vector part* $\mathbf{v}$ and its *point of application* $\mathbf{p}$.

  We shall always picture $\mathbf{v}_p$ as *the arrow from the point* $\mathbf{p}$ *to the point* $\mathbf{p} + \mathbf{v}$. For example, if $\mathbf{p} = (1, 1, 3)$ and $\mathbf{v} = (2, 3, 2)$, then $\mathbf{v}_p$ runs from $(1, 1, 3)$ to $(3, 4, 5)$ as in Fig. 1.1.

  We emphasize that tangent vectors are equal, $\mathbf{v}_p = \mathbf{w}_q$, if and only if they have the same vector part, $\mathbf{v} = \mathbf{w}$, and the same point of application, $\mathbf{p} = \mathbf{q}$. Tangent vectors $\mathbf{v}_p$ and $\mathbf{v}_q$ with the same vector part, but different points of application, are said to be *parallel* (Fig. 1.2). It is essential to recognize that $\mathbf{v}_p$ and $\mathbf{v}_q$ are different tangent vectors if $\mathbf{p} \neq \mathbf{q}$. In physics the concept of moment of a force shows this clearly enough: The same force $\mathbf{v}$ applied at different points $\mathbf{p}$ and $\mathbf{q}$ of a rigid body can produce quite different rotational effects.

---

† A consequence is the identity $f = f(x, y, z)$.

‡ The term "tangent" in this definition will acquire a more direct geometric meaning in Chapter IV.

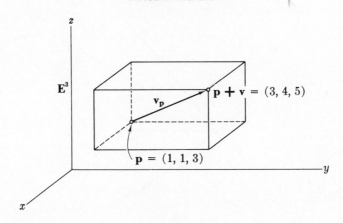

FIG. 1.1

2.2 Definition Let $\mathbf{p}$ be a point of $\mathbf{E}^3$. The set $T_p(\mathbf{E}^3)$ consisting of all tangent vectors that have $\mathbf{p}$ as point of application is called the *tangent space* of $\mathbf{E}^3$ at $\mathbf{p}$ (Fig. 1.3).

We emphasize that $\mathbf{E}^3$ has a different tangent space at each and every one of its points.

Since all the tangent vectors in a given tangent space have the same point of application, we can borrow the vector addition and scalar multiplication of $\mathbf{E}^3$ to turn $T_p(\mathbf{E}^3)$ into a vector space. Explicitly, we define $\mathbf{v}_p + \mathbf{w}_p$ to be $(\mathbf{v} + \mathbf{w})_p$, and if c is a number we define $c(\mathbf{v}_p)$ to be $(c\mathbf{v})_p$. This is just the usual "parallelogram law" for addition of vectors, and scalar multiplication by c merely stretches a tangent vector by factor c—reversing its direction if $c < 0$ (Fig. 1.4).

These operations on each tangent space $T_p(\mathbf{E}^3)$ make it a vector space isomorphic to $\mathbf{E}^3$ itself. Indeed it follows immediately from the definitions above that, for a fixed point $\mathbf{p}$, the function $\mathbf{v} \rightarrow \mathbf{v}_p$ is a linear isomorphism

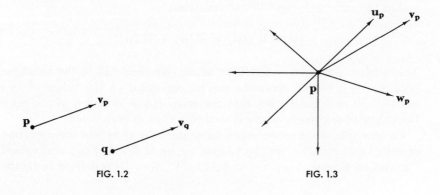

FIG. 1.2   FIG. 1.3

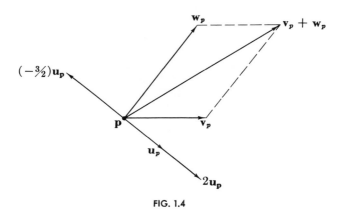

FIG. 1.4

from $\mathbf{E}^3$ to $T_p(\mathbf{E}^3)$—that is, a linear transformation which is one-to-one and onto.

A standard concept in physics and engineering is that of force field. The gravitational force field of the earth, for example, assigns to each point of space a force (vector) directed at the center of the earth.

2.3 Definition A *vector field* V on $\mathbf{E}^3$ is a function that assigns to each point $\mathbf{p}$ of $\mathbf{E}^3$ a tangent vector $V(\mathbf{p})$ to $\mathbf{E}^3$ at $\mathbf{p}$.

Roughly speaking, a vector field is just a big collection of arrows, one at each point of $\mathbf{E}^3$.

There is a natural algebra of vector fields. To describe it, we first re-examine the familar notion of addition of real-valued functions f and g. It is possible to add f and g because it is possible to add their values at each point. The same is true of vector fields V and W. At each point $\mathbf{p}$, their values $V(\mathbf{p})$ and $W(\mathbf{p})$ are in the same vector space—the tangent space $T_p(\mathbf{E}^3)$—hence we can add $V(\mathbf{p})$ and $W(\mathbf{p})$. Consequently we can add V and W by adding their values at each point. The formula for this addition is thus the same as for addition of functions,

$$(V + W)(\mathbf{p}) = V(\mathbf{p}) + W(\mathbf{p}).$$

This scheme occurs over and over again. We shall call it the *pointwise principle:* If a certain operation can be performed on the values of two functions at each point, then that operation can be extended to the functions themselves; simply apply it to their values at each point.

For example, we invoke the pointwise principle to extend the operation of *scalar multiplication* (on the tangent spaces of $\mathbf{E}^3$). If f is a real-valued function on $\mathbf{E}^3$ and V is a vector field on $\mathbf{E}^3$, then fV is defined to be the vector field on $\mathbf{E}^3$ such that

$$(fV)(\mathbf{p}) = f(\mathbf{p})V(\mathbf{p}) \qquad \text{for all } \mathbf{p}.$$

Our aim now is to determine in a concrete way just what vector fields look like. For this purpose we introduce three special vector fields which will serve as a "basis" for all vector fields.

2.4 Definition   Let U_1, U_2, and U_3 be the vector fields on $\mathbf{E}^3$ such that

$$U_1(\mathbf{p}) = (1,0,0)_p$$
$$U_2(\mathbf{p}) = (0,1,0)_p$$
$$U_3(\mathbf{p}) = (0,0,1)_p$$

for each point $\mathbf{p}$ of $\mathbf{E}^3$ (Fig. 1.5). We call U_1, U_2, U_3—collectively—the *natural frame field* on $\mathbf{E}^3$.

Thus U_i $(i = 1,2,3)$ is the unit vector field in the positive x_i direction.

2.5 Lemma   If V is a vector field on $\mathbf{E}^3$, there are three uniquely determined real-valued functions v_1, v_2, v_3 on $\mathbf{E}^3$ such that

$$V = v_1U_1 + v_2U_2 + v_3U_3.$$

The functions v_1, v_2, v_3 are called the *Euclidean coordinate functions* of V.

Proof. By definition, the vector field V assigns to each point $\mathbf{p}$ a tangent vector $V(\mathbf{p})$ at $\mathbf{p}$. Thus the vector part of $V(\mathbf{p})$ depends on $\mathbf{p}$, so we write it $(v_1(\mathbf{p}), v_2(\mathbf{p}), v_3(\mathbf{p}))$. (This defines v_1, v_2, and v_3 as real-valued *functions* on $\mathbf{E}^3$.) Hence

$$V(\mathbf{p}) = (v_1(\mathbf{p}), v_2(\mathbf{p}), v_3(\mathbf{p}))_p$$
$$= v_1(\mathbf{p})(1,0,0)_p + v_2(\mathbf{p})(0,1,0)_p + v_3(\mathbf{p})(0,0,1)_p$$
$$= v_1(\mathbf{p})U_1(\mathbf{p}) + v_2(\mathbf{p})U_2(\mathbf{p}) + v_3(\mathbf{p})U_3(\mathbf{p})$$

for each point $\mathbf{p}$ (Fig. 1.6). By our (pointwise principle) definitions this means that the vector fields V and $\sum v_iU_i$ have the same (tangent vector) value at each point. Hence $V = \sum v_iU_i$.   ∎

This last sentence uses two of our standard conventions: $\sum v_iU_i$ means sum over $i = 1, 2, 3$; the Halmos symbol (∎) indicates the end of a proof.

The tangent-vector identity $(a_1, a_2, a_3)_p = \sum a_iU_i(\mathbf{p})$ appearing in this proof will be used very often.

Computations involving vector fields may always be expressed in terms of their

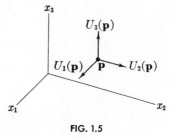

FIG. 1.5

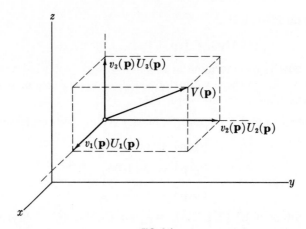

FIG. 1.6

Euclidean coordinate functions. For example, addition, and multiplication by a function, are expressed in terms of coordinates by

$$\sum v_i U_i + \sum w_i U_i = \sum (v_i + w_i) U_i$$
$$f(\sum v_i U_i) = \sum (f v_i) U_i.$$

Since this is differential calculus, we shall naturally require that the various objects we deal with be differentiable. A vector field V is *differentiable* provided its Euclidean coordinate functions are differentiable (in the sense of Definition 1.3). From now on, we shall understand "vector field" to mean "differentiable vector field."

EXERCISES

1. Let $\mathbf{v} = (-2, 1, -1)$ and $\mathbf{w} = (0, 1, 3)$.
   (a) At an arbitrary point $\mathbf{p}$, express the tangent vector $3\mathbf{v}_p - 2\mathbf{w}_p$ as a linear combination of $U_1(\mathbf{p})$, $U_2(\mathbf{p})$, $U_3(\mathbf{p})$.
   (b) For $\mathbf{p} = (1, 1, 0)$, make an accurate sketch showing the four tangent vectors $\mathbf{v}_p$, $\mathbf{w}_p$, $-2\mathbf{v}_p$, and $\mathbf{v}_p + \mathbf{w}_p$.

2. Let $V = xU_1 + yU_2$ and $W = 2x^2U_2 - U_3$. Compute the vector field $W - xV$, and find its value at the point $\mathbf{p} = (-1, 0, 2)$.

3. In each case, express the given vector field V in the standard form $\sum v_i U_i$.
   (a) $2z^2 U_1 = 7V + xy U_3$.
   (b) $V(\mathbf{p}) = (p_1, p_3 - p_1, 0)_p$ for all $\mathbf{p}$.
   (c) $V = 2(xU_1 + yU_2) - x(U_1 - y^2 U_3)$.

(d) At each point $\mathbf{p}$, $V(\mathbf{p})$ is the vector from the point (p_1, p_2, p_3) to the point $(1 + p_1, p_2p_3, p_2)$.

(e) At each point $\mathbf{p}$, $V(\mathbf{p})$ is the vector from $\mathbf{p}$ to the origin.

4. If $V = y^2U_1 - x^2U_3$ and $W = x^2U_1 - zU_2$, find functions f and g such that the vector field $fV + gW$ can be expressed in terms of U_2 and U_3 only.

5. Let $V_1 = U_1 - xU_3$, $V_2 = U_2$, and $V_3 = xU_1 + U_3$.

(a) Prove that the vectors $V_1(\mathbf{p})$, $V_2(\mathbf{p})$, $V_3(\mathbf{p})$ are linearly independent at each point of $\mathbf{E}^3$.

(b) Express the vector field $xU_1 + yU_2 + zU_3$ as a linear combination of V_1, V_2, V_3.

3   Directional Derivatives

Associated with each tangent vector $\mathbf{v}_p$ to $\mathbf{E}^3$ is the straight line $t \to \mathbf{p} + t\mathbf{v}$ (see Example 4.2). If f is a differentiable function on $\mathbf{E}^3$, then $t \to f(\mathbf{p} + t\mathbf{v})$ is an ordinary differentiable function on the real line. Evidently the derivative of this function at $t = 0$ tells the initial rate of change of f as $\mathbf{p}$ moves in the $\mathbf{v}$ direction.

3.1 Definition   Let f be a differentiable real-valued function on $\mathbf{E}^3$, and let $\mathbf{v}_p$ be a tangent vector to $\mathbf{E}^3$. Then the number

$$\mathbf{v}_p[f] = \frac{d}{dt}\,(f(\mathbf{p} + t\mathbf{v}))|_{t=0}$$

is called the *derivative of f with respect* to $\mathbf{v}_p$.

This definition appears in elementary calculus with the additional restriction that $\mathbf{v}_p$ be a unit vector. Even though we do not impose this restriction, we shall nevertheless refer to $\mathbf{v}_p[f]$ as a *directional derivative*.

For example, we compute $\mathbf{v}_p[f]$ for the function $f = x^2yz$, with $\mathbf{p} = (1, 1, 0)$ and $\mathbf{v} = (1, 0, -3)$. Then

$$\mathbf{p} + t\mathbf{v} = (1, 1, 0) + t(1, 0, -3) = (1 + t, 1, -3t)$$

describes the line through $\mathbf{p}$ in the $\mathbf{v}$ direction. Evaluating f along this line, we get

$$f(\mathbf{p} + t\mathbf{v}) = (1 + t)^2 \cdot 1 \cdot (-3t) = -3t - 6t^2 - 3t^3.$$

Now

$$\frac{d}{dt}\,(f(\mathbf{p} + t\mathbf{v})) = -3 - 12t - 9t^2;$$

hence at $t = 0$, we find $\mathbf{v}_p[f] = -3$. Thus, in particular, the function f is (initially) decreasing as $\mathbf{p}$ moves in the $\mathbf{v}$ direction.

The following lemma shows how to compute $\mathbf{v}_p[f]$ in general, in terms of the partial derivatives of f at the point $\mathbf{p}$.

3.2 Lemma   If $\mathbf{v}_p = (v_1, v_2, v_3)_p$ is a tangent vector to $\mathbf{E}^3$, then

$$\mathbf{v}_p[f] = \sum v_i \frac{\partial f}{\partial x_i}(\mathbf{p}).$$

Proof. Let $\mathbf{p} = (p_1, p_2, p_3)$; then

$$\mathbf{p} + t\mathbf{v} = (p_1 + tv_1, p_2 + tv_2, p_3 + tv_3).$$

We use the chain rule to compute the derivative at $t = 0$ of the function

$$f(\mathbf{p} + t\mathbf{v}) = f(p_1 + tv_1, p_2 + tv_2, p_3 + tv_3).$$

Since

$$\frac{d}{dt}(p_i + tv_i) = v_i,$$

we obtain

$$\mathbf{v}_p[f] = \frac{d}{dt}(f(\mathbf{p} + t\mathbf{v}))\,|_{t=0} = \sum \frac{\partial f}{\partial x_i}(\mathbf{p})v_i. \qquad \blacksquare$$

Using this lemma, we recompute $\mathbf{v}_p[f]$ for the example above. Since $f = x^2yz$, we have

$$\frac{\partial f}{\partial x} = 2xyz, \qquad \frac{\partial f}{\partial y} = x^2z, \qquad \frac{\partial f}{\partial z} = x^2y.$$

Thus at the point $\mathbf{p} = (1,1,0)$,

$$\frac{\partial f}{\partial x}(\mathbf{p}) = 0, \qquad \frac{\partial f}{\partial y}(\mathbf{p}) = 0, \qquad \text{and} \qquad \frac{\partial f}{\partial z}(\mathbf{p}) = 1.$$

Then by the lemma,

$$\mathbf{v}_p[f] = 0 + 0 + (-3)1 = -3,$$

as before.

The main properties of this notion of derivative are given in Theorem 3.3.

3.3 Theorem   Let f and g be functions on $\mathbf{E}^3$, $\mathbf{v}_p$ and $\mathbf{w}_p$ tangent vectors, a and b numbers. Then

(1) $(a\mathbf{v}_p + b\mathbf{w}_p)[f] = a\mathbf{v}_p[f] + b\mathbf{w}_p[f]$.
(2) $\mathbf{v}_p[af + bg] = a\mathbf{v}_p[f] + b\mathbf{v}_p[g]$.
(3) $\mathbf{v}_p[fg] = \mathbf{v}_p[f] \cdot g(\mathbf{p}) + f(\mathbf{p}) \cdot \mathbf{v}_p[g]$.

Proof. All three properties may be deduced easily from the preceding lemma. For example, we prove (3). By the lemma, if $\mathbf{v} = (v_1, v_2, v_3)$, then

$$\mathbf{v}_p[fg] = \sum v_i \frac{\partial(fg)}{\partial x_i}(\mathbf{p}).$$

But

$$\frac{\partial(fg)}{\partial x_i} = \frac{\partial f}{\partial x_i} \cdot g + f \cdot \frac{\partial g}{\partial x_i}.$$

Hence

$$\mathbf{v}_p[fg] = \sum v_i \left(\frac{\partial f}{\partial x_i}(\mathbf{p}) \cdot g(\mathbf{p}) + f(\mathbf{p}) \cdot \frac{\partial g}{\partial x_i}(\mathbf{p}) \right)$$

$$= \left(\sum v_i \frac{\partial f}{\partial x_i}(\mathbf{p}) \right) g(\mathbf{p}) + f(\mathbf{p}) \left(\sum v_i \frac{\partial g}{\partial x_i}(\mathbf{p}) \right)$$

$$= \mathbf{v}_p[f] \cdot g(\mathbf{p}) + f(\mathbf{p}) \cdot \mathbf{v}_p[g]. \qquad \blacksquare$$

The first two properties in the preceding theorem may be summarized by saying that $\mathbf{v}_p[f]$ is *linear* in $\mathbf{v}_p$ and in f. The third property, as its proof makes clear, is essentially just the usual Leibniz rule for differentiation of a product. *No matter what form differentiation may take, it will always have suitable* linear *and* Leibnizian *properties.*

We now use the pointwise principal to define the *operation of a vector field V on a function f.* The result is the real-valued function $V[f]$ whose value at each point $\mathbf{p}$ is the number $V(\mathbf{p})[f]$—that is, the derivative of f with respect to the tangent vector $V(\mathbf{p})$ at $\mathbf{p}$. This process should be no surprise, since for a function f on the real line, one begins by defining the derivative of f *at a point*—then the derivative *function* df/dx is the function whose value at each point is the derivative at that point. Evidently the definition of $V[f]$ is strictly analogous to this familiar process. In particular, if U_1, U_2, U_3 is the standard frame field on $\mathbf{E}^3$, then $U_i[f] = \partial f/\partial x_i$. This is an immediate consequence of Lemma 3.2. For example, $U_1(\mathbf{p}) = (1,0,0)_p$; hence

$$U_1(\mathbf{p})[f] = \frac{d}{dt}\left(f(p_1 + t, p_2, p_3) \right)\big|_{t=0},$$

which is precisely the definition of $(\partial f/\partial x_1)(\mathbf{p})$. This is true for all points $\mathbf{p} = (p_1, p_2, p_3)$; hence $U_1[f] = \partial f/\partial x_1$.

We shall use this notion of directional derivative more in the case of vector fields than for individual tangent vectors.

3.4 Corollary If V and W are vector fields on $\mathbf{E}^3$, and f, g, h are real-valued functions, then

(1) $(fV + gW)[h] = fV[h] + gW[h]$.

(2) $V[af + bg] = aV[f] + bV[g]$, for all real numbers a and b.

(3) $V[fg] = V[f] \cdot g + fV[g]$.

Proof. The pointwise principle guarantees that to derive these properties from Theorem 3.3 we need only be careful about the placement of parentheses. For example, we prove the third formula. By definition, the value of the function $V[fg]$ at $\mathbf{p}$ is $V(\mathbf{p})[fg]$. But by Theorem 3.3 this is

$$V(\mathbf{p})[f] \cdot g(\mathbf{p}) + f(\mathbf{p})V(\mathbf{p})[g] = V[f](\mathbf{p}) \cdot g(\mathbf{p}) + f(\mathbf{p})V[g](\mathbf{p})$$

$$= (V[f] \cdot g + f \cdot V[g])(\mathbf{p}). \qquad \blacksquare$$

If the use of parentheses here seems extravagant, we remind the reader that a meticulous proof of Leibniz' formula

$$\frac{d}{dx}(fg) = \frac{df}{dx} \cdot g + f \cdot \frac{dg}{dx}$$

must consist of exactly the same shifting of parentheses.

Note that the linearity of $V[f]$ in V and f is for *functions* as "scalars" in the first formula in Corollary 3.4 but only for *numbers* as "scalars" in the second. This stems from the fact that fV signifies merely multiplication, but $V[f]$ is differentiation.

The identity $U_i[f] = \partial f/\partial x_i$ makes it a simple matter to carry out explicit computations. For example, if $V = xU_1 - y^2U_3$ and $f = x^2y + z^3$, then

$$V[f] = xU_1[x^2y] + xU_1[z^3] - y^2U_3[x^2y] - y^2U_3[z^3]$$

$$= x(2xy) + 0 - 0 - y^2(3z^2) = 2x^2y - 3y^2z^2.$$

3.5 Remark Since the subscript notation $\mathbf{v}_p$ for a tangent vector is somewhat cumbersome, from now on we shall frequently omit the point of application $\mathbf{p}$ from the notation. This can cause no confusion, since $\mathbf{v}$ and $\mathbf{w}$ will always denote tangent vectors, and $\mathbf{p}$ and $\mathbf{q}$ points of $\mathbf{E}^3$. In many situations (for example, Definition 3.1) the point of application is crucial, and will be indicated by using either the old notation $\mathbf{v}_p$ or the phrase "a tangent vector $\mathbf{v}$ to $\mathbf{E}^3$ at $\mathbf{p}$."

EXERCISES

1. Let $\mathbf{v}_p$ be the tangent vector to $\mathbf{E}^3$ for which $\mathbf{v} = (2, -1, 3)$ and $\mathbf{p} = (2, 0, -1)$. Working directly from the definition, compute the directional derivative $\mathbf{v}_p[f]$, where

   (a) $f = y^2z$.   (b) $f = x^7$.   (c) $f = e^x \cos y$.

2. Compute the derivatives in Ex. 1 using Lemma 3.2.

3. Let $V = y^2 U_1 - x U_3$, and let $f = xy$, $g = z^3$. Compute the functions
   (a) $V[f]$.   (c) $V[fg]$.   (e) $V[f^2 + g^2]$.
   (b) $V[g]$.   (d) $fV[g] - gV[f]$. (f) $V[V[f]]$.

4. Prove the identity $V = \sum V[x_i]U_i$, where x_1, x_2, x_3 are the natural coordinate functions. (*Hint:* Evaluate $V = \sum v_i U_i$ on x_j.)

5. If $V[f] = W[f]$ for every function f on $\mathbf{E}^3$, prove that $V = W$.

4   Curves in E³

Let I be an open interval in the real line R. We shall interpret this liberally to include not only the usual type of open interval $a < t < b$ (a, b real numbers), but also the types $a < t$ (a half-line to $+\infty$), $t < b$ (a half-line to $-\infty$), and also the whole real line.

One can picture a curve in $\mathbf{E}^3$ as a trip taken by a moving point α. At each "time" t in some open interval, α is located at the point

$$\alpha(t) = (\alpha_1(t), \alpha_2(t), \alpha_3(t))$$

in $\mathbf{E}^3$. In rigorous terms then, α is a function from I to $\mathbf{E}^3$, and the real-valued functions α_1, α_2, α_3 are its *Euclidean coordinate functions*. Thus we write $\alpha = (\alpha_1, \alpha_2, \alpha_3)$, meaning, of course, that

$$\alpha(t) = (\alpha_1(t), \alpha_2(t), \alpha_3(t))$$

for all t in the interval I. We define the function α to be *differentiable* provided its (real-valued) coordinate functions are differentiable in the usual sense.

4.1 Definition   A *curve* in $\mathbf{E}^3$ is a differentiable function $\alpha: I \to \mathbf{E}^3$ from an open interval I into $\mathbf{E}^3$.

We shall give several examples of curves, which will be used in Chapter II to experiment with results on the *geometry* of curves.

4.2 Example

(1) *Straight line.*   A line is the simplest type of curve in Euclidean space; its coordinate functions are linear (in the sense $t \to at + b$, not in the homogeneous sense $t \to at$). Explicitly, the curve $\alpha: \mathbf{R} \to \mathbf{E}^3$, such that

$$\alpha(t) = \mathbf{p} + t\mathbf{q} = (p_1 + tq_1, p_2 + tq_2, p_3 + tq_3) \qquad (\mathbf{q} \neq 0)$$

is the *straight line* through the point $\mathbf{p} = \alpha(0)$ in the $\mathbf{q}$ direction.

(2) *Helix.*   (Fig. 1.7). The curve $t \to (a \cos t, a \sin t, 0)$ travels around

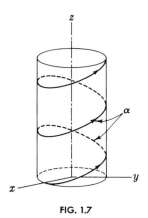

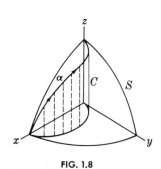

FIG. 1.7   FIG. 1.8

a circle of radius $a > 0$ in the xy plane of $\mathbf{E}^3$. If we allow this curve to rise (or fall) at a constant rate, we obtain a *helix* $\alpha: \mathbf{R} \to \mathbf{E}^3$, given by the formula

$$\alpha(t) = (a \cos t,\, a \sin t,\, bt)$$

where $a > 0$, $b \neq 0$. (We shall always use the term *helix* to mean *right circular helix*.)

(3) Let

$$\alpha(t) = (2 \cos^2 t,\, \sin 2t,\, 2 \sin t) \qquad \text{for} \qquad 0 < t < \pi/2.$$

This curve α has a noteworthy property: Let C be the cylinder in $\mathbf{E}^3$ constructed on the circle in the xy plane with center at $(1,0,0)$ and radius 1. Then α follows the route sliced from C by the sphere S with radius 2 and center at the origin (Fig. 1.8).

(4) The curve $\alpha: \mathbf{R} \to \mathbf{E}^3$ such that

$$\alpha(t) = (e^t,\, e^{-t},\, \sqrt{2}\, t)$$

shares with the helix in (2) the property of rising constantly. However, it lies over the hyperbola $xy = 1$ in the xy plane instead of a circle.

(5) The curve $\alpha: \mathbf{R} \to \mathbf{E}^3$ such that

$$\alpha(t) = (3t - t^3,\, 3t^2,\, 3t + t^3).$$

If the coordinate functions of a curve are simple enough, the shape the curve has in $\mathbf{E}^3$ can be found, at least approximately, by the brute-force procedure of plotting points. We could get a reasonable picture of this curve for $0 \leq t \leq 1$ by computing $\alpha(t)$ for $t = 0, \frac{1}{10}, \frac{1}{2}, \frac{9}{10}, 1$.

If we visualize a curve α in $\mathbf{E}^3$ as a moving point, then at every time t

there is a tangent vector at the point $\alpha(t)$ which gives the instantaneous velocity of α at that time.

4.3 Definition   Let $\alpha\colon I \to \mathbf{E}^3$ be a curve in $\mathbf{E}^3$ with $\alpha = (\alpha_1, \alpha_2, \alpha_3)$. For each number t in I, the *velocity vector of α at t* is the tangent vector

$$\alpha'(t) = \left(\frac{d\alpha_1}{dt}(t), \frac{d\alpha_2}{dt}(t), \frac{d\alpha_3}{dt}(t) \right)_{\alpha(t)}$$

at the point $\alpha(t)$ in $\mathbf{E}^3$ (Fig. 1.9).

We interpret this definition geometrically as follows. The derivative at t of a real-valued function f on $\mathbf{R}$ is given by

$$\frac{df}{dt}(t) = \lim_{\Delta t \to 0} \frac{f(t + \Delta t) - f(t)}{\Delta t}.$$

This formula still makes sense if f is replaced by a curve $\alpha = (\alpha_1, \alpha_2, \alpha_3)$. In fact,

$$\frac{1}{\Delta t}(\alpha(t + \Delta t) - \alpha(t)) = \left(\frac{\alpha_1(t + \Delta t) - \alpha_1(t)}{\Delta t}, \right.$$

$$\left. \frac{\alpha_2(t + \Delta t) - \alpha_2(t)}{\Delta t}, \frac{\alpha_3(t + \Delta t) - \alpha_3(t)}{\Delta t} \right).$$

This is the vector from $\alpha(t)$ to $\alpha(t + \Delta t)$, scalar multiplied by $1/\Delta t$ (Fig. 1.10).

Now as Δt gets smaller, $\alpha(t + \Delta t)$ approaches $\alpha(t)$, and in the limit as $\Delta t \to 0$, we get a vector *tangent* to the curve α at the point $\alpha(t)$, namely, $(d\alpha_1/dt(t),\ d\alpha_2/dt(t),\ d\alpha_3/dt(t))$. As the figure suggests, the point of application of this vector must be the point $\alpha(t)$. Thus the standard limit operation for derivatives gives rise to our definition of the velocity of a curve.

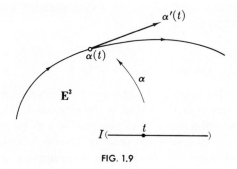

FIG. 1.9

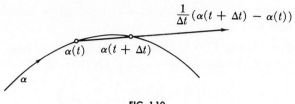

FIG. 1.10

An application of the identity

$$(v_1, v_2, v_3)_p = \sum v_i U_i(\mathbf{p})$$

to the velocity vector $\alpha'(t)$ at t yields the alternative formula

$$\alpha'(t) = \sum \frac{d\alpha_i}{dt}(t) U_i(\alpha(t)).$$

For example, the velocity of the straight line $\alpha(t) = \mathbf{p} + t\mathbf{q}$ is

$$\alpha'(t) = (q_1, q_2, q_3)_{\alpha(t)} = \mathbf{q}_{\alpha(t)}.$$

The fact that α is straight is reflected in the fact that all its velocity vectors are parallel; only the point of application changes as t changes.

For the helix

$$\alpha(t) = (a \cos t, a \sin t, bt),$$

the velocity is

$$\alpha'(t) = (-a \sin t, a \cos t, b)_{\alpha(t)}.$$

The fact that the helix rises constantly is shown by the constancy of the z coordinate of $\alpha'(t)$.

Given a curve α, one can construct many new curves which follow the same route as α, but travel at different speeds.

4.4 Definition Let I and J be open intervals in the real line $\mathbf{R}$. Let α: $I \to \mathbf{E}^3$ be a curve and let $h: J \to I$ be a differentiable (real-valued) function. Then the composite function

$$\beta = \alpha(h): J \to \mathbf{E}^3$$

is a curve called the *reparametrization* of α by h.

At each time s in the interval J, the curve β is at the point $\beta(s) = \alpha(h(s))$ reached by the curve α at time $h(s)$ in the interval I (Fig. 1.11). Thus β does follow the route of α, but β generally reaches a given point on the route at a different time than α does. In practice, to compute the coordinates of β, one simply substitutes $t = h(s)$ in the coordinates $\alpha_1(t)$, $\alpha_2(t)$, $\alpha_3(t)$ of

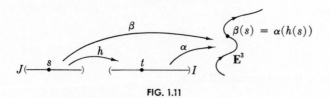

FIG. 1.11

α. For example, suppose $\alpha(t) = (\sqrt{t}, t\sqrt{t}, 1 - t)$ on $I: 0 < t < 4$. If $h(s) = s^2$ on $J: 0 < s < 2$, then

$$\beta(s) = \alpha(h(s)) = \alpha(s^2) = (s, s^3, 1 - s^2).$$

Thus the curve $\alpha: I \to \mathbf{E}^3$ has been reparametrized by h to yield the curve $\beta: J \to \mathbf{E}^3$.

The following lemma relates the velocities of a curve and of a reparametrization.

4.5 Lemma If β is the reparametrization of α by h, then

$$\beta'(s) = (dh/ds)(s)\, \alpha'(h(s)).$$

Proof. If $\alpha = (\alpha_1, \alpha_2, \alpha_3)$, then

$$\beta(s) = \alpha(h(s)) = (\alpha_1(h(s)), \alpha_2(h(s)), \alpha_3(h(s))).$$

Using the "prime" notation for derivatives, the chain rule for a composition of real-valued functions f and g reads $(g(f))' = g'(f) \cdot f'$. Thus in the case at hand, we obtain

$$\alpha_i(h)'(s) = \alpha_i'(h(s)) \cdot h'(s).$$

By the definition of velocity, this yields

$$\begin{aligned}
\beta'(s) &= \alpha(h)'(s) \\
&= (\alpha_1'(h(s)) \cdot h'(s), \alpha_2'(h(s)) \cdot h'(s), \alpha_3'(h(s)) \cdot h'(s)) \\
&= h'(s)\alpha'(h(s)).
\end{aligned}$$ ∎

According to this lemma, to obtain the velocity of a reparametrization of α by h, first reparametrize α' by h, then scalar multiply by the derivative of h.

Since velocities are tangent vectors, we can take the derivative of a function with respect to a velocity.

4.6 Lemma Let α be a curve in $\mathbf{E}^3$ and let f be a differentiable function on $\mathbf{E}^3$. Then

$$\alpha'(t)[f] = \frac{d(f(\alpha))}{dt}(t).$$

Proof. Since

$$\alpha'(t) = \left(\frac{d\alpha_1}{dt}, \frac{d\alpha_2}{dt}, \frac{d\alpha_3}{dt}\right)_\alpha,$$

we conclude from Lemma 3.2 that

$$\alpha'(t)[f] = \sum \frac{\partial f}{\partial x_i}(\alpha(t)) \frac{d\alpha_i}{dt}(t).$$

But the composite function $f(\alpha)$ may be written $f(\alpha_1, \alpha_2, \alpha_3)$, and the chain rule then gives exactly the same result for the derivative of $f(\alpha)$. ∎

By definition, $\alpha'(t)[f]$ is the rate of change of f along the line through $\alpha(t)$ in the $\alpha'(t)$ direction (Fig. 1.12). (If $\alpha'(t) \neq 0$, this is the tangent line to α at $\alpha(t)$; see Exercise 9.) The lemma shows that this rate of change is the same as that of f along the curve α itself.

Since a curve $\alpha: I \to \mathbf{E}^3$ is a function, it makes sense to say that α is one-to-one; that is, $\alpha(t) = \alpha(t_1)$ only if $t = t_1$. Another special property of curves is periodicity: A curve $\alpha: \mathbf{R} \to \mathbf{E}^3$ is *periodic* if there is a number $p > 0$ such that $\alpha(t + p) = \alpha(t)$ for all t—and the smallest such number p is then called the *period* of α.

From the viewpoint of calculus, the most important condition on a curve α is that it be *regular*, that is, have all velocity vectors different from zero. Such a curve can have no corners or cusps.

The following remarks about curves (offered without proof) are not an essential part of our exposition, but will be of use in Chapter IV. We consider, in the case of the plane $\mathbf{E}^2$, another familiar way to formulate the concept of "curve." If f is a differentiable real-valued function on $\mathbf{E}^2$, let

$$C: f = a$$

be the set of all points $\mathbf{p}$ in $\mathbf{E}^2$ such that $f(\mathbf{p}) = a$. Now if the partial derivatives $\partial f/\partial x$ and $\partial f/\partial y$ are never simultaneously zero at any point of C, then C consists of one or more separate "components" which we shall

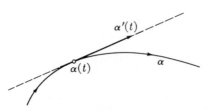

FIG. 1.12

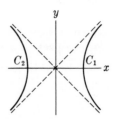

FIG. 1.13

call *Curves.*† Thus, for example, $C: x^2 + y^2 = r^2$ is the circle of radius r centered at the origin of $\mathbf{E}^2$, and the hyperbola $C: x^2 - y^2 = r^2$ splits into two Curves ("branches") C_1 and C_2 as in Fig. 1.13.

Every Curve C is the route of many regular curves α, called *parametrizations* of C. If C is a *closed* Curve, then it has a periodic parametrization $\alpha: \mathbf{R} \to C$. For example, the curve

$$\alpha(t) = (r \cos t, r \sin t)$$

is a well-known periodic parametrization of the circle given above. If C is a Curve that is not closed (sometimes called an *arc*), then every parametrization $\beta: I \to C$ is one-to-one. For example,

$$\beta(t) = (r \cosh t, r \sinh t)$$

parametrizes the branch, $x > 0$, of the hyperbola given above.

EXERCISES

1. Compute the velocity vector of curve (3) in Example 4.2 for arbitrary t, and at $t = \pi/4$.

2. Plot curve (5) in Example 4.2 by the method there suggested. On the sketch show the velocity vectors at $t = 0, \frac{1}{2}, 1$.

3. Find the coordinate functions of the curve $\beta = \alpha(h)$, where α is curve (3) in Example 4.2 and h is the function on $J: 0 < s < 1$ such that $h(s) = \sin^{-1}s$.

4. Find the (unique) curve α such that $\alpha(0) = (1, 0, -5)$ and $\alpha'(t) = (t^2, t, e^t)$.

5. Find a straight line passing through the points $(1, -3, -1)$ and $(6, 2, 1)$. Does this line meet the line through the points $(-1, 1, 0)$ and $(-5, -1, -1)$?

6. Deduce from Lemma 4.6 that in the definition of directional derivative (Definition 3.1) the straight line $t \to \mathbf{p} + t\mathbf{v}$ may be replaced by *any* curve α with *initial velocity* $\mathbf{v}_p$, that is, such that $\alpha(0) = \mathbf{p}$ and $\alpha'(0) = \mathbf{v}_p$.

7. *(Continuation).* Show that the curves given by $(t, 1 + t^2, t)$, $(\sin t, \cos t, t)$, and $(\sinh t, \cosh t, t)$ all have the same initial velocity $\mathbf{v}_p$. If $f = x^2 - y^2 + z^2$, compute $\mathbf{v}_p[f]$ by evaluating f on each of the curves.

8. Let $h(s) = \log s$ on $J: s > 0$. Reparametrize curve (4) in Example 4.2

† In this section (only) we use a capital C to distinguish this notion from a curve $\alpha: I \to \mathbf{E}^3$.

using h. Check the equation in Lemma 4.5 in this case by computing each side separately.

9. For a fixed t, the *tangent line* to a regular curve α at $\alpha(t)$ is the straight line $u \rightarrow \alpha(t) + u\alpha'(t)$, where we delete the point of application of $\alpha'(t)$. Find the tangent line to the helix $\alpha(t) = (2 \cos t, 2 \sin t, t)$ at the points $\alpha(0)$ and $\alpha(\pi/4)$.

10. Sketch the following Curves in $\mathbf{E}^2$ and find parametrizations for each.
    (a) $C: 4x^2 + y^2 = 1$.   (c) $C: y = e^x$.
    (b) $C: 3x + 4y = 1$.   (d) $C: x^{2/3} + y^{2/3} = 1, x > 0, y > 0$.

5 1-Forms

If f is a real-valued function on $\mathbf{E}^3$, then in elementary calculus one defines the differential of f to be

$$df = \frac{\partial f}{\partial x} \, dx + \frac{\partial f}{\partial y} \, dy + \frac{\partial f}{\partial z} \, dz.$$

It is not always made clear exactly what this formal expression means. In this section we give a rigorous treatment using the notion of 1-form, and these will tend to appear at crucial moments in our later work.

5.1 Definition   A *1-form* ϕ on $\mathbf{E}^3$ is a real-valued function on the set of all tangent vectors to $\mathbf{E}^3$ such that ϕ is linear at each point, that is,

$$\phi(a\mathbf{v} + b\mathbf{w}) = a\phi(\mathbf{v}) + b\phi(\mathbf{w})$$

for any numbers a, b and tangent vectors $\mathbf{v}$, $\mathbf{w}$ at the same point of $\mathbf{E}^3$.

We emphasize that for every tangent vector $\mathbf{v}$ to $\mathbf{E}^3$, a 1-form ϕ defines a real number $\phi(\mathbf{v})$; and for each point $\mathbf{p}$ in $\mathbf{E}^3$, the resulting function $\phi_p: T_p(\mathbf{E}^3) \rightarrow \mathbf{R}$ is linear. [Thus at each point $\mathbf{p}$, ϕ_p is an element of the *dual space* of $T_p(\mathbf{E}^3)$. In this sense the notion of 1-form is dual to that of vector field.]

The sum of 1-forms ϕ and ψ is defined in the usual pointwise fashion

$$(\phi + \psi)(\mathbf{v}) = \phi(\mathbf{v}) + \psi(\mathbf{v}) \qquad \text{for all tangent vectors } \mathbf{v}.$$

Similarly if f is a real-valued function on $\mathbf{E}^3$ and ϕ is a 1-form, then $f\phi$ is the 1-form such that

$$(f\phi)(\mathbf{v}_p) = f(\mathbf{p}) \, \phi(\mathbf{v}_p)$$

for all tangent vectors $\mathbf{v}_p$.

There is also a natural way to *evaluate a 1-form ϕ on a vector field V* to obtain a real-valued function $\phi(V)$: At each point $\mathbf{p}$ the value of $\phi(V)$

is the number $\phi(V(\mathbf{p}))$. Thus a 1-form may also be viewed as a machine which converts vector fields into real-valued functions. If $\phi(V)$ is differentiable whenever V is, we say that ϕ is *differentiable*. As with vector fields, we shall always assume that the 1-forms we deal with are differentiable.

A routine check of definitions shows that $\phi(V)$ is linear in both ϕ and V; that is,

$$\phi(fV + gW) = f\,\phi(V) + g\,\phi(W)$$

and

$$(f\phi + g\psi)(V) = f\,\phi(V) + g\,\psi(V)$$

where f and g are functions.

Using the notion of directional derivative, we now define a most important way to convert functions into 1-forms.

5.2 Definition If f is a differentiable real-valued function on $\mathbf{E}^3$, the *differential df* of f is the 1-form such that

$$df(\mathbf{v}_p) = \mathbf{v}_p[f] \qquad \text{for all tangent vectors } \mathbf{v}_p.$$

In fact, *df is* a 1-form, since by definition it is a real-valued function on tangent vectors, and by (1) of Theorem 3.3 it is linear at each point $\mathbf{p}$. Clearly df knows all rates of change of f in all directions on $\mathbf{E}^3$, so it is not surprising that differentials are fundamental to the calculus on $\mathbf{E}^3$.

Our task now is to show that these rather abstract definitions lead to familiar results when expressed in terms of coordinates.

5.3 Example 1-Forms on $\mathbf{E}^3$. (1) The differentials dx_1, dx_2, dx_3 of the natural coordinate functions. Using Lemma 3.2 we find

$$dx_i(\mathbf{v}_p) \;=\; \mathbf{v}_p[x_i] \;=\; \sum_j v_j \frac{\partial x_i}{\partial x_j}(\mathbf{p}) \;=\; \sum_j v_j \,\delta_{ij} \;=\; v_i$$

where δ_{ij} is the Kronecker delta (0 if $i \neq j$, 1 if $i = j$). Thus *the value of dx_i on an arbitrary tangent vector $\mathbf{v}_p$ is the ith coordinate v_i of its vector part*—and does not depend at all on the point of application $\mathbf{p}$.

(2) The 1-form $\psi = f_1\,dx_1 + f_2\,dx_2 + f_3\,dx_3$. Since dx_i is a 1-form, our definitions show that ψ is also a 1-form for any functions f_1, f_2, f_3. The value of ψ on an arbitrary tangent vector $\mathbf{v}_p$ is

$$\psi(\mathbf{v}_p) \;=\; \Big(\sum f_i\,dx_i\Big)(\mathbf{v}_p) \;=\; \sum f_i(\mathbf{p})\,dx_i(\mathbf{v}) \;=\; \sum f_i(\mathbf{p})v_i.$$

The first of these examples shows that the 1-forms dx_1, dx_2, dx_3 are the analogues for tangent vectors of the natural coordinate functions x_1, x_2, x_3 for points. Alternatively, we can view dx_1, dx_2, dx_3 as the "duals" of the natural unit vector fields U_1, U_2, U_3. In fact, it follows immediately from (1) above that the function $dx_i(U_j)$ has the constant value δ_{ij}.

We shall now show that every 1-form can be written in the concrete manner given in (2) above.

5.4 Lemma   If ϕ is a 1-form on $\mathbf{E}^3$, then $\phi = \sum f_i \, dx_i$, where $f_i = \phi(U_i)$. These functions f_1, f_2, f_3 are called the *Euclidean coordinate functions* of ϕ.

Proof. By definition a 1-form is a function on tangent vectors; thus ϕ and $\sum f_i \, dx_i$ are equal if and only if they have the same value on every tangent vector $\mathbf{v}_p = \sum v_i U_i(\mathbf{p})$. In (2) of Example 5.3 we saw that

$$\left(\sum f_i \, dx_i\right)(\mathbf{v}_p) = \sum f_i(\mathbf{p}) v_i.$$

On the other hand,

$$\phi(\mathbf{v}_p) = \phi\left(\sum v_i U_i(\mathbf{p})\right) = \sum v_i \phi(U_i(\mathbf{p})) = \sum v_i f_i(\mathbf{p})$$

since $f_i = \phi(U_i)$. Thus ϕ and $\sum f_i \, dx_i$ do have the same value on every tangent vector. ∎

This lemma shows that a 1-form on $\mathbf{E}^3$ is nothing more than an expression $f \, dx + g \, dy + h \, dz$, and such expressions are now rigorously defined as functions on tangent vectors. Let us now show that the definition of differential of a function (Definition 5.2) agrees with the informal definition given at the start of this section.

5.5 Corollary   If f is a differentiable function on $\mathbf{E}^3$, then

$$df = \sum \frac{\partial f}{\partial x_i} \, dx_i \, .$$

Proof. The value of $\sum (\partial f/\partial x_i) \, dx_i$ on an arbitrary tangent vector $\mathbf{v}_p$ is $\sum (\partial f/\partial x_i)(\mathbf{p}) v_i$. By Lemma 3.2 $df(\mathbf{v}_p) = \mathbf{v}_p[f]$ is the same. Thus the 1-forms df and $\sum (\partial f/\partial x_i) \, dx_i$ are equal. ∎

Using either this result or the definition of d, it is immediate that

$$d(f + g) = df + dg.$$

Finally we determine the effect of d on *products* of functions and on *compositions* of functions.

5.6 Lemma   Let fg be the product of differentiable functions f and g on $\mathbf{E}^3$. Then

$$d(fg) = g \, df + f \, dg.$$

Proof. Using Corollary 5.5, we obtain

$$d(fg) = \sum \frac{\partial(fg)}{\partial x_i} \, dx_i = \sum \left(\frac{\partial f}{\partial x_i} g + f \frac{\partial g}{\partial x_i}\right) dx_i$$

$$= g\left(\sum \frac{\partial f}{\partial x_i} \, dx_i\right) + f\left(\sum \frac{\partial g}{\partial x_i} \, dx_i\right) = g \, df + f \, dg. \qquad ∎$$

5.7 Lemma   Let $f: \mathbf{E}^3 \to \mathbf{R}$ and $h: \mathbf{R} \to \mathbf{R}$ be differentiable functions, so the composite function $h(f): \mathbf{E}^3 \to \mathbf{R}$ is also differentiable. Then

$$d(h(f)) = h'(f)\, df.$$

Proof. (The prime here is just the ordinary derivative, so $h'(f)$ is again a composite function, from $\mathbf{E}^3$ to $\mathbf{R}$.) The usual chain rule for a composite function such as $h(f)$ reads

$$\frac{\partial(h(f))}{\partial x_i} = h'(f)\frac{\partial f}{\partial x_i}.$$

Hence

$$d(h(f)) = \sum \frac{\partial(h(f))}{\partial x_i}\, dx_i = \sum h'(f)\,\frac{\partial f}{\partial x_i}\, dx_i = h'(f)df. \qquad \blacksquare$$

To compute df for a given function f it is almost always simpler to use these properties of d rather than substitute in the formula of Corollary 5.5. Then from df we immediately get the partial derivatives of f and, in fact, *all its directional derivatives.* For example, suppose

$$f = (x^2 - 1)y + (y^2 + 2)z.$$

Then by Lemmas 5.6 and 5.7,

$$df = (2x\, dx)y + (x^2 - 1)\, dy + (2y\, dy)z + (y^2 + 2)\, dz$$

$$= \underbrace{2xy\, dx}_{\partial f/\partial x} + \underbrace{(x^2 + 2yz - 1)}_{\partial f/\partial y}\, dy + \underbrace{(y^2 + 2)}_{\partial f/\partial z}\, dz$$

Now use the rules above to evaluate this expression on a tangent vector $\mathbf{v}$. The result is

$$\mathbf{v}_p[f] = df(\mathbf{v}_p) = 2p_1 p_2 v_1 + (p_1{}^2 + 2p_2 p_3 - 1)v_2 + (p_2{}^2 + 1)v_3.$$

EXERCISES

1. Let $\mathbf{v} = (1, 2, -3)$ and $\mathbf{p} = (0, -2, 1)$. Evaluate the following 1-forms on the tangent vector $\mathbf{v}_p$.
   (a) $y^2\, dx$.   (b) $z\, dy - y\, dz$.   (c) $(z^2 - 1)dx - dy + x^2\, dz$.

2. If $\phi = \sum f_i\, dx_i$ and $V = \sum v_i U_i$, show that the 1-form ϕ evaluated on the vector field V is the function $\phi(V) = \sum f_i v_i$.

3. Evaluate the 1-form $\phi = x^2\, dx - y^2\, dz$ on the vector fields
   $V = xU_1 + yU_2 + zU_3$,
   $W = xy(U_1 - U_3) + yz(U_1 - U_2)$,   and $(1/x)V + (1/y)W$.

4. Express the following differentials in terms of df:

   (a) $d(f^5)$.   (b) $d(\sqrt{f})$,   where $f > 0$.   (c) $d(\log(1 + f^2))$.

5. Express the differentials of the following functions in the standard form $\sum f_i \, dx_i$.

   (a) $(x^2 + y^2 + z^2)^{1/2}$.   (b) $\tan^{-1}(y/x)$.

6. In each case compute the differential of f and find the directional derivative $\mathbf{v}_p[f]$, for $\mathbf{v}_p$ as in Ex. 1.

   (a) $f = xy^2 - yz^2$.   (b) $f = xe^{yz}$.   (c) $f = \sin(xy)\cos(xz)$.

7. Which of the following are 1-forms? In each case ϕ is the function on tangent vectors such that the value of ϕ on $(v_1, v_2, v_3)_p$ is

   (a) $v_1 - v_3$.   (c) $v_1 p_3 + v_2 p_1$.   (e) 0.

   (b) $p_1 - p_3$.   (d) $v_p[x^2 + y^2]$.   (f) $(p_1)^2$.

   In case ϕ is a 1-form, express it as $\sum f_i \, dx_i$.

8. Prove Lemma 5.6 directly from the definition of d—without using Corollary 5.5.

9. A 1-form ϕ is *zero* at a point $\mathbf{p}$ provided $\phi(\mathbf{v}_p) = 0$ for all tangent vectors at $\mathbf{p}$. A point at which its differential df is zero is called a *critical point* of the function f. Prove that $\mathbf{p}$ is a critical point of f if and only if

$$\frac{\partial f}{\partial x}(\mathbf{p}) = \frac{\partial f}{\partial y}(\mathbf{p}) = \frac{\partial f}{\partial z}(\mathbf{p}) = 0.$$

Find all critical points of $f = (1 - x^2)y + (1 - y^2)z$.

(*Hint:* Find the partial derivatives of f by computing df.)

10. (*Continuation*). Prove that the local maxima and local minima of f are critical points of f. (f has a *local maximum* at $\mathbf{p}$ if $f(\mathbf{q}) \leq f(\mathbf{p})$ for all $\mathbf{q}$ near $\mathbf{p}$.)

11. It is sometimes asserted that df is the linear approximation of Δf.

    (a) Explain the sense in which $(df)(\mathbf{v}_p)$ is the linear approximation of $f(\mathbf{p} + \mathbf{v}) - f(\mathbf{p})$.

    (b) Compute the exact and approximate values of $f(0.9, 1.6, 1.2) - f(1, 1.5, 1)$, where $f = x^2 y/z$.

6 Differential Forms

The 1-forms on $\mathbf{E}^3$ are part of a larger system called the *differential forms* on $\mathbf{E}^3$. We shall not give as rigorous an account of differential forms as we did of 1-forms, for our use of the full system will be limited to Section 8 of Chapter II. Roughly speaking, a *differential form* on $\mathbf{E}^3$ is an expression obtained by adding and multiplying real-valued functions and the differentials dx_1, dx_2, dx_3 of the natural coordinate functions of $\mathbf{E}^3$. These two operations obey the usual associative and distributive laws; however, the multiplication is not commutative. Instead it obeys the

> *alternation rule:* $dx_i\,dx_j \;=\; -dx_j\,dx_i$ $(1 \leqq i, j \leqq 3)$.

This rule appears—although rather inconspicuously—in elementary calculus (see Exercise 9).

A consequence of the alternation rule is the fact that "repeats are zero," that is, $dx_i\,dx_i = 0$, since if $i = j$ the alternation rule reads

$$dx_i\,dx_i \;=\; -dx_i\,dx_i.$$

If each summand of a differential form contains $p\,dx_i$'s $(p = 0, 1, 2, 3)$, the form is called a *p-form*, and is said to have *degree p*. Thus, shifting to dx,dy,dz, we find

A 0-form is just a differentiable function f.

A 1-form is an expression $f\,dx + g\,dy + h\,dz$, just as in the preceding section.

A 2-form is an expression $f\,dx\,dy + g\,dx\,dz + h\,dy\,dz$.

A 3-form is an expression $f\,dx\,dy\,dz$.

We already know how to add 1-forms: simply add corresponding coefficient functions. Thus, in index notation,

$$\sum f_i\,dx_i + \sum g_i\,dx_i \;=\; \sum (f_i + g_i)\,dx_i.$$

The corresponding rule holds for 2-forms or 3-forms.

On three-dimensional Euclidean space, all p-forms with $p > 3$ are zero. This is a consequence of the alternation rule, for a product of more than three dx_i's must contain some dx_i twice, but repeats are zero, as noted above. For example, $dx\,dy\,dx\,dz = -dx\,dx\,dy\,dz = 0$, since $dx\,dx = 0$. As a reminder that the alternation rule is to be used, we denote this multiplication of forms by a *wedge* $\wedge$. (However, we do not bother with the wedge when only products of dx, dy, dz are involved.)

6.1 Example   Computation of wedge products. (1) Let

$$\phi = x\,dx - y\,dy \qquad \text{and} \qquad \psi = z\,dx + x\,dz.$$

Then

$$\phi \wedge \psi = (x\,dx - y\,dy) \wedge (z\,dx + x\,dz)$$

$$= xz\,dx\,dx + x^2\,dx\,dz - yz\,dy\,dx - yx\,dy\,dz.$$

But

$$dx\, dx = 0, \quad \text{and} \quad dy\, dx = -dx\, dy.$$

Thus

$$\phi \wedge \psi = yz\, dx\, dy + x^2\, dx\, dz - xy\, dy\, dz.$$

In general, the product of two 1-forms is a 2-form.

(2) Let ϕ and ψ be the 1-forms given above and let

$$\theta = z\, dy.$$

Then

$$\theta \wedge \phi \wedge \psi = yz^2\, dy\, dx\, dy + x^2z\, dy\, dx\, dz - xyz\, dy\, dy\, dz.$$

Since $dy\, dx\, dy$ and $dy\, dy\, dz$ each contain repeats, both are zero. Thus

$$\theta \wedge \phi \wedge \psi = -x^2z\, dx\, dy\, dz.$$

(3) Let ϕ be as above, and let η be the 2-form $y\, dx\, dz + x\, dy\, dz$. Omitting forms containing repeats, we find

$$\phi \wedge \eta = x^2\, dx\, dy\, dz - y^2\, dy\, dx\, dz = (x^2 + y^2)\, dx\, dy\, dz.$$

It should be clear from these examples that the wedge product of a p-form and a q-form is a $(p + q)$-form. Thus such a product is automatically zero whenever $p + q > 3$.

6.2 Lemma If ϕ and ψ are 1-forms, then

$$\phi \wedge \psi = -\psi \wedge \phi.$$

Proof. Write

$$\phi = \sum f_i\, dx_i, \quad \psi = \sum g_i\, dx_i.$$

Then by the alternation rule,

$$\phi \wedge \psi = \sum f_i g_j\, dx_i\, dx_j = -\sum g_j f_i\, dx_j\, dx_i = -\psi \wedge \phi. \quad \blacksquare$$

In the language of differential forms, the operator d of Definition 5.2 converts a 0-form f into a 1-form df. It is easy to generalize to an operator (also denoted by d) which converts a p form η into a $(p + 1)$-form $d\eta$: One simply applies d (of Definition 5.2) to the coefficient functions of η. For example, here is the case $p = 1$.

6.3 Definition If $\phi = \sum f_i\, dx_i$ is a 1-form on $\mathbf{E}^3$, the *exterior derivative* of ϕ is the 2-form $d\phi = \sum df_i \wedge dx_i$.

If we expand the preceding definition using Corollary 5.5, we obtain the following interesting formula for the exterior derivative of

$$\phi = f_1 \, dx_1 + f_2 \, dx_2 + f_3 \, dx_3:$$

$$d\phi = \left(\frac{\partial f_2}{\partial x_1} - \frac{\partial f_1}{\partial x_2}\right) dx_1 \, dx_2 + \left(\frac{\partial f_3}{\partial x_1} - \frac{\partial f_1}{\partial x_3}\right) dx_1 \, dx_3 + \left(\frac{\partial f_3}{\partial x_2} - \frac{\partial f_2}{\partial x_3}\right) dx_2 \, dx_3.$$

There is no need to memorize this formula; it is more reliable to simply apply the definition in each case. For example, suppose

$$\phi = xy \, dx + x^2 \, dz.$$

Then

$$
\begin{aligned}
d\phi &= d(xy) \wedge dx + d(x^2) \, dz \\
&= (y \, dx + x \, dy) \wedge dx + 2x \, dx \, dz \\
&= -x \, dx \, dy + 2x \, dx \, dz.
\end{aligned}
$$

It is easy to check that the general exterior derivative enjoys the same linearity property as the particular case in Definition 5.2; that is,

$$d(a\phi + b\psi) = a \, d\phi + b \, d\psi,$$

where ϕ and ψ are arbitrary forms and a and b are numbers.

The exterior derivative and the wedge product work together nicely.

6.4 Theorem   Let f and g be functions, ϕ and ψ 1-forms. Then
(1) $d(fg) = df \, g + f \, dg$.
(2) $d(f\phi) = df \wedge \phi + f \, d\phi$.
(3) $d(\phi \wedge \psi) = d\phi \wedge \psi - \phi \wedge d\psi.$ †

Proof. The first formula is just Lemma 5.6. We include it to show the family resemblance of all three formulas. The proof of the second formula is a simpler variant of that of the third, so we prove only the latter.

Case 1. $\phi = f \, dx, \psi = g \, dx$. Since

$$\phi \wedge \psi = fg \, dx \, dx = 0,$$

we must show that the right side of the equation is also zero. Now

$$d\phi = df \wedge dx = \frac{\partial f}{\partial y} \, dy \, dx + \frac{\partial f}{\partial z} \, dz \, dx;$$

hence each term of $d\varphi \wedge \psi$ has a repeated dx. Thus $d\phi \wedge \psi = 0$, and similarly $\phi \wedge d\psi = 0$.

Case 2. $\phi = f \, dx, \psi = g \, dy$. Using the formula for $d\phi$ computed above, we get

† As usual, multiplication takes precedence over addition or subtraction, so this expression should be read as $(d\phi \wedge \psi) - (\phi \wedge d\psi)$.

$$d\phi \wedge \psi = \left(\frac{\partial f}{\partial y}\, dy\, dx + \frac{\partial f}{\partial z}\, dz\, dx\right) \wedge g\, dy$$

$$= 0 + \frac{\partial f}{\partial z}\, g\, dz\, dx\, dy = g\, \frac{\partial f}{\partial z}\, dx\, dy\, dz.$$

Similarly,

$$\phi \wedge d\psi = f\, dx \wedge \left(\frac{\partial g}{\partial x}\, dx\, dy + \frac{\partial g}{\partial z}\, dz\, dy\right)$$

$$= f\, \frac{\partial g}{\partial z}\, dx\, dz\, dy = -f\, \frac{\partial g}{\partial z}\, dx\, dy\, dz.$$

Thus

$$d\phi \wedge \psi - \phi \wedge d\psi = \left(g\, \frac{\partial f}{\partial z} + \frac{\partial g}{\partial z}\, f\right) dx\, dy\, dz.$$

But

$$\phi \wedge \psi = fg\, dx\, dy,$$

so we get

$$d(\phi \wedge \psi) = d(fg)\, dx\, dy = \frac{\partial (fg)}{\partial z}\, dz\, dx\, dy$$

$$= \left(\frac{\partial f}{\partial z}\, g + f\, \frac{\partial g}{\partial z}\right) dx\, dy\, dz.$$

Hence the formula is proved in this case.

Case 3. The general case. From cases 1 and 2 we know that the formula is true whenever ϕ and ψ are "simple," that is, of the form $f\, du$, where u is x, y, or z. Since every 1-form is a sum of simple 1-forms, the general case follows from the linearity of d and the distributive law for the wedge product. ∎

One way to remember the minus sign which occurs in formula (3) of Theorem 6.4 is to pretend that d is a 1-form. To reach ψ, d must change places with ϕ; hence the minus sign is consistent with Lemma 6.2.

Differential forms, and the associated concepts of wedge product and exterior derivative, provide a means of expressing rather complicated relationships in a simple, methodical way. For example, as its proof shows, the tidy formula

$$d(\phi \wedge \psi) = d\phi \wedge \psi - \phi \wedge d\psi$$

involves some rather tricky relations among partial derivatives. Before forms were invented, it was necessary to struggle through these relations in many a separate problem, but now we simply apply the general formula.

A variety of interesting applications is given in Flanders [1]. Later on we shall use differential forms to express the fundamental equations of geometry.

EXERCISES

1. Let $\phi = yz\,dx + dz$, $\psi = \sin z\,dx + \cos z\,dy$, $\xi = dy + z\,dz$. Find the standard expressions (in terms of $dx\,dy, \cdots,$) for
   (a) $\phi \wedge \psi$, $\psi \wedge \xi$, $\xi \wedge \phi$.   (b) $d\phi$, $d\psi$, $d\xi$.

2. Let $\phi = dx/y$ and $\psi = z\,dy$. Check the Leibnizian formula (3) of Theorem 6.4 in this case by computing each term separately.

3. For any function f show that $d\,(df) = 0$. Deduce that $d\,(f\,dg) = df \wedge dg$.

4. Simplify the following forms:
   (a) $d\,(f\,dg + g\,df)$.   (c) $d\,(f\,dg \wedge g\,df)$.
   (b) $d\{\,(f - g)\,(df + dg)\}$.   (d) $d\,(gf\,df) + d\,(f\,dg)$.

5. For any three 1-forms $\phi_i = \sum_j f_{ij}\,dx_j$ $(1 \leqq i \leqq 3)$, prove

$$\phi_1 \wedge \phi_2 \wedge \phi_3 = \begin{vmatrix} f_{11} & f_{12} & f_{13} \\ f_{21} & f_{22} & f_{23} \\ f_{31} & f_{32} & f_{33} \end{vmatrix} dx_1\,dx_2\,dx_3.$$

6. If r, ϑ, and z are the cylindrical coordinate functions on $\mathbf{E}^3$, then $x = r\cos\vartheta$, $y = r\sin\vartheta$, $z = z$. Compute the *volume element* $dx\,dy\,dz$ of $\mathbf{E}^3$ in cylindrical coordinates. (That is, express $dx\,dy\,dz$ in terms of the functions r, φ, z and their differentials.)

7. For a 2-form

$$\eta = f\,dx\,dy + g\,dx\,dz + h\,dy\,dz,$$

the *exterior derivative* $d\eta$ is defined to be the 3-form obtained by replacing f, g, and h by their differentials. Prove that for any 1-form ϕ, $d\,(d\phi) = 0$.

Exercises 3 and 7 show that $d^2 = 0$, that is, for any form ξ, $d\,(d\xi) = 0$. (If ξ is a 2-form, then $d\,(d\xi) = 0$, since its degree exceeds 3.)

8. Classical *vector analysis* avoids the use of differential forms on $\mathbf{E}^3$ by converting 1-forms and 2-forms into vector fields by means of the following one-to-one correspondences:

$$\sum f_i\,dx_i \xrightarrow{\ (1)\ } \sum f_i U_i \xrightarrow{\ (2)\ } f_3\,dx_1\,dx_2 - f_2\,dx_1\,dx_3 + f_1\,dx_2\,dx_3.$$

Vector analysis uses three basic operations based on partial differentiation:

Gradient of a function f:

$$\text{grad } f = \sum \frac{\partial f}{\partial x_i} U_i.$$

Curl of a vector field $V = \sum f_i U_i$:

$$\text{curl } V = \left(\frac{\partial f_3}{\partial x_2} - \frac{\partial f_2}{\partial x_3} \right) U_1 + \left(\frac{\partial f_1}{\partial x_3} - \frac{\partial f_3}{\partial x_1} \right) U_2 + \left(\frac{\partial f_2}{\partial x_1} - \frac{\partial f_1}{\partial x_2} \right) U_3.$$

Divergence of a vector field $V = \sum f_i U_i$:

$$\text{div } V = \sum \frac{\partial f_i}{\partial x_i}.$$

Prove that all three operations may be expressed by exterior derivatives as follows:

(a) $df \xrightarrow{(1)} \text{grad } f.$
(b) If $\phi \xrightarrow{(1)} V$, then $d\phi \xrightarrow{(2)} \text{curl } V.$
(c) If $\eta \xrightarrow{(2)} V$, then $d\eta = (\text{div } V)\, dx\, dy\, dz.$

9. Let f and g be real-valued functions on $\mathbf{E}^2$. Prove that

$$df \wedge dg = \begin{vmatrix} \dfrac{\partial f}{\partial x} & \dfrac{\partial f}{\partial y} \\[2mm] \dfrac{\partial g}{\partial x} & \dfrac{\partial g}{\partial y} \end{vmatrix} dx\, dy.$$

This formula appears in elementary calculus; show that it implies the alternation rule.

7 Mappings

In this section we discuss functions from $\mathbf{E}^n$ to $\mathbf{E}^m$. If $n = 3$ and $m = 1$, then such a function is just a real-valued function on $\mathbf{E}^3$. If $n = 1$ and $m = 3$, then such a function is a curve in $\mathbf{E}^3$. Although our results will necessarily be stated for arbitrary m and n, we are primarily interested only in the three cases:

$$\mathbf{E}^2 \to \mathbf{E}^2 \qquad \mathbf{E}^2 \to \mathbf{E}^3 \qquad \mathbf{E}^3 \to \mathbf{E}^3.$$

The fundamental observation about a function $F: \mathbf{E}^n \to \mathbf{E}^m$ is that it can be completely described by m real-valued functions on $\mathbf{E}^n$. (We saw this already in Section 4 for $n = 1$, $m = 3$.)

7.1 Definition   Given a function $F: \mathbf{E}^n \to \mathbf{E}^m$, let $f_1, f_2, \cdots, f_m$ denote the real-valued functions on $\mathbf{E}^n$ such that

$$F(\mathbf{p}) = (f_1(\mathbf{p}), f_2(\mathbf{p}), \cdots, f_m(\mathbf{p}))$$

for all points $\mathbf{p}$ in $\mathbf{E}^n$. These functions are called the *Euclidean coordinate functions* of F, and we write $F = (f_1, f_2, \cdots, f_m)$.

The function F is *differentiable* provided its coordinate functions are differentiable in the usual sense. A differentiable function $F: \mathbf{E}^n \to \mathbf{E}^m$ is called a *mapping* from $\mathbf{E}^n$ to $\mathbf{E}^m$.

Note that the coordinate functions of F are the composite functions $f_i = x_i(F)$, where $x_1, \cdots, x_m$ are the coordinate functions of $\mathbf{E}^m$.

Mappings may be described in many different ways. For example, suppose $F: \mathbf{E}^3 \to \mathbf{E}^3$ is the mapping $F = (x^2, yz, xy)$. Thus

$$F(\mathbf{p}) = (x(\mathbf{p})^2, y(\mathbf{p})z(\mathbf{p}), x(\mathbf{p})y(\mathbf{p})) \qquad \text{for all } \mathbf{p}.$$

Now $\mathbf{p} = (p_1, p_2, p_3)$, and, by definition of the coordinate functions,

$$x(\mathbf{p}) = p_1, \; y(\mathbf{p}) = p_2, \; z(\mathbf{p}) = p_3.$$

Hence we obtain the following *pointwise* formula for F:

$$F(p_1, p_2, p_3) = (p_1^2, p_2 p_3, p_1 p_2) \qquad \text{for all } p_1, p_2, p_3.$$

In particular,

$$F(1, -2, 0) = (1, 0, -2), \qquad F(-3, 1, 3) = (9, 3, -3),$$

and so on.

In principal, one could deduce the theory of curves from the general theory of mappings. But curves are reasonably simple, while a mapping, even in the case $\mathbf{E}^2 \to \mathbf{E}^2$, can be quite complicated. Hence we reverse this process and use curves, at every stage, to gain an understanding of mappings.

7.2 Definition   If $\alpha: I \to \mathbf{E}^n$ is a curve in $\mathbf{E}^n$ and $F: \mathbf{E}^n \to \mathbf{E}^m$ is a mapping, then the composite function $\beta = F(\alpha): I \to \mathbf{E}^m$ is a curve in $\mathbf{E}^m$ called the *image of α under F* (Fig. 1.14).

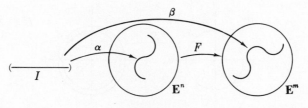

FIG. 1.14

7.3 Example Mappings. (1) Consider the mapping $F: \mathbf{E}^3 \to \mathbf{E}^3$ such that

$$F = (x - y, x + y, 2z).$$

In pointwise terms then,

$$F(p_1, p_2, p_3) = (p_1 - p_2, p_1 + p_2, 2p_3) \qquad \text{for all } p_1, p_2, p_3.$$

Only when a mapping is quite simple can one hope to get a good idea of its behavior by merely computing its values on some finite number of points. But this function *is* quite simple—it is a *linear* transformation from $\mathbf{E}^3$ to $\mathbf{E}^3$. Thus by a well-known theorem of linear algebra, F is completely determined by its values on three (linearly independent) points, say the *unit points*

$$\mathbf{u}_1 = (1, 0, 0) \qquad \mathbf{u}_2 = (0, 1, 0) \qquad \mathbf{u}_3 = (0, 0, 1).$$

(2) The mapping $F: \mathbf{E}^2 \to \mathbf{E}^2$ such that $F(u, v) = (u^2 - v^2, 2uv)$. (Here u and v are the coordinate functions of $\mathbf{E}^2$.) To analyze this mapping, we examine its effect on the curve $\alpha(t) = (r \cos t, r \sin t)$, where $0 \leq t \leq 2\pi$. This curve takes one counterclockwise trip around the circle of radius r (center at the origin.) The image curve is

$$\beta(t) = F(\alpha(t)) = F(r \cos t, r \sin t) = (r^2 \cos^2 t - r^2 \sin^2 t, 2r^2 \cos t \sin t),$$

with $0 \leq t \leq 2\pi$. Using the trigonometric identities

$$\cos 2t = \cos^2 t - \sin^2 t, \qquad \sin 2t = 2 \sin t \cos t,$$

we find for $\beta = F(\alpha)$ the formula

$$\beta(t) = (r^2 \cos 2t, r^2 \sin 2t),$$

with $0 \leq t \leq 2\pi$. This curve takes *two* counterclockwise trips around the circle of radius r^2 (center at origin) (Fig. 1.15).

Thus the effect of F is to wrap the plane $\mathbf{E}^2$ smoothly around itself twice—leaving the origin fixed, since $F(0, 0) = (0, 0)$. In this process, each circle of radius r is wrapped twice around the circle of radius r^2.

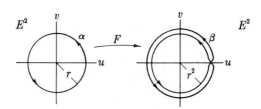

FIG. 1.15

Each time we have defined a new object in this chapter we have proceeded to define a suitable notion of derivative of that object. For example, the "derivative" of a curve α is its velocity α'. Using the notion of velocity of a curve, we shall now define the derivative F_* of a mapping F: $\mathbf{E}^n \to \mathbf{E}^m$. F_* is going to be a function that assigns to each tangent vector $\mathbf{v}$ to $\mathbf{E}^n$ at $\mathbf{p}$ a tangent vector $F_*(\mathbf{v})$ to $\mathbf{E}^m$ at $F(\mathbf{p})$. We get $F_*(\mathbf{v})$ by the following process: The tangent vector $\mathbf{v}$ is the initial velocity of the curve $\alpha(t) = \mathbf{p} + t\mathbf{v}$, where by Remark 3.5 we are consistently abbreviating $\mathbf{v}_p$ to simply $\mathbf{v}$. Now the image of α under the mapping F is the curve β such that

$$\beta(t) = F(\alpha(t)) = F(\mathbf{p} + t\mathbf{v}).$$

We define $F_*(\mathbf{v})$ to be the initial velocity $\beta'(0)$ of β (Fig. 1.16).

Summarizing this process, we obtain the following definition.

7.4 Definition Let $F: \mathbf{E}^n \to \mathbf{E}^m$ be a mapping. If $\mathbf{v}$ is a tangent vector to $\mathbf{E}^n$ at $\mathbf{p}$, let $F_*(\mathbf{v})$ be the initial velocity of the curve $t \to F(\mathbf{p} + t\mathbf{v})$ in $\mathbf{E}^m$. The resulting function F_* (from tangent vectors of $\mathbf{E}^n$ to tangent vectors of $\mathbf{E}^m$) is called the *derivative map* F_* of F.

Note that the initial position $t = 0$ of the curve $t \to F(\mathbf{p} + t\mathbf{v})$ is $F(\mathbf{p})$. Thus by Definition 4.3, the point of application of its initial velocity is $F(\mathbf{p})$. It follows from the definition, then, that F_* transforms a tangent vector to $\mathbf{E}^n$ at $\mathbf{p}$ into a tangent vector to $\mathbf{E}^m$ *at* $F(\mathbf{p})$.

For example, let us compute the derivative map of the mapping

$$F(u,v) = (u^2 - v^2, 2uv)$$

in (2) of Example 7.3. For a tangent vector $\mathbf{v}$ at $\mathbf{p}$, we have

$$\mathbf{p} + t\mathbf{v} = (p_1 + tv_1, p_2 + tv_2);$$

thus

$$F(\mathbf{p} + t\mathbf{v}) = ((p_1 + tv_1)^2 - (p_2 + tv_2)^2, 2(p_1 + tv_1)(p_2 + tv_2)).$$

As t varies, this formula describes that curve in $\mathbf{E}^2$ which, by definition,

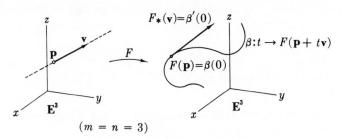

$$F_*(\mathbf{v}) = \beta'(0)$$

$$\beta : t \to F(\mathbf{p} + t\mathbf{v})$$

$$F(\mathbf{p}) = \beta(0)$$

$$(m = n = 3)$$

FIG. 1.16

has initial velocity $F_*(\mathbf{v})$. Differentiating the coordinates above with respect to t (Definition 4.3), we obtain.

$$F_*(\mathbf{v}) = F(\mathbf{p} + t\mathbf{v})'(0) = 2(p_1v_1 - p_2v_2, v_1p_2 + p_1v_2) \qquad \text{at } F(\mathbf{p}).$$

7.5 Theorem Let $F = (f_1, f_2, \cdots, f_m)$ be a mapping from $\mathbf{E}^n$ to $\mathbf{E}^m$. If $\mathbf{v}$ is a tangent vector to $\mathbf{E}^n$ at $\mathbf{p}$, then

$$F_*(\mathbf{v}) = (\mathbf{v}[f_1], \cdots, \mathbf{v}[f_m]) \qquad \text{at } F(\mathbf{p}).$$

Thus $F_*(\mathbf{v})$ *is determined by the derivatives* $\mathbf{v}[f_i]$ *of the coordinate functions of F with respect to* $\mathbf{v}$.

Proof. For the sake of concreteness we take $m = 3$. Given $\mathbf{v}$ at $\mathbf{p}$, we refer to the definition (7.4) of F_* and let β be the curve

$$\beta(t) = F(\mathbf{p} + t\mathbf{v}) = (f_1(\mathbf{p} + t\mathbf{v}), f_2(\mathbf{p} + t\mathbf{v}), f_3(\mathbf{p} + t\mathbf{v})).$$

By definition, $\beta'(0) = F_*(\mathbf{v})$. According to Definition 4.3, to get the velocity vector $\beta'(0)$ we must take the derivatives at $t = 0$ of the coordinate functions $f_i(\mathbf{p} + t\mathbf{v})$ of β. But $(d/dt)(f_i(\mathbf{p} + t\mathbf{v}))\,|_{t=0}$ is precisely $\mathbf{v}[f_i]$, where as usual the point of application $\mathbf{p}$ is now omitted from the notation. Thus

$$F_*(\mathbf{v}) = (\mathbf{v}[f_1], \mathbf{v}[f_2], \mathbf{v}[f_3])_{\beta(0)}.$$

But by definition of β,

$$\beta(0) = F(\mathbf{p}). \qquad\blacksquare$$

Fix a particular point $\mathbf{p}$ in $\mathbf{E}^n$. As noted above, each tangent vector $\mathbf{v}$ to $\mathbf{E}^n$ at $\mathbf{p}$ is transformed by F_* into a tangent vector $F_*(v)$ to $\mathbf{E}^m$ at $F(\mathbf{p})$. Thus for each point $\mathbf{p}$ in $\mathbf{E}^n$, the derivative map F_* gives rise to a function

$$F_{*p} \colon T_p(\mathbf{E}^n) \to T_{F(p)}(\mathbf{E}^m)$$

which we call derivative map of F *at* $\mathbf{p}$. Compare the corresponding situation in elementary calculus where a differentiable function $f\colon \mathbf{R} \to \mathbf{R}$ has a derivative function $f'\colon \mathbf{R} \to \mathbf{R}$ which at each point t of $\mathbf{R}$ gives the derivative $f'(t)$ of f at t.

The links between calculus and linear algebra are tighter than one might expect from a conventional calculus course. A most significant link is provided by

7.6 Corollary Let F be a mapping from $\mathbf{E}^n$ to $\mathbf{E}^m$. Then at each point $\mathbf{p}$ of $\mathbf{E}^n$, the derivative map $F_{*p}\colon T_p(\mathbf{E}^n) \to T_{F(p)}(\mathbf{E}^m)$ is a linear transformation.

Proof. If $\mathbf{v}$ and $\mathbf{w}$ are tangent vectors at $\mathbf{p}$, and a and b are numbers, we must show that

$$F_*(a\mathbf{v} + b\mathbf{w}) = aF_*(\mathbf{v}) + bF_*(\mathbf{w}).$$

Using the first assertion in Theorem 3.3, this follows easily from the preceding theorem. ∎

The linearity of F_{*p} is a generalization of the fact that the derivative $f'(t)$ of $f\colon \mathbf{R} \to \mathbf{R}$ is the slope of the tangent *line* to the graph of f at t. Indeed *for each point* $\mathbf{p}$, F_{*p} *is the linear transformation which best approximates the behavior of F near* $\mathbf{p}$. This idea is fully developed in advanced calculus, where it is used to prove Theorem 7.10.

Since $F_{*p}\colon T_p(\mathbf{E}^n) \to T_{F(p)}(\mathbf{E}^m)$ is a linear transformation, it is reasonable to compute its matrix with respect to the natural bases

$U_1(\mathbf{p}),\ \cdots,\ U_n(\mathbf{p}) \quad \text{for } T_p(\mathbf{E}^n)$

$$U_1(F(\mathbf{p})),\ \cdots,\ U_m(F(\mathbf{p})) \quad \text{for } T_{F(p)}(\mathbf{E}^m).$$

This matrix is called the *Jacobian matrix* of F at $\mathbf{p}$.

7.7 Corollary If $F = (f_1,\ \cdots,\ f_m)$ is a mapping from $\mathbf{E}^n$ to $\mathbf{E}^m$, then

$$F_*(U_j(\mathbf{p})) = \sum_{i=1}^{m} \frac{\partial f_i}{\partial x_j}(\mathbf{p})\, U_i(F(\mathbf{p})) \qquad (1 \le j \le n).$$

Hence the Jacobian matrix of F at $\mathbf{p}$ is $((\partial f_i/\partial x_j)(\mathbf{p}))_{1 \le i \le m,\, 1 \le j \le n}$.

Proof. Set $\mathbf{v} = U_j(\mathbf{p})$ in Corollary 7.6. Since the unit vector $U_j(\mathbf{p})$ applied to f_i is just $(\partial f_i/\partial x_j)(\mathbf{p})$, we get

$$F_*(U_j(\mathbf{p})) = \left(\frac{\partial f_1}{\partial x_j}(\mathbf{p}),\ \cdots,\ \frac{\partial f_m}{\partial x_j}(\mathbf{p}) \right) = \sum_{i=1}^{m} \frac{\partial f_i}{\partial x_j}(\mathbf{p})\, U_i(F(\mathbf{p})). \quad \blacksquare$$

Standard abbreviation:

$$F_*(U_j) = \sum_i \frac{\partial f_i}{\partial x_j}\, \bar{U}_i,$$

where U_j and $\partial f_i/\partial x_j$ are evaluated at $\mathbf{p}$, and $\bar{U}_j$ is evaluated at $F(\mathbf{p})$. This result shows that the derivative map of F is completely determined by the partial derivatives of its coordinate functions. For example, consider the second mapping in Example 7.3. Its coordinate functions are $f = u^2 - v^2$ and $g = 2uv$. Hence

$$\begin{pmatrix} \dfrac{\partial f}{\partial u} & \dfrac{\partial f}{\partial v} \\[2mm] \dfrac{\partial g}{\partial u} & \dfrac{\partial g}{\partial v} \end{pmatrix} = \begin{pmatrix} 2u & -2v \\ 2v & 2u \end{pmatrix}.$$

Thus the Jacobian matrix of this mapping at the point $\mathbf{p} = (p_1,\ p_2)$ is

$$\begin{pmatrix} 2p_1 & -2p_2 \\ 2p_2 & 2p_1 \end{pmatrix}.$$

7.8 Theorem   Let $F: \mathbf{E}^n \to \mathbf{E}^m$ be a mapping. If $\beta = F(\alpha)$ is the image in $\mathbf{E}^m$ of the curve α in $\mathbf{E}^n$, then $\beta' = F_*(\alpha')$.

This theorem asserts that F_* *preserves velocities* of curves, since for each t, the velocity $\beta'(t)$ of the image curve is the image, under F_*, of the velocity $\alpha'(t)$ of α.

Proof. For definiteness, set $m = 3$. Now if $F = (f_1, f_2, f_3)$, then

$$\beta = F(\alpha) = (f_1(\alpha), f_2(\alpha), f_3(\alpha)).$$

Thus the coordinate functions of β are $\beta_i = f_i(\alpha)$. By Theorem 7.5,

$$F_*(\alpha'(t)) = (\alpha'(t)[f_1], \quad \alpha'(t)[f_2], \quad \alpha'(t)[f_3]).$$

But applying Lemma 4.6, we find that

$$\alpha'(t)[f_i] = \frac{d(f_i(\alpha))}{dt}(t) = \frac{d\beta_i}{dt}(t).$$

Hence

$$F_*(\alpha'(t)) = \left(\frac{d\beta_1}{dt}(t), \frac{d\beta_2}{dt}(t), \frac{d\beta_3}{dt}(t)\right).$$

Furthermore, this tangent vector has point of application $F(\alpha(t)) = \beta(t)$; hence it is precisely $\beta'(t)$.   ∎

Just as one uses the derivative of a function $f: \mathbf{R} \to \mathbf{R}$ to gain information about the function f, one can use the derivative map F_* in the study of a mapping F. A detailed investigation of this matter belongs in advanced calculus; we shall give only one or two basic definitions needed in later work.

7.9 Definition   A mapping $F: \mathbf{E}^n \to \mathbf{E}^m$ is *regular* provided that for each point $\mathbf{p}$ of $\mathbf{E}^n$ the derivative map F_{*p} is one-to-one.

Since each F_{*p} is a linear transformation, we can apply standard results of linear algebra to conclude that the following conditions are equivalent:

(1) F_{*p} is one-to-one.

(2) If $F_*(\mathbf{v}_p) = 0$, then $\mathbf{v}_p = 0$.

(3) The Jacobian matrix of F at $\mathbf{p}$ has rank n (dimension of the domain $\mathbf{E}^n$ of F).

For example, the second mapping in Example 7.3 is not regular. But the one-to-one condition fails at only a single point, the origin. In fact, the computation immediately preceding Theorem 7.8 shows that its Jacobian matrix has rank 2 at $\mathbf{p} \neq 0$, rank 0 at $\mathbf{0}$.

A mapping that has an inverse mapping is called a *diffeomorphism*. A diffeomorphism is thus necessarily both one-to-one and onto, but a mapping which is one-to-one and onto need not be a diffeomorphism (Exercise 11). The results of this section apply equally well to mappings defined only on open sets of $\mathbf{E}^n$. In particular, we may speak of a diffeomorphism from one open set of $\mathbf{E}^n$ to another.

We state, without proof, one of the basic results of advanced calculus.

7.10 Theorem   Let $F: \mathbf{E}^n \to \mathbf{E}^n$ be a mapping such that F_{*p} is one-to-one at some point $\mathbf{p}$. Then there is an open set $\mathcal{U}$ containing $\mathbf{p}$ such that the restriction of F to $\mathcal{U}$ is a diffeomorphism $\mathcal{U} \to \mathcal{V}$ onto an open set $\mathcal{V}$.

This is called the *inverse function theorem*, because it asserts that the restricted mapping $\mathcal{U} \to \mathcal{V}$ has an inverse mapping $\mathcal{V} \to \mathcal{U}$. The proof is based on the idea that at points $\mathbf{p} + \Delta\mathbf{p}$ very near $\mathbf{p}$, $F(\mathbf{p} + \Delta\mathbf{p})$ is approximately $F(\mathbf{p}) + F_*(\Delta\mathbf{p})$. Since the tangent spaces at $\mathbf{p}$ and $F(\mathbf{p})$ have the same dimension, it follows that the one-to-one linear transformation F_{*p} has an inverse; hence so does F—near $\mathbf{p}$.

EXERCISES

1. If F is the mapping $F = (u^2 - v^2, 2uv)$ in Example 7.3, find all points $\mathbf{p}$ such that
   (a) $F(\mathbf{p}) = (0,0)$. (b) $F(\mathbf{p}) = (8, -6)$. (c) $F(\mathbf{p}) = \mathbf{p}$.

2. The mapping F in Exercise 1 carries the horizontal line $v = 1$ to the parabola $u \to F(u,1) = (u^2 - 1, 2u)$. Sketch the lines $u = 1$ and $v = 1$, and their images under F.

3. The image $F(S)$ of a set S under a mapping F consists of all points $F(\mathbf{p})$ with $\mathbf{p}$ in S. For F as in Ex. 1, find the image of each of the following sets:
   (a) The horizontal strip $S: 1 \leq v \leq 2$.
   (b) The half-disc $S: u^2 + v^2 \leq 1, v \geq 0$.
   (c) The wedge $S: -u \leq v \leq u, u \geq 0$.
   In each case, show the set S and its image $F(S)$ on a single sketch. (*Hint:* Begin by finding the image of the boundary curves of S.)

4. (a) Show that the derivative map of the mapping (1) in Example 7.3 is given by

   $$F_*(\mathbf{v}_p) = (v_1 - v_2, v_1 + v_2, 2v_3)_{F(p)}.$$

   (*Hint:* Work directly from the definition of derivative map.)
   (b) In general, if $F: \mathbf{E}^n \to \mathbf{E}^m$ is a linear transformation, prove that

   $$F_*(\mathbf{v}_p) = F(\mathbf{v})_{F(p)}.$$

5. If $F = (f_1, \cdots, f_m)$ is a mapping from E^n to E^m, we write

$$F_* = (df_1, \cdots, df_m),$$

since by Theorem 7.5,

$$F_*(\mathbf{v}_p) = (df_1(\mathbf{v}_p), \cdots, df_m(\mathbf{v}_p))_{F(p)}.$$

Find F_* for the mapping $F = (x \cos y, x \sin y, z)$ from $\mathbf{E}^3$ to $\mathbf{E}^3$, and compute $F_*(\mathbf{v}_p)$ if
(a) $\mathbf{v} = (2, -1, 3)$, $\mathbf{p} = (0, 0, 0)$.
(b) $\mathbf{v} = (2, -1, 3)$, $\mathbf{p} = (2, \pi/2, \pi)$.

6. Is the mapping in the preceding exercise regular?

7. Let $F = (f_1, f_2)$ and $G = (g_1, g_2)$ be mappings from $\mathbf{E}^2$ to $\mathbf{E}^2$. Compute the Euclidean coordinate functions of the composite function GF: $\mathbf{E}^2 \to \mathbf{E}^2$ and show that it is a mapping.

8. In the definition (7.4) of $F_*(\mathbf{v}_p)$, show that the straight line may be replaced by any curve α with initial velocity $\mathbf{v}_p$.

9. Prove that a mapping $F: \mathbf{E}^n \to \mathbf{E}^m$ preserves directional derivatives in this sense: If $\mathbf{v}_p$ is a tangent vector to $\mathbf{E}^n$ and g is a differentiable function on $\mathbf{E}^m$, then $F_*(\mathbf{v}_p)[g] = \mathbf{v}_p[g(F)]$.

10. Let $F = (f_1, f_2)$ be a mapping from $\mathbf{E}^2$ to $\mathbf{E}^2$. If for *every* point $\mathbf{q}$ of $\mathbf{E}^2$ the equations

$$\begin{cases} q_1 = f_1(p_1, p_2) \\ q_2 = f_2(p_1, p_2) \end{cases}$$

have a *unique* solution

$$\begin{cases} p_1 = g_1(q_1, q_2) \\ p_2 = g_2(q_1, q_2) \end{cases}$$

prove that F is one-to-one and onto, and that $F^{-1} = (g_1, g_2)$.

11. (*Continuation*). In each case, show that F is one-to-one and onto, compute the inverse function F^{-1}, and decide whether F is a diffeomorphism (that is, whether F^{-1} is differentiable).
(a) $F = (ve^u, u)$.
(b) $F = (u^3, v - u)$.
(c) $F = (1 + 2u - 2v, 4 - 2u + v)$.

12. Let $F: \mathbf{E}^n \to \mathbf{E}^m$ and $G: \mathbf{E}^m \to \mathbf{E}^p$ be mappings.
(a) Generalize the results of Exercise 7 to this case.
(b) If α is a curve in $\mathbf{E}^n$, show that $(GF)_*(\alpha') = G_*(F_*(\alpha'))$. [*Hint*: $(GF)(\alpha) = G(F(\alpha))$.]

(c) Deduce that $(GF)_* = G_*F_*$: The derivative map of a composition of mappings is the composition of their derivative maps.

13. If $f: \mathbf{R} \to \mathbf{R}$ is a differentiable real-valued function on the real line $\mathbf{R}$, prove that $f_*(\mathbf{v}_p)$ is the tangent vector $f'(\mathbf{p})\,\mathbf{v}$ at the point $f(\mathbf{p})$.

8 Summary

Starting from the familiar notion of real-valued functions, and using linear algebra at every stage, we have constructed a variety of mathematical objects. The basic notion of tangent vector led to vector fields, which dualized to 1-forms—which in turn led to arbitrary differential forms. The notions of curve and differentiable function were generalized to that of a mapping $F: \mathbf{E}^n \to \mathbf{E}^m$.

Then starting from the usual notion of the derivative of a real-valued function, we proceeded to construct appropriate differentiation operations for these objects: the directional derivative of a function, the exterior derivative of a form, the velocity of a curve, the derivative map of a mapping. These operations all reduced to (ordinary or partial) derivatives of real-valued coordinate functions, but it is noteworthy that in most cases the *definitions* of these operations did not involve coordinates. (This could be achieved in all cases.) Generally speaking, the differentiation operations all exhibited in one form or another the characteristic linear and Leibnizian properties of ordinary differentiation.

Most of these concepts are probably already familiar to the reader, at least in special cases. But we now have careful definitions and a catalogue of basic properties which will enable us to begin our exploration of differential geometry.

CHAPTER II

Frame Fields

Roughly speaking, geometry begins with the measurement of distances and angles. We shall see that the geometry of Euclidean space can be derived from the *dot product*, the natural inner product on Euclidean space.

Much of this chapter is devoted to the geometry of curves in $\mathbf{E}^3$. We emphasize this topic not only because of its intrinsic importance, but also because the basic method used to investigate curves has proved effective throughout differential geometry. A curve in $\mathbf{E}^3$ is studied by assigning at each point a certain *frame*—that is, set of three orthogonal unit vectors. The rate of change of these vectors along the curve is then expressed in terms of the vectors themselves by the celebrated *Frenet formulas* (Theorem 3.2). In a real sense the theory of curves in $\mathbf{E}^3$ is merely a corollary of these fundamental formulas.

Later on we shall use this "method of moving frames" to study a *surface* in $\mathbf{E}^3$. The general idea is to think of a surface as a kind of two-dimensional curve and follow the Frenet approach as closely as possible. To carry out this scheme we shall need the generalization (Theorem 7.2) of the Frenet formulas devised by E. Cartan. It was Cartan who, at the beginning of this century, first realized the full power of this method not only in differential geometry but also in a variety of related fields.

1 Dot Product

We begin by reviewing some basic facts about the natural inner product on the vector space $\mathbf{E}^3$.

1.1 Definition The dot product of points $\mathbf{p} = (p_1, p_2, p_3)$ and $\mathbf{q} = (q_1, q_2, q_3)$ in $\mathbf{E}^3$ is the number

$$\mathbf{p \cdot q} = p_1 q_1 + p_2 q_2 + p_3 q_3.$$

The dot product *is* an inner product, that is, it has three properties
(1) Bilinearity:

$$(a\mathbf{p} + b\mathbf{q})\cdot\mathbf{r} = a\mathbf{p}\cdot\mathbf{r} + b\mathbf{q}\cdot\mathbf{r}$$

$$\mathbf{r}\cdot(a\mathbf{p} + b\mathbf{q}) = a\mathbf{r}\cdot\mathbf{p} + b\mathbf{r}\cdot\mathbf{q}.$$

(2) Symmetry: $\mathbf{p}\cdot\mathbf{q} = \mathbf{q}\cdot\mathbf{p}$.
(3) Positive definiteness: $\mathbf{p}\cdot\mathbf{p} \geq 0$, and $\mathbf{p}\cdot\mathbf{p} = 0$ if and only if $\mathbf{p} = 0$.
(Here $\mathbf{p}$, $\mathbf{q}$, and $\mathbf{r}$ are arbitrary points of $\mathbf{E}^3$, and a and b are numbers.)

The *norm* of a point $\mathbf{p} = (p_1, p_2, p_3)$ is the number

$$\| \mathbf{p} \| = (\mathbf{p}\cdot\mathbf{p})^{1/2} = (p_1^2 + p_2^2 + p_3^2)^{1/2}.$$

The norm is thus a real-valued function on $\mathbf{E}^3$; it has the fundamental properties $\| \mathbf{p} + \mathbf{q} \| \leq \| \mathbf{p} \| + \| \mathbf{q} \|$ and $\| a\mathbf{p} \| = |a| \| \mathbf{p} \|$, where $|a|$ is the absolute value of the number a.

In terms of the norm we get a compact version of the usual distance formula in $\mathbf{E}^3$.

1.2 Definition   If $\mathbf{p}$ and $\mathbf{q}$ are points of $\mathbf{E}^3$, the *Euclidean distance* from $\mathbf{p}$ to $\mathbf{q}$ is the number

$$\mathrm{d}(\mathbf{p}, \mathbf{q}) = \| \mathbf{p} - \mathbf{q} \|.$$

In fact, since

$$\mathbf{p} - \mathbf{q} = (p_1 - q_1, p_2 - q_2, p_3 - q_3),$$

expansion of the norm gives the well-known formula (Fig. 2.1)

$$d(\mathbf{p}, \mathbf{q}) = ((p_1 - q_1)^2 + (p_2 - q_2)^2 + (p_3 - q_3)^2)^{1/2}.$$

Euclidean distance may be used to give a more precise definition of open sets (Chapter 1, Section 1). First, if $\mathbf{p}$ is a point of $\mathbf{E}^3$ and $\epsilon > 0$ is a number, the ϵ-*neighborhood* $\mathfrak{N}_\epsilon$ of $\mathbf{p}$ in $\mathbf{E}^3$ is the set of all points $\mathbf{q}$ of $\mathbf{E}^3$ such that $d(\mathbf{p}, \mathbf{q}) < \epsilon$. Then a subset $\mathcal{O}$ of $\mathbf{E}^3$ is *open* provided that each point of $\mathcal{O}$ has an ϵ-neighborhood which is entirely contained in $\mathcal{O}$. In short, all points near enough to a point of an open set are also in the set. This definition is

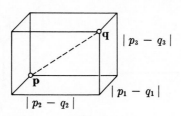

FIG. 2.1

valid with $\mathbf{E}^3$ replaced by $\mathbf{E}^n$—or indeed any set furnished with a reasonable distance function.

We saw in Chapter I that for each point $\mathbf{p}$ of $\mathbf{E}^3$ there is a *canonical isomorphism* $\mathbf{v} \to \mathbf{v}_p$ from $\mathbf{E}^3$ onto the tangent space $T_p(\mathbf{E}^3)$ at $\mathbf{p}$. These isomorphisms lie at the heart of Euclidean geometry—using them, the dot product on $\mathbf{E}^3$ itself may be transferred to each of its tangent spaces.

1.3 Definition The dot product of tangent vectors $\mathbf{v}_p$ and $\mathbf{w}_p$ at the same point of $\mathbf{E}^3$ is the number $\mathbf{v}_p \cdot \mathbf{w}_p = \mathbf{v} \cdot \mathbf{w}$.

For example, $(1, 0, -1)_p \cdot (3, -3, 7)_p = 1(3) + 0(-3) + (-1)7 = -4$. Evidently this definition provides a dot product on each tangent space $T_p(\mathbf{E}^3)$ with the same properties as the original dot product on $\mathbf{E}^3$. In particular, each tangent vector $\mathbf{v}_p$ to $\mathbf{E}^3$ has *norm* (or *length*) $\| \mathbf{v}_p \| = \| \mathbf{v} \|$.

A fundamental result of linear algebra is the Schwarz inequality $| \mathbf{v} \cdot \mathbf{w} | \leq \| \mathbf{v} \| \, \| \mathbf{w} \|$. This permits us to define the cosine of the angle ϑ between $\mathbf{v}$ and $\mathbf{w}$ by the equation (Fig. 2.2).

$$\mathbf{v} \cdot \mathbf{w} = \| \mathbf{v} \| \, \| \mathbf{w} \| \cos \vartheta.$$

Thus the dot product of two vectors is the product of their lengths times the cosine of the angle between them. (The angle ϑ is not uniquely determined unless further restrictions are imposed, say $0 \leq \vartheta \leq \pi$.)

In particular, if $\vartheta = \pi/2$, then $\mathbf{v} \cdot \mathbf{w} = 0$. Thus we shall define two vectors to be *orthogonal* provided their dot product is zero. A vector of length 1 is called a *unit vector*.

1.4 Definition A set $\mathbf{e}_1, \mathbf{e}_2, \mathbf{e}_3$ of three mutually orthogonal unit vectors, tangent to $\mathbf{E}^3$ at $\mathbf{p}$, is called a *frame* at the point $\mathbf{p}$.

Thus $\mathbf{e}_1, \mathbf{e}_2, \mathbf{e}_3$ is a frame if and only if

$$\mathbf{e}_1 \cdot \mathbf{e}_1 = \mathbf{e}_2 \cdot \mathbf{e}_2 = \mathbf{e}_3 \cdot \mathbf{e}_3 = 1$$

$$\mathbf{e}_1 \cdot \mathbf{e}_2 = \mathbf{e}_1 \cdot \mathbf{e}_3 = \mathbf{e}_2 \cdot \mathbf{e}_3 = 0.$$

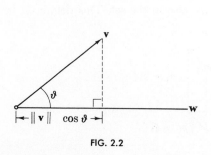

FIG. 2.2

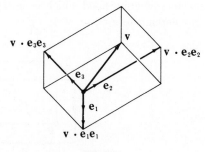

FIG. 2.3

By the symmetry of the dot product, the second row of equations is, of course, the same as

$$\mathbf{e}_2 \cdot \mathbf{e}_1 = \mathbf{e}_3 \cdot \mathbf{e}_1 = \mathbf{e}_3 \cdot \mathbf{e}_2 = 0.$$

Using index notation, all nine equations may be concisely expressed as $\mathbf{e}_i \cdot \mathbf{e}_j = \delta_{ij}$ for $1 \leq i,j \leq 3$, where δ_{ij} is the Kronecker delta (0 if $i \neq j$, 1 if $i = j$). For example, at each point $\mathbf{p}$ of $\mathbf{E}^3$, the vectors $U_1(\mathbf{p})$, $U_2(\mathbf{p})$, $U_3(\mathbf{p})$ of Definition 2.4 in Chapter I constitute a frame at $\mathbf{p}$.

1.5 Theorem Let $\mathbf{e}_1$, $\mathbf{e}_2$, $\mathbf{e}_3$ be a frame at a point $\mathbf{p}$ of $\mathbf{E}^3$. If $\mathbf{v}$ is any tangent vector to $\mathbf{E}^3$ at $\mathbf{p}$, then (Fig. 2.3)

$$\mathbf{v} = (\mathbf{v} \cdot \mathbf{e}_1)\mathbf{e}_1 + (\mathbf{v} \cdot \mathbf{e}_2)\mathbf{e}_2 + (\mathbf{v} \cdot \mathbf{e}_3)\mathbf{e}_3.$$

Proof. First we show that the vectors $\mathbf{e}_1$, $\mathbf{e}_2$, $\mathbf{e}_3$ are linearly independent. Suppose

$$\sum a_i \mathbf{e}_i = 0.$$

Then

$$0 = \left(\sum a_i \mathbf{e}_i\right) \cdot \mathbf{e}_j = \sum a_i \mathbf{e}_i \cdot \mathbf{e}_j = \sum a_i\, \delta_{ij} = a_j,$$

where all sums are over $i = 1, 2, 3$. Thus

$$a_1 = a_2 = a_3 = 0,$$

as required. Now the tangent space $T_p(\mathbf{E}^3)$ has dimension 3, since it is linearly isomorphic to $\mathbf{E}^3$. Thus by a well-known theorem of linear algebra, the three independent vectors $\mathbf{e}_1$, $\mathbf{e}_2$, $\mathbf{e}_3$ form a basis for $T_p(\mathbf{E}^3)$. Hence for each vector $\mathbf{v}$ there are three (unique) numbers c_1, c_2, c_3 such that

$$\mathbf{v} = \sum c_i \mathbf{e}_i.$$

But

$$\mathbf{v} \cdot \mathbf{e}_j = \left(\sum c_i \mathbf{e}_i\right) \cdot \mathbf{e}_j = \sum c_i \delta_{ij} = c_j,$$

and thus

$$\mathbf{v} = \sum (\mathbf{v} \cdot \mathbf{e}_i)\mathbf{e}_i. \qquad\blacksquare$$

This result (valid in any inner-product space) is one of the great labor-saving devices in mathematics. For to find the coordinates of a vector $\mathbf{v}$ with respect to an *arbitrary* basis, one must in general solve a set of nonhomogeneous linear equations, a task which even in dimension 3 is not always entirely trivial. But the theorem shows that to find the coordinates of $\mathbf{v}$ with respect to a frame (that is, an *orthonormal* basis) it suffices merely to compute the three dot products $\mathbf{v} \cdot \mathbf{e}_1$, $\mathbf{v} \cdot \mathbf{e}_2$, $\mathbf{v} \cdot \mathbf{e}_3$. We call this process *orthonormal expansion* of $\mathbf{v}$ in terms of the frame $\mathbf{e}_1$, $\mathbf{e}_2$, $\mathbf{e}_3$.

In the special case of the natural frame $U_1(\mathbf{p})$, $U_2(\mathbf{p})$, $U_3(\mathbf{p})$ the identity

$$\mathbf{v} = (v_1, v_2, v_3) = \sum v_i U_i(\mathbf{p})$$

is an orthonormal expansion, and the dot product is defined in terms of these *Euclidean coordinates* by $\mathbf{v} \cdot \mathbf{w} = \sum v_i w_i$. If we use instead an arbitrary frame $\mathbf{e}_1$, $\mathbf{e}_2$, $\mathbf{e}_3$, then each vector $\mathbf{v}$ has new coordinates $a_i = \mathbf{v} \cdot \mathbf{e}_i$ relative to this frame, but *the dot product is still given by the same simple formula*

$$\mathbf{v} \cdot \mathbf{w} = \sum a_i b_i$$

since

$$\mathbf{v} \cdot \mathbf{w} = \left(\sum a_i \mathbf{e}_i\right) \cdot \left(\sum b_i \mathbf{e}_i\right) = \sum_{i,j} a_i b_j \, \mathbf{e}_i \cdot \mathbf{e}_j$$

$$= \sum_{i,j} a_i b_j \, \delta_{ij} = \sum a_i b_i.$$

When applied to more complicated geometric situations, the advantage of using frames becomes enormous, and this is why they appear so frequently throughout this book.

The notion of frame is very close to that of orthogonal matrix.

1.6 Definition Let $\mathbf{e}_1$, $\mathbf{e}_2$, $\mathbf{e}_3$ be a frame at a point $\mathbf{p}$ of $\mathbf{E}^3$. The 3×3 matrix A whose rows are the Euclidean coordinates of these three vectors is called the *attitude matrix* of the frame.

Explicitly, if

$$\mathbf{e}_1 = (a_{11}, a_{12}, a_{13})_p$$

$$\mathbf{e}_2 = (a_{21}, a_{22}, a_{23})_p$$

$$\mathbf{e}_3 = (a_{31}, a_{32}, a_{33})_p$$

then

$$A = \begin{pmatrix} a_{11} & a_{12} & a_{13} \\ a_{21} & a_{22} & a_{23} \\ a_{31} & a_{32} & a_{33} \end{pmatrix}.$$

Thus A does describe the "attitude" of the frame in $\mathbf{E}^3$, although not its point of application.

Evidently the rows of A are orthonormal, since

$$\sum_k a_{ik} a_{jk} = \mathbf{e}_i \cdot \mathbf{e}_j = \delta_{ij} \qquad \text{for } 1 \leq i,j \leq 3.$$

By definition, this means that A is an *orthogonal* matrix.

In terms of matrix multiplication, these equations may be written

$A \, {}^{t}A = I$, where I is the 3×3 identity matrix, and ${}^{t}A$ is the *transpose* of A:

$$ {}^{t}A = \begin{pmatrix} a_{11} & a_{21} & a_{31} \\ a_{12} & a_{22} & a_{32} \\ a_{13} & a_{23} & a_{33} \end{pmatrix}. $$

It follows, by a standard theorem of linear algebra, that ${}^{t}AA = I$, so that ${}^{t}A = A^{-1}$, the *inverse* of A.

There is another product on $\mathbf{E}^{3}$, closely related to the wedge product of 1-forms, and second in importance only to the dot product. We shall transfer it immediately to each tangent space of $\mathbf{E}^{3}$.

1.7 Definition If $\mathbf{v}$ and $\mathbf{w}$ are tangent vectors to $\mathbf{E}^{3}$ at the same point $\mathbf{p}$, then the *cross product* of $\mathbf{v}$ and $\mathbf{w}$ is the tangent vector

$$ \mathbf{v} \times \mathbf{w} = \begin{vmatrix} U_{1}(\mathbf{p}) & U_{2}(\mathbf{p}) & U_{3}(\mathbf{p}) \\ v_{1} & v_{2} & v_{3} \\ w_{1} & w_{2} & w_{3} \end{vmatrix}. $$

This formal determinant is to be expanded along its first row. For example, if $\mathbf{v} = (1, 0, -1)_{p}$ and $\mathbf{w} = (2, 2, -7)_{p}$, then

$$ \mathbf{v} \times \mathbf{w} = \begin{vmatrix} U_{1}(\mathbf{p}) & U_{2}(\mathbf{p}) & U_{3}(\mathbf{p}) \\ 1 & 0 & -1 \\ 2 & 2 & -7 \end{vmatrix} $$

$$ = 2U_{1}(\mathbf{p}) + 5U_{2}(\mathbf{p}) + 2U_{3}(\mathbf{p}) = (2, 5, 2)_{p}. $$

Using familiar properties of determinants, we see that the cross product $\mathbf{v} \times \mathbf{w}$ is *linear* in $\mathbf{v}$ and in $\mathbf{w}$, and satisfies the *alternation rule*

$$ \mathbf{v} \times \mathbf{w} = -\mathbf{w} \times \mathbf{v} $$

(hence, in particular, $\mathbf{v} \times \mathbf{v} = 0$). The geometric usefulness of the cross product is based mostly on

1.8 Lemma The cross product $\mathbf{v} \times \mathbf{w}$ is orthogonal to both $\mathbf{v}$ and $\mathbf{w}$, and has length such that

$$ \| \mathbf{v} \times \mathbf{w} \|^{2} = \mathbf{v} \cdot \mathbf{v} \, \mathbf{w} \cdot \mathbf{w} - (\mathbf{v} \cdot \mathbf{w})^{2}. $$

Proof. Let $\mathbf{v} \times \mathbf{w} = \sum c_{i} U_{i}(\mathbf{p})$. Then the dot product $\mathbf{v} \cdot (\mathbf{v} \times \mathbf{w})$ is just $\sum v_{i} c_{i}$. But by the definition of cross product, the Euclidean coordinates c_{1}, c_{2}, c_{3} of $\mathbf{v} \times \mathbf{w}$ are such that

$$ \mathbf{v} \cdot (\mathbf{v} \times \mathbf{w}) = \begin{vmatrix} v_{1} & v_{2} & v_{3} \\ v_{1} & v_{2} & v_{3} \\ w_{1} & w_{2} & w_{3} \end{vmatrix}. $$

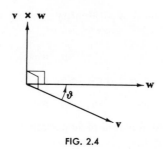

FIG. 2.4

This determinant is zero, since two of its rows are the same; thus $\mathbf{v} \times \mathbf{w}$ is orthogonal to $\mathbf{v}$— and, similarly, to $\mathbf{w}$.

Rather than use tricks to prove the length formula, we give a brute-force computation. Now

$$
\begin{aligned}
\mathbf{v}\cdot\mathbf{v}\ \mathbf{w}\cdot\mathbf{w} - (\mathbf{v}\cdot\mathbf{w})^2 &= \left(\sum v_i^2\right)\left(\sum w_j^2\right) - \left(\sum v_i w_i\right)^2 \\
&= \sum_{i,j} v_i^2 w_j^2 - \left\{\sum v_i^2 w_i^2 + 2\sum_{i<j} v_i w_i v_j w_j\right\} \\
&= \sum_{i\neq j} v_i^2 w_j^2 - 2\sum_{i<j} v_i w_i v_j w_j.
\end{aligned}
$$

On the other hand,

$$
\begin{aligned}
\| \mathbf{v} \times \mathbf{w} \|^2 &= (\mathbf{v} \times \mathbf{w})\cdot(\mathbf{v} \times \mathbf{w}) = \sum c_i^2 \\
&= (v_2 w_3 - v_3 w_2)^2 + (v_3 w_1 - v_1 w_3)^2 + (v_1 w_2 - v_2 w_1)^2
\end{aligned}
$$

and expanding these squares gives the same result as above. ∎

A more intuitive description of the length of a cross product is

$$
\| \mathbf{v} \times \mathbf{w} \| = \| \mathbf{v} \| \ \| \mathbf{w} \| \sin \vartheta,
$$

where $0 \leq \vartheta \leq \pi$ is the smaller of the two angles from $\mathbf{v}$ to $\mathbf{w}$. The direction of $\mathbf{v} \times \mathbf{w}$ on the line orthogonal to $\mathbf{v}$ and $\mathbf{w}$ is given, for practical purposes, by this "right-hand rule": If the fingers of the right hand point in the direction of the shortest rotation of $\mathbf{v}$ to $\mathbf{w}$, then the thumb points in the direction of $\mathbf{v} \times \mathbf{w}$ (Fig. 2.4).

Combining the dot and cross product, we get the *triple scalar product*, which assigns to any three vectors $\mathbf{u}, \mathbf{v}, \mathbf{w}$ the number $\mathbf{u}\cdot\mathbf{v} \times \mathbf{w}$ (Exercise 4). Parentheses are unnecessary: $\mathbf{u}\cdot(\mathbf{v} \times \mathbf{w})$ is the only possible meaning.

EXERCISES

1. Let $\mathbf{v} = (1, 2, -1)$ and $\mathbf{w} = (-1, 0, 3)$ be tangent vectors at a point of E^3. Compute   (a) $\mathbf{v}\cdot\mathbf{w}$.   (b) $\mathbf{v} \times \mathbf{w}$.

(c) $\mathbf{v}/\|\,\mathbf{v}\,\|$, $\mathbf{w}/\|\,\mathbf{w}\,\|$. (d) $\|\,\mathbf{v}\times\mathbf{w}\,\|$.

(e) the cosine of the angle between $\mathbf{v}$ and $\mathbf{w}$.

2. Prove that Euclidean distance has the properties

(a) $d(\mathbf{p},\mathbf{q}) \geqq 0$; $d(\mathbf{p},\mathbf{q}) = 0$ if and only if $\mathbf{p} = \mathbf{q}$,

(b) $d(\mathbf{p},\mathbf{q}) = d(\mathbf{q},\mathbf{p})$,

(c) $d(\mathbf{p},\mathbf{q}) + d(\mathbf{q},\mathbf{r}) \geqq d(\mathbf{p},\mathbf{r})$,

for any points $\mathbf{p}, \mathbf{q}, \mathbf{r}$ in $\mathbf{E}^3$.

3. Prove that the tangent vectors

$$\mathbf{e}_1 = \frac{(1, 2, 1)}{\sqrt{6}}, \qquad \mathbf{e}_2 = \frac{(-2, 0, 2)}{\sqrt{8}}, \qquad \mathbf{e}_3 = \frac{(1, -1, 1)}{\sqrt{3}}$$

constitute a frame. Express $\mathbf{v} = (6, 1, -1)$ as a linear combination of these vectors. (Check the result by direct computation.)

4. Let $\mathbf{u} = (u_1, u_2, u_3)$, $\mathbf{v} = (v_1, v_2, v_3)$, $\mathbf{w} = (w_1, w_2, w_3)$. Prove that

(a) $\mathbf{u}\cdot\mathbf{v}\times\mathbf{w} = \begin{vmatrix} u_1 & u_2 & u_3 \\ v_1 & v_2 & v_3 \\ w_1 & w_2 & w_3 \end{vmatrix}$.

(b) $\mathbf{u}\cdot\mathbf{v}\times\mathbf{w} \neq 0$ if and only if $\mathbf{u}$, $\mathbf{v}$, and $\mathbf{w}$ are linearly independent.

(c) If any two vectors in $\mathbf{u}\cdot\mathbf{v}\times\mathbf{w}$ are reversed, the product changes sign. Explicitly,

$$\mathbf{u}\cdot\mathbf{v}\times\mathbf{w} = \mathbf{v}\cdot\mathbf{w}\times\mathbf{u} = \mathbf{w}\cdot\mathbf{u}\times\mathbf{v}$$

$$= -\mathbf{w}\cdot\mathbf{v}\times\mathbf{u} = -\mathbf{v}\cdot\mathbf{u}\times\mathbf{w} = -\mathbf{u}\cdot\mathbf{w}\times\mathbf{v}.$$

(d) $\mathbf{u}\cdot\mathbf{v}\times\mathbf{w} = \mathbf{u}\times\mathbf{v}\cdot\mathbf{w}$.

5. Prove that $\mathbf{v}\times\mathbf{w} \neq 0$ if and only if $\mathbf{v}$ and $\mathbf{w}$ are linearly independent, and show that $\|\,\mathbf{v}\times\mathbf{w}\,\|$ is the area of the parallelogram with sides $\mathbf{v}$ and $\mathbf{w}$.

6. If $\mathbf{e}_1, \mathbf{e}_2, \mathbf{e}_3$ is a frame, show that

$$\mathbf{e}_1\cdot\mathbf{e}_2\times\mathbf{e}_3 = \pm 1.$$

Deduce that any 3×3 orthogonal matrix has determinant ± 1.

7. If $\mathbf{u}$ is a unit vector, then the *component* of $\mathbf{v}$ in the $\mathbf{u}$ direction is

$$\mathbf{v}\cdot\mathbf{u}\,\mathbf{u} = \|\,\mathbf{v}\,\| \cos\vartheta\,\mathbf{u}.$$

Show that $\mathbf{v}$ has a unique expression $\mathbf{v} = \mathbf{v}_1 + \mathbf{v}_2$, where $\mathbf{v}_1\cdot\mathbf{v}_2 = 0$ and $\mathbf{v}_1$ is the component of $\mathbf{v}$ in the $\mathbf{u}$ direction.

8. Show that the volume of the parallelopiped with sides $\mathbf{u}$, $\mathbf{v}$, $\mathbf{w}$ is $\pm\mathbf{u}\cdot\mathbf{v}\times\mathbf{w}$ (Fig. 2.5). (*Hint:* Use the indicated unit vector $\mathbf{e} = \mathbf{v}\times\mathbf{w}/\|\,\mathbf{v}\times\mathbf{w}\,\|$.)

9. Give rigorous proofs, using ϵ-neighborhoods, that each of the following subsets of $\mathbf{E}^3$ is open:
   (a) All points $\mathbf{p}$ such that $\| \mathbf{p} \| < 1$.
   (b) All $\mathbf{p}$ such that $p_3 > 0$. (*Hint:* $| p_i - q_i | \leqq d(\mathbf{p},\mathbf{q})$.)

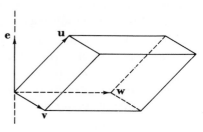

FIG. 2.5

10. In each case, let S be the set of all points $\mathbf{p}$ that satisfy the given condition. Describe S, and decide whether it is *open*.
    (a) $p_1^2 + p_2^2 + p_3^2 = 1$.　　(c) $p_1 = p_2 \neq p_3$.
    (b) $p_3 \neq 0$.　　(d) $p_1^2 + p_2^2 < 9$.

11. If f is a differentiable function on $\mathbf{E}^3$, show that the gradient

$$\nabla f = \sum (\partial f/\partial x_i) U_i$$

(Ex. 8 of I.6) has the following properties:
   (a) $\mathbf{v}[f] = (df)(\mathbf{v}) = \mathbf{v} \cdot (\nabla f)(\mathbf{p})$ for any tangent vector at $\mathbf{p}$.
   (b) The norm $\|(\nabla f)(\mathbf{p})\| = \big[\sum (\partial f/\partial x_i)^2(\mathbf{p}) \big]^{1/2}$ of $(\nabla f)(\mathbf{p})$ is the maximum of the directional derivatives $\mathbf{u}[f]$ for all *unit* vectors at $\mathbf{p}$. Furthermore, if $(\nabla f)(\mathbf{p}) \neq 0$ the unit vector for which the maximum occurs is

$$(\nabla f)(\mathbf{p})/\|(\nabla f)(\mathbf{p})\|.$$

   The notations grad f, curl V, and div V (in the exercise referred to) are often replaced by ∇f, $\nabla \times V$, and $\nabla \cdot V$, respectively.

12. *Angle functions.* Let f and g be differentiable real-valued functions on an interval I. Suppose that $f^2 + g^2 = 1$ and that ϑ_0 is a number such that $f(0) = \cos \vartheta_0$, $g(0) = \sin \vartheta_0$. If ϑ is the function such that

$$\vartheta(t) = \vartheta_0 + \int_0^t (fg' - gf') \, dt,$$

prove that

$$f = \cos \vartheta, \, g = \sin \vartheta.$$

(*Hint:* We want $(f - \cos \vartheta)^2 + (g - \sin \vartheta)^2 = 0$; show that

$$(f \cos \vartheta + g \sin \vartheta)' = 0.)$$

   The point of this exercise is that ϑ is a differentiable function, unambiguously defined on the whole interval I.

2 Curves

We begin the geometric study of curves by reviewing some familiar defini-tions. Let $\alpha \colon I \to \mathbf{E}^3$ be a curve. In Chapter I, Section 4, we defined the velocity vector $\alpha'(t)$ of α at t. Now we define the *speed* of α at t to be the length $v(t) = \| \alpha'(t) \|$ of the velocity vector. Thus speed is a real-valued function on the interval I. In terms of Euclidean coordinates $\alpha = (\alpha_1, \alpha_2, \alpha_3)$, we have

$$\alpha'(t) = \left(\frac{d\alpha_1}{dt}(t), \frac{d\alpha_2}{dt}(t), \frac{d\alpha_3}{dt}(t) \right).$$

Hence the speed function v of α is given by the usual formula

$$v = \| \alpha' \| = \left(\left(\frac{d\alpha_1}{dt} \right)^2 + \left(\frac{d\alpha_2}{dt} \right)^2 + \left(\frac{d\alpha_3}{dt} \right)^2 \right)^{1/2}.$$

In physics, the distance traveled by a moving point is determined by integrating its speed with respect to time. Thus we define the *arc length* of α from $t = a$ to $t = b$ to be the number

$$\int_a^b \| \alpha'(t) \| \, dt.$$

Substituting the formula for $\| \alpha' \|$ given above, we get the usual formula for arc length.

Sometimes one is interested only in the route followed by a curve and not in the particular speed at which it traverses its route. One way to ig-nore the speed of a curve α is to reparametrize to a curve β which has *unit speed* $\| \beta' \| = 1$. Then β represents a "standard trip" along the route of α.

2.1 Theorem If α is a regular curve in $\mathbf{E}^3$, then there exists a reparame-trization β of α such that β has unit speed.

Proof. Fix a number a in the domain I of $\alpha \colon I \to \mathbf{E}^3$, and consider the *arc-length function*

$$s(t) = \int_a^t \| \alpha'(u) \| \, du.$$

(The resulting reparametrization is said to be *based at* $t = a$.) Thus the derivative ds/dt of the function $s = s(t)$ is the speed function $v = \| \alpha' \|$ of α. Since α is regular, by definition α' is never zero; hence $ds/dt > 0$. By a standard theorem of calculus, the function s has an inverse function $t = t(s)$, whose derivative dt/ds at $s = s(t)$ is the reciprocal of ds/dt at $t = t(s)$. In particular, $dt/ds > 0$.

Now let β be the reparametrization $\beta(s) = \alpha(t(s))$ of α. We assert that β has unit speed. In fact, by Lemma 4.5 of Chapter I,

$$\beta'(s) = (dt/ds)(s)\alpha'(t(s)).$$

Hence, by the preceding remarks, the speed of β is

$$\| \beta'(s) \| = \frac{dt}{ds}(s) \| \alpha'(t(s)) \| = \frac{dt}{ds}(s) \frac{ds}{dt}(t(s)) = 1. \qquad \blacksquare$$

We shall use the notation of this proof frequently in later work. The unit-speed curve β is sometimes said to have *arc-length parametrization*, since the arc length of β from $s = a$ to $s = b$ $(a < b)$ is just $b - a$.

For example, consider the helix α in Example 4.2 of Chapter I. Since $\alpha(t) = (a \cos t, a \sin t, bt)$, the velocity α' is given by the formula

$$\alpha'(t) = (-a \sin t, a \cos t, b).$$

Hence

$$\| \alpha'(t) \|^2 = \alpha'(t) \cdot \alpha'(t) = a^2 \sin^2 t + a^2 \cos^2 t + b^2 = a^2 + b^2.$$

Thus α has *constant* speed $c = \| \alpha' \| = (a^2 + b^2)^{1/2}$. If we measure arc length from $t = 0$, then

$$s(t) = \int_0^t c \, dt = ct.$$

Hence, $t(s) = s/c$. Substituting in the formula for α, we get the unit-speed reparametrization

$$\beta(s) = \alpha\left(\frac{s}{c}\right) = \left(a \cos \frac{s}{c}, a \sin \frac{s}{c}, \frac{bs}{c}\right).$$

It is easy to check directly that $\| \beta'(s) \| = 1$ for all s.

A reparametrization $\alpha(h)$ of a curve α is said to be *orientation-preserving* if $h' \geq 0$, *orientation-reversing* if $h' \leq 0$. In the latter case, α and $\alpha(h)$ traverse their common route in opposite directions. By the conventions above, a unit-speed reparametrization is always orientation-preserving since $ds/dt > 0$ for a regular curve α.

We now define a variant of the general notion of vector field (Definition 2.3 of Chapter I) which is adapted to the study of curves. Roughly speaking, a vector field on a curve consists of a vector at each point of the curve.

2.2 Definition A *vector field* Y on a curve $\alpha: I \to \mathbf{E}^3$ is a function that assigns to each number t in I a tangent vector $Y(t)$ to $\mathbf{E}^3$ at the point $\alpha(t)$.

We have already used such vector fields, since for any curve α, its ve-

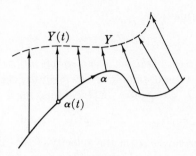

FIG. 2.6

locity α' evidently satisfies this definition. Note that, unlike α', arbitrary vector fields on α need not be tangent to α, but may point in any direction (Fig. 2.6).

The properties of vector fields on curves are analogous to those of vector fields on $\mathbf{E}^3$. For example, if Y is a vector field on $\alpha \colon I \to \mathbf{E}^3$, then for each t in I we can write

$$Y(t) = (y_1(t), y_2(t), y_3(t))_{\alpha(t)} = \sum y_i(t) U_i(\alpha(t)).$$

We have thus defined real-valued functions y_1, y_2, y_3 on I called the *Euclidean coordinate functions* of Y. These will always be assumed to be differentiable. Note that the composite function $t \to U_i(\alpha(t))$ is a vector field on α. Where it seems safe to do so, we shall often write merely U_i instead of $U_i(\alpha(t))$.

The operations of addition, scalar multiplication, dot product, and cross product of vector fields (on the same curve) are all defined in the usual pointwise fashion. Thus if

$$Y(t) = t^2 U_1 - t U_3, \quad Z(t) = (1 - t^2) U_2 + t U_3,$$

and

$$f(t) = \frac{t + 1}{t},$$

we obtain the vector fields

$$(Y + Z)(t) = t^2 U_1 + (1 - t^2) U_2$$

$$(fY)(t) = t(t + 1) U_1 - (t + 1) U_3$$

$$(Y \times Z)(t) = \begin{vmatrix} U_1 & U_2 & U_3 \\ t^2 & 0 & -t \\ 0 & 1 - t^2 & t \end{vmatrix}$$

$$= t(1 - t^2) U_1 - t^3 U_2 + t^2(1 - t^2) U_3$$

and the real-valued function

$$(Y \cdot Z)(t) = -t^2.$$

To differentiate a vector field on α one simply differentiates its Euclidean coordinate functions, thus obtaining a new vector field on α. Explicitly, if $Y = \sum y_i U_i$, then $Y' = \sum (dy_i/dt) U_i$. Thus, for Y as above, we get

$$Y' = 2tU_1 - U_3, \qquad Y'' = 2U_1, \qquad \text{and} \qquad Y''' = 0.$$

In particular, the derivative α'' of the velocity α' of α is called the *acceleration* of α. Thus if $\alpha = (\alpha_1, \alpha_2, \alpha_3)$, the acceleration α'' is the vector field

$$\alpha'' = \left(\frac{d^2\alpha_1}{dt^2}, \frac{d^2\alpha_1}{dt^2}, \frac{d^2\alpha_3}{dt^2} \right)_\alpha$$

on α. By contrast with velocity, acceleration is generally not tangent to the curve.

As we mentioned earlier, in whatever form it appears, differentiation always has suitable linearity and Leibnizian properties. In the case of vector fields on a curve, it is easy to prove the linearity property

$$(aY + bZ)' = aY' + bZ'$$

(a and b numbers) and the Leibnizian properties

$$(fY)' = \frac{df}{dt} Y + fY' \qquad \text{and} \qquad (Y \cdot Z)' = Y' \cdot Z + Y \cdot Z'.$$

If the function $Y \cdot Z$ is constant, the last formula shows that

$$Y' \cdot Z + Y \cdot Z' = 0.$$

This observation will be used frequently in later work. In particular, if Y has constant length $\| Y \|$, then Y and Y' are orthogonal at each point, since $\| Y \|^2 = Y \cdot Y$ constant implies $2Y \cdot Y' = 0$.

Recall that tangent vectors are parallel if they have the same vector parts. We say that a vector field Y on a curve is *parallel* provided all its (tangent vector) values are parallel. In this case, if the common vector part is (c_1, c_2, c_3), then

$$Y(t) = (c_1, c_2, c_3)_{\alpha(t)} = \sum c_i U_i \qquad \text{for all } t.$$

Thus parallelism for a vector field is equivalent to the constancy of its Euclidean coordinate functions.

Vanishing of derivatives is always important in calculus; here are three simple cases.

2.3 Lemma   (1) A curve α is constant if and only if its velocity is zero, $\alpha' = 0$.

(2) A nonconstant curve α is a straight line if and only if its acceleration is zero, $\alpha'' = 0$.

(3) A vector field Y on a curve is parallel if and only if its derivative is zero, $Y' = 0$.

Proof. In each case it suffices to look at the Euclidean coordinate functions. For example, we shall prove (2). If $\alpha = (\alpha_1, \alpha_2, \alpha_3)$, then

$$\alpha'' = \left(\frac{d^2\alpha_1}{dt^2}, \frac{d^2\alpha_2}{dt^2}, \frac{d^2\alpha_3}{dt^2}\right)$$

Thus $\alpha'' = 0$ if and only if each $d^2\alpha_i/dt^2 = 0$. By elementary calculus, this is equivalent to the existence of constants p_i and q_i such that

$$\alpha_i(t) = p_i + tq_i, \quad \text{for} \quad i = 1, 2, 3.$$

Thus $\alpha(t) = \mathbf{p} + t\mathbf{q}$, and α is a straight line as defined in Example 4.2 of Chapter I. (Note that nonconstancy implies $\mathbf{q} \neq 0$.) ∎

EXERCISES

1. For the curve $\alpha(t) = (2t, t^2, t^3/3)$,
   (a) find the velocity, speed, and acceleration for arbitrary t, and at $t = 1$;
   (b) find the arc-length function $s = s(t)$ (based at $t = 0$), and determine the arc length of α from $t = -1$ to $t = +1$.

2. Show that the curve $\alpha(t) = (t\cos t, t\sin t, t)$ lies on a cone in $\mathbf{E}^3$. Find the velocity, speed, and acceleration of α at the vertex of the cone.

3. Show that the curve $\alpha(t) = (\cosh t, \sinh t, t)$ has arc-length function $s(t) = \sqrt{2} \sinh t$, and find a unit-speed reparametrization of α.

4. Consider the curve $\alpha(t) = (2t, t^2, \log t)$ on $I: t > 0$. Show that this curve passes through the points $\mathbf{p} = (2,1,0)$ and $\mathbf{q} = (4, 4, \log 2)$, and find its arc length between these points.

5. Suppose that β_1 and β_2 are unit-speed reparametrizations of the same curve α. Show that there is a number s_0 such that $\beta_2(s) = \beta_1(s + s_0)$ for all s. What is the geometric significance of s_0?

6. Let Y be a vector field on the helix $\alpha(t) = (\cos t, \sin t, t)$. In each of the following cases, express Y in the form $\sum y_i U_i$:
   (a) $Y(t)$ is the vector from $\alpha(t)$ to the origin of $\mathbf{E}^3$.
   (b) $Y(t) = \alpha'(t) - \alpha''(t)$.
   (c) $Y(t)$ has unit length and is orthogonal to both $\alpha'(t)$ and $\alpha''(t)$.
   (d) $Y(t)$ is the vector from $\alpha(t)$ to $\alpha(t + \pi)$.

7. Let Y be a vector field on a curve α. If $\alpha(h)$ is a reparametrization of α, show that $Y(h)$ is a vector field on $\alpha(h)$, and prove the chain rule $Y(h)' = h' \, Y'(h)$.

8. Let $\alpha, \beta : I \to \mathbf{E}^3$ be curves such that $\alpha'(t)$ and $\beta'(t)$ are parallel (same Euclidean coordinates) for each t. Prove that α and β are *parallel* in the sense that there is a point $\mathbf{p}$ in $\mathbf{E}^3$ such that $\beta(t) = \alpha(t) + \mathbf{p}$ for all t.

9. If α is a regular curve show that
   (a) α has constant speed if and only if the acceleration α'' is always orthogonal to α (that is, to α').
   (b) α is a reparametrization of a straight line $t \to \mathbf{p} + t\mathbf{q}$ if and only if α'' is always tangent to α (that is, α'' and α' are collinear).

10. A portion of a curve defined on a closed interval $[a,b]: a \leq t \leq b$, is called a *curve segment*. A reparametrization $\alpha(h): [a,b] \to \mathbf{E}^3$ of a curve segment $\alpha: [c,d] \to \mathbf{E}^3$ is *monotone* provided either
    (a) $h' \geq 0$, $h(a) = c$, $h(b) = d$, or (b) $h' \leq 0$, $h(a) = d$, $h(b) = c$.
    Prove that monotone reparametrization does not change arc length.

11. Prove that a straight line is the shortest distance between two points in $\mathbf{E}^3$. Use the following scheme; let $\alpha: [a,b] \to \mathbf{E}^3$ be an arbitrary curve segment from $\mathbf{p} = \alpha(a)$ to $\mathbf{q} = \alpha(b)$. Let $\mathbf{u} = (\mathbf{q} - \mathbf{p})/\| \mathbf{q} - \mathbf{p} \|$.
    (a) If σ is a straight-line segment from $\mathbf{p}$ to $\mathbf{q}$, say

    $$\sigma(t) = (1 - t)\mathbf{p} + t\mathbf{q} \quad (0 \leq t \leq 1),$$

    show that $L(\sigma) = d(\mathbf{p}, \mathbf{q})$.
    (b) From $\| \alpha' \| \geq \alpha' \cdot \mathbf{u}$, deduce $L(\alpha) \geq d(\mathbf{p}, \mathbf{q})$, where $L(\alpha)$ is the length of α and d is Euclidean distance.
    (c) Furthermore, show that if $L(\alpha) = d(\mathbf{p}, \mathbf{q})$, then (but for parametrization) α is a straight line segment. (*Hint*: write $\alpha' = (\alpha' \cdot \mathbf{u})\mathbf{u} + Y$, where $Y \cdot \mathbf{u} = 0$.)

3 The Frenet Formulas

We now derive mathematical measurements of the turning and twisting of a curve in $\mathbf{E}^3$. Throughout this section we deal only with *unit-speed* curves; in the next we extend the results to arbitrary regular curves.

Let $\beta: I \to \mathbf{E}^3$ be a unit-speed curve, so $\| \beta'(s) \| = 1$ for each s in I. Then $T = \beta'$ is called the *unit tangent* vector field on β. Since T has constant length 1, its derivative $T' = \beta''$ measures the way the curve is turning in $\mathbf{E}^3$. We call T' the *curvature* vector field of β. Differentiation of $T \cdot T = 1$ gives $2T' \cdot T = 0$, so T' is always orthogonal to T, that is, *normal* to β.

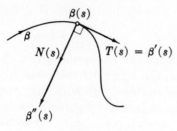

FIG. 2.7

The length of the curvature vector field T' gives a numerical measurement of the turning of β. The real-valued function κ such that $\kappa(s) = \| T'(s) \|$ for all s in I is called the *curvature* function of β. Thus $\kappa \geq 0$, and the larger κ is, the sharper the turning of β.

To carry this analysis further, we impose the restriction that κ is never zero, so $\kappa > 0$.† Then the unit-vector field $N = T'/\kappa$ on β tells the *direction* in which β is turning at each point. N is called the *principal normal* vector field of β (Fig. 2.7). The vector field $B = T \times N$ on β is then called the *binormal* vector field of β.

3.1 Lemma   Let β be a unit-speed curve in $\mathbf{E}^3$ with $\kappa > 0$. Then the three vector fields T, N, and B on β are unit vector fields which are mutually orthogonal at each point. We call T, N, B the *Frenet frame field* on β.

Proof.   By definition $\| T \| = 1$. Since $\kappa = \| T' \| > 0$,

$$\| N \| = (1/\kappa) \| T' \| = 1.$$

We saw above that T and N are orthogonal—that is, $T \cdot N = 0$. Then by applying Lemma 1.8 at each point, we conclude that $\| B \| = 1$, and B is orthogonal to both T and N. ∎

In summary, we have $T = \beta'$, $N = T'/\kappa$, and $B = T \times N$, satisfying $T \cdot T = N \cdot N = B \cdot B = 1$, with all other dot products zero.

The key to the successful study of the geometry of a curve β is to use its Frenet frame field T, N, B whenever possible, instead of the natural frame field U_1, U_2, U_3. For the Frenet frame field of β is full of information about β, whereas the natural frame field contains none at all.

The first and most important use of this idea is to express the *derivatives* T', N', B' *in terms of* T, N, B. Since $T = \beta'$, we have $T' = \beta'' = \kappa N$. Next consider B'. We claim that B' is, at each point, a scalar multiple of N. To prove this, it suffices by orthonormal expansion to show that $B' \cdot B = 0$ and $B' \cdot T = 0$. The former holds since B is a unit vector. To prove the

† For an *arbitrary* unit-speed curve, this means that we must make a separate study of each segment on which $\kappa > 0$; see Exercise 19 of Section 4.

latter, differentiate $B \cdot T = 0$, obtaining $B' \cdot T + B \cdot T' = 0$; then

$$B' \cdot T = -B \cdot T' = -B \cdot \kappa N = 0.$$

Thus we can now define the *torsion* function τ of the curve β to be the real-valued function on the interval I such that $B' = -\tau N$. (The minus sign is traditional.) By contrast with curvature, there is no restriction on the values of τ—it may be positive, negative, or zero at various points of I. (Indeed the sign of τ, at each point, turns out to have an interesting geometric significance.) We shall presently show that τ does measure the torsion, or twisting, of the curve β.

3.2 Theorem (Frenet formulas)   If $\beta \colon I \to \mathbf{E}^3$ is a unit-speed curve with curvature $\kappa > 0$ and torsion τ, then

$$
\begin{aligned}
T' &= & \kappa N & \\
N' &= -\kappa T & & + \tau B. \\
B' &= & -\tau N &
\end{aligned}
$$

Proof. As we saw above, the first and third formulas are essentially just the definitions of curvature and torsion. To prove the second, we use orthonormal expansion to express N' in terms of T, N, B:

$$N' = N' \cdot T \, T + N' \cdot N \, N + N' \cdot B \, B.$$

These coefficients are easily found. Differentiating $N \cdot T = 0$, we get $N' \cdot T + N \cdot T' = 0$; hence

$$N' \cdot T = -N \cdot T' = -N \cdot \kappa N = -\kappa.$$

As usual, $N' \cdot N = 0$, since N is a unit vector field. Finally,

$$N' \cdot B = -N \cdot B' = -N \cdot (-\tau N) = \tau. \qquad \blacksquare$$

3.3 Example   We compute the Frenet frame T, N, B and the curvature and torsion functions of the unit-speed helix

$$\beta(s) = \left(a \cos \frac{s}{c}, a \sin \frac{s}{c}, \frac{bs}{c} \right),$$

where $c = (a^2 + b^2)^{1/2}$ and $a > 0$. Now

$$T(s) = \beta'(s) = \left(-\frac{a}{c} \sin \frac{s}{c}, \frac{a}{c} \cos \frac{s}{c}, \frac{b}{c} \right)$$

Hence

$$T'(s) = \left(-\frac{a}{c^2} \cos \frac{s}{c}, -\frac{a}{c^2} \sin \frac{s}{c}, 0 \right).$$

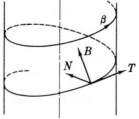

FIG. 2.8   Thus

$$\kappa(s) = \| T'(s) \| = \frac{a}{c^2} = \frac{a}{a^2 + b^2} > 0.$$

Since $T' = \kappa N$, we get

$$N(s) = \left(-\cos \frac{s}{c}, \ -\sin \frac{s}{c}, \ 0 \right)$$

Note that regardless of what values a and b have, N always points straight in toward the axis of the cylinder on which β lies (Fig. 2.8).

Applying the definition of cross product to $B = T \times N$, we get

$$B(s) = \left(\frac{b}{c} \sin \frac{s}{c}, \ -\frac{b}{c} \cos \frac{s}{c}, \ \frac{a}{c} \right).$$

It remains to compute torsion. Now

$$B'(s) = \left(\frac{b}{c^2} \cos \frac{s}{c}, \ \frac{b}{c^2} \sin \frac{s}{c}, \ 0 \right),$$

and by definition, $B' = -\tau N$. Comparing the formulas for B' and N, we conclude that

$$\tau(s) = \frac{b}{c^2} = \frac{b}{a^2 + b^2}.$$

So the torsion of the helix is also constant.

Note that when the parameter b is zero, the helix reduces to a circle of radius a. The curvature of this circle is $\kappa = 1/a$ (so the smaller the radius, the larger the curvature), and the torsion is identically zero.

This example is a very special one—in general (as the examples in the exercises show) neither the curvature nor the torsion functions of a curve need be constant.

3.4 Remark We have emphasized all along the distinction between a tangent vector and a point of $\mathbf{E}^3$. However, Euclidean space has, as we have seen, the remarkable property that given a point $\mathbf{p}$, there is a natural one-to-one correspondence between points (v_1, v_2, v_3) and tangent vectors $(v_1, v_2, v_3)_p$ at $\mathbf{p}$. Thus one can transform points into tangent vectors (and vice versa) by means of this canonical isomorphism. In the next two sections particularly, it will often be convenient to switch quietly from one to the other without change of notation. Since *corresponding objects have the same Euclidean coordinates*, this switching can have no effect on scalar multiplication, addition, dot products, differentiation—or any other operation defined in terms of Euclidean coordinates.

Thus a vector field $Y = (y_1, y_2, y_3)_\beta$ on a curve β becomes itself a curve (y_1, y_2, y_3) in $\mathbf{E}^3$. In particular, if Y is parallel, its Euclidean coordinate functions are constant, so Y is identified with a single point of $\mathbf{E}^3$.

In solid geometry one describes a *plane* in $\mathbf{E}^3$ as being composed of all perpendiculars to a given line at a given point. In vector language then, the *plane through* $\mathbf{p}$ *orthogonal to* $\mathbf{q} \neq 0$ consists of all points $\mathbf{r}$ in $\mathbf{E}^3$ such that $(\mathbf{r} - \mathbf{p}) \cdot \mathbf{q} = 0$. By the remark above, we may picture $\mathbf{q}$ as a tangent vector at $\mathbf{p}$ as shown in Fig. 2.9.

We can now give an informative approximation of a given curve near an arbitrary point on the curve. The goal is to show how curvature and torsion influence the shape of the curve. To derive this approximation we use a Taylor approximation of the curve—and express this in terms of the Frenet frame at the selected point.

For simplicity, we shall consider the unit-speed curve $\beta = (\beta_1, \beta_2, \beta_3)$ near the point $\beta(0)$. For s small, each coordinate $\beta_i(s)$ is closely approximated by the initial terms of its Taylor series:

$$\beta_i(s) \sim \beta_i(0) + \frac{d\beta_i}{ds}(0)\, s + \frac{d^2\beta_i}{ds^2}(0)\, \frac{s^2}{2} + \frac{d^3\beta_i}{ds^3}(0)\, \frac{s^3}{6}.$$

Hence

$$\beta(s) \sim \beta(0) + s\beta'(0) + \frac{s^2}{2}\beta''(0) + \frac{s^3}{6}\beta'''(0).$$

But $\beta'(0) = T_0$, and $\beta''(0) = \kappa_0 N_0$, where the subscript indicates evaluation at $s = 0$, and we assume $\kappa_0 \neq 0$. Now

$$\beta''' = (\kappa N)' = \frac{d\kappa}{ds} N + \kappa N'.$$

Thus by the Frenet formula for N', we get

$$\beta'''(0) = -\kappa_0^2 T_0 + \frac{d\kappa}{ds}(0)\, N_0 + \kappa_0 \tau_0 B_0.$$

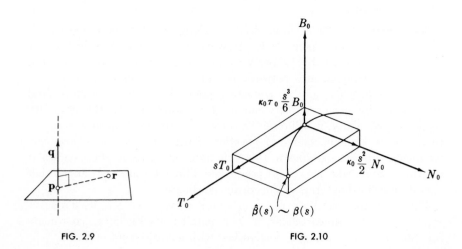

FIG. 2.9   FIG. 2.10

Finally, substitute these derivatives into the approximation of $\beta(s)$ given above, and keep only the dominant term in each component (that is, the one containing the smallest power of s). The result is

$$\beta(s) \sim \beta(0) + sT_0 + \kappa_0 \frac{s^2}{2} N_0 + \kappa_0 \tau_0 \frac{s^3}{6} B_0.$$

Denoting the right side by $\hat{\beta}(s)$, we obtain a curve $\hat{\beta}$ called the *Frenet approximation* of β near $s = 0$. We emphasize that β has a different Frenet approximation near each of its points; if 0 is replaced by an arbitrary number s_0, then s is replaced by $s - s_0$, as usual in Taylor expansions.

Let us now examine the Frenet approximation given above. The first term in the expression for $\hat{\beta}$ is just the point $\beta(0)$. The first two terms give the *tangent line* $s \to \beta(0) + sT_0$ of β at $\beta(0)$—the best linear approximation of β near $\beta(0)$. The first three terms give the parabola

$$s \to \beta(0) + sT_0 + \kappa_0(s^2/2)N_0,$$

which is the best quadratic approximation of β near $\beta(0)$. Note that this parabola lies in the plane through $\beta(0)$ orthogonal to B_0, the *osculating plane* of β at $\beta(0)$. This parabola has the same shape as the parabola $y = \kappa_0 x^2/2$ in the xy plane, and is completely determined by the curvature κ_0 of β at $s = 0$.

Finally, the torsion τ_0, which appears in the last and smallest term of $\hat{\beta}$, controls the motion of β orthogonal to its osculating plane at $\beta(0)$, as shown in Fig. 2.10.

On the basis of this discussion, it is a reasonable guess that *if a unit-speed curve has curvature identically zero, then it is a straight line*. In fact, this follows immediately from (2) of Lemma 2.3, since $\kappa = \| T' \| = \| \beta'' \|$, so that $\kappa = 0$ if and only if $\beta'' = 0$. Thus curvature does measure deviation from straightness.

A *plane curve* in $\mathbf{E}^3$ is a curve that lies in a single plane of $\mathbf{E}^3$. Evidently a plane curve does not twist in as interesting a way as even the simple helix in Example 3.3. The discussion above shows that for s small the curve β tends to stay in its osculating plane at $\beta(0)$; it is $\tau_0 \neq 0$ which causes β to twist out of the osculating plane. Thus if the torsion of β is identically zero, we may well suspect that β *never* leaves this plane.

3.5 Corollary Let β be a unit-speed curve in $\mathbf{E}^3$ with $\kappa > 0$. Then β is a plane curve if and only if $\tau = 0$.

Proof. Suppose β is a plane curve. Then by the remarks above, there exist points $\mathbf{p}$ and $\mathbf{q}$ such that $(\beta(s) - \mathbf{p}) \cdot \mathbf{q} = 0$ for all s. Differentiation yields

$$\beta'(s) \cdot \mathbf{q} = \beta''(s) \cdot \mathbf{q} = 0 \qquad \text{for all } s.$$

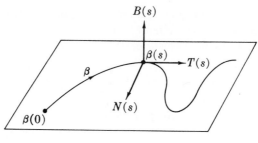

FIG. 2.11

Thus $\mathbf{q}$ is always orthogonal to $T = \beta'$ and $N = \beta''/\kappa$. But B is also orthogonal to T and N, so, since B has unit length, $B = \pm\mathbf{q}/\|\,\mathbf{q}\,\|$. Thus $B' = 0$, and by definition $\tau = 0$ (Fig. 2.11).

Conversely, suppose $\tau = 0$. Thus $B' = 0$; that is, B is parallel and may thus be identified (by Remark 3.4) with a *point* of $\mathbf{E}^3$. We assert that β lies in the plane through $\beta(0)$ orthogonal to B. To prove this, consider the real-valued function

$$f(s) = (\beta(s) - \beta(0))\cdot B \qquad \text{for all } s.$$

Then

$$\frac{df}{ds} = \beta'\cdot B = T\cdot B = 0.$$

But obviously, $f(0) = 0$, so f is identically zero. Thus

$$(\beta(s) - \beta(0))\cdot B = 0 \qquad \text{for all } s,$$

which shows that β lies entirely in this plane orthogonal to the (parallel) binormal of β.   ∎

We saw at the end of Example 3.3 that a circle of radius a has curvature $1/a$ and torsion zero. Furthermore the formula given there for the principal normal shows that for a circle, N always points toward its center. This suggests how to prove the following converse.

3.6 Lemma If β is a unit-speed curve with constant curvature $\kappa > 0$ and torsion zero, then β is part of a circle of radius $1/\kappa$.

Proof. Since $\tau = 0$, β is a plane curve. What we must now show is that every point of β is at distance $1/\kappa$ from some fixed point—which will thus be the center of the circle. Consider the curve $\gamma = \beta + (1/\kappa)N$. Using the hypothesis on β, and (as usual) a Frenet formula, we find

$$\gamma' = \beta' + \frac{1}{\kappa}N' = T + \frac{1}{\kappa}(-\kappa T) = 0.$$

Hence the curve γ is constant; that is, $\beta(s) + (1/\kappa)N(s)$ has the same value, say $\mathbf{c}$, for all s (see Fig. 2.12). But the distance from $\mathbf{c}$ to $\beta(s)$ is

$$d(\mathbf{c}, \beta(s)) = \| \mathbf{c} - \beta(s) \| = \left\| \frac{1}{\kappa} N(s) \right\| = \frac{1}{\kappa}. \quad \blacksquare$$

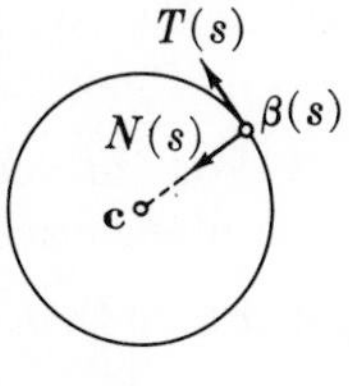

FIG. 2.12

In principle, every geometric problem about curves can be solved by means of the Frenet formulas. In simple cases it may be just enough to record the data of the problem in convenient form, differentiate, and use the Frenet formulas. For example, suppose β is a unit-speed curve that lies entirely in the sphere Σ of radius a centered at the origin of $\mathbf{E}^3$. To stay in the sphere, β must curve; in fact it is a reasonable guess that the minimum possible curvature occurs when β is on a great circle of Σ. Such a circle has radius a, so we conjecture that *a spherical curve β has curvature $\kappa \geq 1/a$, where a is the radius of its sphere.*

To prove this, observe that since every point of Σ has distance a from the origin, we have $\beta \cdot \beta = a^2$. Differentiation yields $2\beta' \cdot \beta = 0$, that is, $\beta \cdot T = 0$. Another differentiation gives $\beta' \cdot T + \beta \cdot T' = 0$, and by using a Frenet formula we get $T \cdot T + \kappa\beta \cdot N = 0$; hence

$$\kappa\beta \cdot N = -1.$$

By the Schwarz inequality,

$$| \beta \cdot N | \leq \| \beta \| \, \| N \| = a,$$

and since $\kappa \geq 0$ we obtain the required result:

$$\kappa = | \kappa | = \frac{1}{| \beta \cdot N |} \geq \frac{1}{a}.$$

Continuation of this procedure leads to a necessary and sufficient condition (expressed in terms of curvature and torsion) for a curve to be *spherical*, that is, lie on some sphere in $\mathbf{E}^3$ (Exercise 10).

EXERCISES

1. Compute the *Frenet apparatus* κ, τ, T, N, B of the unit-speed curve $\beta(s) = (\frac{4}{5} \cos s, 1 - \sin s, -\frac{3}{5} \cos s)$. Show that this curve is a circle; find its center and radius.

2. Consider the curve

$$\beta(s) = \left(\frac{(1 + s)^{3/2}}{3}, \frac{(1 - s)^{3/2}}{3}, \frac{s}{\sqrt{2}} \right)$$

defined on $I: -1 < s < 1$. Show that β has unit speed, and compute its Frenet apparatus.

3. For the helix in Example 3.3, check the Frenet formulas by direct substitution of the computed values of κ, τ, T, N, B.

4. Prove that

$$T = N \times B = -B \times N$$

$$N = B \times T = -T \times B$$

$$B = T \times N = -N \times T.$$

(A formal proof uses properties of the cross product established in the Exercises of Section 1—but one can recall these formulas by using the right-hand rule given on p. 48.)

5. If A is the vector field $\tau T + \kappa B$ on a unit-speed curve β, show that the Frenet formulas become

$$T' = A \times T$$

$$N' = A \times N$$

$$B' = A \times B.$$

6. A unit-speed parametrization of a circle may be written

$$\gamma(s) = \mathbf{c} + r \cos s/r \; \mathbf{e}_1 + r \sin s/r \; \mathbf{e}_2,$$

where $\mathbf{e}_i \cdot \mathbf{e}_j = \delta_{ij}$.

If β is a unit-speed curve with $\kappa(0) > 0$, prove that there is one and only one circle γ which approximates β near $\beta(0)$ in the sense that

$$\gamma(0) = \beta(0), \qquad \gamma'(0) = \beta'(0), \qquad \text{and} \qquad \gamma''(0) = \beta''(0).$$

Show that γ lies in the osculating plane of β at $\beta(0)$ and find its center $\mathbf{c}$ and radius r. The circle γ is called the *osculating circle* and $\mathbf{c}$ the

FIG. 2.13   FIG. 2.14

center of curvature of β at $\beta(0)$. (The same results hold when 0 is replaced by any number s.)

7. If α and a reparametrization $\bar{\alpha} = \alpha(h)$ are both unit-speed curves, show that
   (a) $h(s) = \pm s + s_0$ for some number s_0
   (b) $\bar{T} = \pm T(h)$
   $\bar{N} = \quad N(h) \qquad \bar{\kappa} = \kappa(h) \qquad \bar{\tau} = \tau(h)$
   $\bar{B} = \pm B(h)$
   where the sign $(\pm)$ is the same as that in (a), and we assume $\kappa > 0$ (Fig. 2.14). Thus even in the orientation-reversing case, the principal normals N and $\bar{N}$ still point in the same direction.

8. *Curves in the plane.* For a unit-speed curve $\beta(s) = (x(s), y(s))$ in $\mathbf{E}^2$, the *unit tangent* is $T = \beta' = (x', y')$, but the *unit normal* N is defined by rotating T through $+90°$, so $N = (-y', x')$. Thus T' and N are collinear, and the *curvature* of β is defined by the Frenet equation $T' = \kappa N$. Prove
   (a) $N' = -\kappa T$.
   (b) If φ is the slope angle† of β, then $\kappa = \varphi'$.

   This procedure differs from that for $\mathbf{E}^3$, since κ need not be positive—indeed its sign tells which way β is turning. Furthermore, N is defined without assuming $\kappa > 0$.

9. Let $\hat{\beta}$ be the Frenet approximation of an arbitrary unit-speed curve β near $s = 0$. If, say, the B_0 component of $\hat{\beta}$ is removed, the resulting curve is the *orthogonal projection* of $\hat{\beta}$ in the $T_0 N_0$ plane. It is the view of $\hat{\beta} \sim \beta$ one gets by looking toward $\hat{\beta}(0) = \beta(0)$ directly along the vector B_0. Sketch the general shape of the orthogonal projections of $\hat{\beta}$ on each of the planes $T_0 N_0$, $T_0 B_0$, $N_0 B_0$, assuming $\tau > 0$. (These views of β may be confirmed experimentally using a bent piece of wire.)

10. *Spherical curves.* Let α be a unit-speed curve with $\kappa > 0$, $\tau \neq 0$.
   (a) If α lies on a sphere of center $\mathbf{c}$ and radius r, show that

$$\alpha - \mathbf{c} = -\rho N - \rho'\sigma B,$$

   where $\rho = 1/\kappa$ and $\sigma = 1/\tau$. Thus $r^2 = \rho^2 + (\rho'\sigma)^2$.
   (b) Conversely, if $\rho^2 + (\rho'\sigma)^2$ has constant value r^2 and $\rho' \neq 0$, show that α lies on a sphere of radius r.
   (*Hint:* For (b), show that the "center curve" $\gamma = \alpha + \rho N + \rho'\sigma B$—suggested by (a)—is constant.)

---

† The existence of φ as a differentiable function with $T = \cos \varphi\, U_1 + \sin \varphi\, U_2$ derives from Exercise 12 of Section 1.

11. Let $\beta, \bar{\beta}: I \to \mathbf{E}^3$ be unit-speed curves with nonvanishing curvature and torsion. If $T = \bar{T}$, then β and $\bar{\beta}$ are parallel (Ex. 8 of II.2). If $B = \bar{B}$, prove that $\bar{\beta}$ is parallel to either β or the curve $s \to -\beta(s)$.

4 Arbitrary-Speed Curves

It is a simple matter to adapt the results of the previous section to the study of a regular curve $\alpha: I \to \mathbf{E}^3$ which does not necessarily have unit speed. We merely transfer to α the Frenet apparatus of a unit-speed re-parametrization $\bar{\alpha}$ of α. Explicitly, if s is an arc-length function for α as in Theorem 2.1, then

$$\alpha(t) = \bar{\alpha}(s(t)) \qquad \text{for all } t$$

or, in functional notation, $\alpha = \bar{\alpha}(s)$. Now if $\bar{\kappa} > 0$, $\bar{\tau}$, $\bar{T}$, $\bar{N}$, and $\bar{B}$ are defined for $\bar{\alpha}$ as in Section 3, we define for α the

Curvature function: $\kappa = \bar{\kappa}(s)$
Torsion function: $\tau = \bar{\tau}(s)$
Unit tangent vector field: $T = \bar{T}(s)$
Principal normal vector field: $N = \bar{N}(s)$
Binormal vector field: $B = \bar{B}(s)$

In general κ and $\bar{\kappa}$ are different functions, defined on different intervals. But *they give exactly the same description of the turning of the common route of α and $\bar{\alpha}$*, since at any point $\alpha(t) = \bar{\alpha}(s(t))$ the numbers $\kappa(t)$ and $\bar{\kappa}(s(t))$ are by definition the same. Similarly with the rest of the Frenet apparatus; since only a change of parametrization is involved, its fundamental geometric meaning is the same as before. In particular, T, N, B is again a frame field on α linked to the shape of α as indicated in the discussion of Frenet approximations.

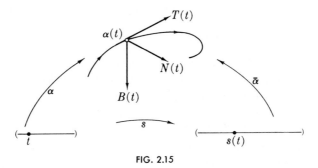

FIG. 2.15

For purely theoretical work this simple transference is often all that is needed. Data about α converts into data about the unit-speed reparametrization $\bar{\alpha}$; results about $\bar{\alpha}$ convert to results about α. For example, if α is a regular curve with $\tau = 0$, then by the definition above $\bar{\alpha}$ has $\bar{\tau} = 0$; by Corollary 3.5, $\bar{\alpha}$ is a plane curve, so obviously α is, too.

However, for explicit numerical computations—and occasionally for the theory as well—this transference is impractical, since it is rarely possible to find explicit formulas for $\bar{\alpha}$. (For example, try to find a unit-speed parametrization for the curve $\alpha(t) = (t, t^2, t^3)$.)

The Frenet formulas are valid only for unit-speed curves; they tell the rate of change of the frame field T, N, B *with respect to arc length*. However, the speed v of the curve is the proper correction factor in the general case.

4.1 Lemma   If α is a regular curve in $\mathbf{E}^3$ with $\kappa > 0$, then

$$
\begin{aligned}
T' &= & \kappa v N \\
N' &= -\kappa v T & & + \tau v B. \\
B' &= & -\tau v N
\end{aligned}
$$

Proof. Let $\bar{\alpha}$ be a unit-speed reparametrization of α. Then by definition, $T = \bar{T}(s)$, where s is an arc-length function for α. The chain rule as applied to differentiation of vector fields (Exercise 7 of Section 2) gives

$$ T' = \bar{T}'(s)\, \frac{ds}{dt}. $$

By the usual Frenet equations, $\bar{T}' = \bar{\kappa}\bar{N}$. Substituting the function s in this equation yields

$$ \bar{T}'(s) = \bar{\kappa}(s)\bar{N}(s) = \kappa N $$

by the definition of κ and N in the arbitrary-speed case. Since ds/dt is the speed function v of α, these two equations combine to yield $T' = \kappa v N$. The formulas for N' and B' are derived in the same way.   ∎

There is a commonly used notation for the calculus that completely ignores change of parametrization. For example, the same letter would designate both a curve α and its unit-speed parametrization $\bar{\alpha}$, and similarly with the Frenet apparatus of these two curves. Differences in derivatives are handled by writing, say, dT/dt for T', but dT/ds for either $\bar{T}'$ or its reparametrization $\bar{T}'(s)$. With these conventions, the proof above would combine the chain rule $dT/dt = (dT/ds)(ds/dt)$ and the Frenet formula $dT/ds = \kappa N$ to give $dT/dt = \kappa v N$.

Only for a *constant-speed* curve is acceleration orthogonal to velocity, since $\beta' \cdot \beta'$ constant is equivalent to $(\beta' \cdot \beta')' = 2\beta' \cdot \beta'' = 0$. In the general

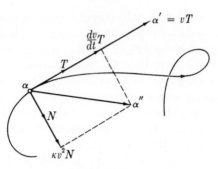

FIG. 2.16

case, we analyze velocity and acceleration by expressing them in terms of the Frenet frame field.

4.2 Lemma   If α is a regular curve with speed function v, then the velocity and acceleration of α are given by (Fig. 2.16)

$$\alpha' = vT \qquad \alpha'' = \frac{dv}{dt}\, T + \kappa v^2 N$$

Proof. Since $\alpha = \bar{\alpha}(s)$, where s is the arc-length function of α, we find, using Lemma 4.5 of Chapter I that

$$\alpha' = \bar{\alpha}'(s)\, \frac{ds}{dt} = v\bar{T}(s) = vT.$$

Then a second differentiation yields

$$\alpha'' = \frac{dv}{dt}\, T + vT' = \frac{dv}{dt}\, T + \kappa v^2 N$$

where we use Lemma 4.1.   ∎

The formula $\alpha' = vT$ is to be expected—α' and T are each tangent to the curve, and T has a unit length while $\| \alpha' \| = v$. The formula for acceleration is more interesting. By definition, α'' is the rate of change of the velocity α', and in general both the length and the direction of α' are changing. The *tangential component* $(dv/dt)T$ of α'' measures the rate of change of the length of α' (that is, of the speed of α). The *normal component* $\kappa v^2 N$ measures the rate of change of the direction of α'. Newton's laws of motion show that these components may be experienced as forces. For example, in a car that is speeding up or slowing down on a straight road, the only force one feels is due to $(dv/dt)T$. If one takes an unbanked curve at speed v, the sideways force one feels is due to $\kappa v^2 N$. Here κ measures how sharply the *road* turns; the effect of speed is given by v^2, so 60 miles per hour is four times as unsettling as 30.

We now find effectively computable expressions for the Frenet apparatus.

4.3 Theorem   Let α be a regular curve in $\mathbf{E}^3$. Then

$$T = \alpha'/\|\,\alpha'\,\|$$

$$N = B \times T \qquad\qquad \kappa = \|\,\alpha' \times \alpha''\,\|/\|\,\alpha'\,\|^3$$

$$B = \alpha' \times \alpha''/\|\,\alpha' \times \alpha''\,\| \qquad \tau = (\alpha' \times \alpha'') \cdot \alpha'''/\|\,\alpha' \times \alpha''\,\|^2.$$

Proof. Since $v = \|\,\alpha'\,\| > 0$, the formula $T = \alpha'/\|\,\alpha'\,\|$ is equivalent to $\alpha' = vT$. From the preceding lemma we get

$$\alpha' \times \alpha'' = (vT) \times \left(\frac{dv}{dt}\,T + \kappa v^2 N\right)$$

$$= v\,\frac{dv}{dt}\,T \times T + \kappa v^3 T \times N = \kappa v^3 B$$

since $T \times T = 0$. Taking norms we find

$$\|\,\alpha' \times \alpha''\,\| = \|\,\kappa v^3 B\,\| = \kappa v^3$$

because $\|\,B\,\| = 1$, $\kappa \geq 0$, and $v > 0$. Indeed *this equation shows that for regular curves, $\|\,\alpha' \times \alpha''\,\| > 0$ is equivalent to the usual condition $\kappa > 0$.* (Thus for $\kappa > 0$, α' and α'' are linearly independent and determine the osculating plane at each point, as do T and N.) Then

$$B = \frac{\alpha' \times \alpha''}{\kappa v^3} = \frac{\alpha' \times \alpha''}{\|\,\alpha' \times \alpha''\,\|}\,.$$

Now $N = B \times T$ is true for any Frenet frame field (Exercise 4 of Section 3); thus only the formula for torsion remains to be proved. To find the dot product $(\alpha' \times \alpha'') \cdot \alpha'''$ we express everything in terms of T, N, B. We already know that $\alpha' \times \alpha'' = \kappa v^3 B$. Thus, since $0 = T \cdot B = N \cdot B$, *we need only find the B component of α'''*. But

$$\alpha''' = \left(\frac{dv}{dt}\,T + \kappa v^2 N\right)' = \kappa v^2 N' + \cdots$$

$$= \kappa v^3 \tau B + \cdots$$

where we use Lemma 4.1. Consequently $(\alpha' \times \alpha'') \cdot \alpha''' = \kappa^2 v^6 \tau$, and since $\|\,\alpha' \times \alpha''\,\| = \kappa v^3$, we have the required formula for τ.   ∎

The triple scalar product in this formula for τ could (by Exercise 4 of Section 1) also be written $\alpha' \cdot \alpha'' \times \alpha'''$. But we need $\alpha' \times \alpha''$ anyway, so it is usually easier to find $(\alpha' \times \alpha'') \cdot \alpha'''$.

4.4 Example   We compute the Frenet apparatus of the curve

$$\alpha(t) = (3t - t^3,\, 3t^2,\, 3t + t^3).$$

The derivatives are

$$\alpha'(t) = 3(1 - t^2, 2t, 1 + t^2)$$
$$\alpha''(t) = 6(-t, 1, t)$$
$$\alpha'''(t) = 6(-1, 0, 1).$$

Now

$$\alpha'(t) \cdot \alpha'(t) = 18(1 + 2t^2 + t^4),$$

so

$$v(t) = \| \alpha'(t) \| = \sqrt{18}(1 + t^2).$$

Applying the definition of cross product, we find

$$\alpha'(t) \times \alpha''(t) = 18 \begin{vmatrix} U_1 & U_2 & U_3 \\ 1 - t^2 & 2t & 1 + t^2 \\ -t & 1 & t \end{vmatrix} = 18(-1 + t^2, -2t, 1 + t^2).$$

Dotting this vector with itself, we get

$$(18)^2 \{(-1 + t^2)^2 + 4t^2 + (1 + t^2)^2\} = 2(18)^2 (1 + t^2)^2.$$

Hence

$$\| \alpha'(t) \times \alpha''(t) \| = 18\sqrt{2} (1 + t^2).$$

The expressions above for $\alpha' \times \alpha''$ and α''' yield

$$(\alpha' \times \alpha'') \cdot \alpha''' = 6 \cdot 18 \cdot 2.$$

It remains only to substitute this data into the formulas in Theorem 4.3, with N being computed by another cross product. The final results are

$$T = \frac{(1 - t^2, 2t, 1 + t^2)}{\sqrt{2}(1 + t^2)}$$

$$N = \frac{(-2t, 1 - t^2, 0)}{1 + t^2}$$

$$B = \frac{(-1 + t^2, -2t, 1 + t^2)}{\sqrt{2}(1 + t^2)}$$

$$\kappa = \tau = \frac{1}{3(1 + t^2)^2}$$

Alternatively, we could use the identity in Lemma 1.8 to compute $\| \alpha' \times \alpha'' \|$, and express

$$(\alpha' \times \alpha'') \cdot \alpha''' = \alpha' \cdot (\alpha'' \times \alpha''')$$

as a determinant by Exercise 4 of Section 1.

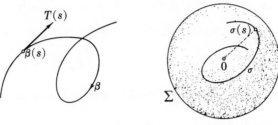

FIG. 2.17

Let us summarize the situation. We now have the Frenet apparatus for an arbitrary-speed curve α. This apparatus satisfies the extended Frenet formulas (with factor v) and may be computed by Theorem 4.3. If $v = 1$, that is, if α is a unit-speed curve, the Frenet formulas in Lemma 4.1 simplify slightly (to Theorem 3.2), but Theorem 4.3 may be replaced by the *much* simpler definitions in Section 3.

Let us consider some applications of the results in this section. There are a number of interesting ways in which one can assign to a given curve β a new curve $\tilde{\beta}$ whose geometric properties illuminate some aspects of the behavior of β. For example, if β is a unit-speed curve, the curve $\sigma = T$ is the *spherical image* of β. According to Remark 3.4, σ is the curve such that each point $\sigma(s)$ has the same Euclidean coordinates as the unit-tangent vector $T(s)$ (Fig. 2.17). Roughly speaking, $\sigma(s)$ is gotten by moving $T(s)$ to the origin. The spherical image lies entirely on the unit sphere Σ of $\mathbf{E}^3$, since $\| \sigma \| = \| T \| = 1$, and *the motion of σ represents the curving of β.*

For example, if β is the helix in Example 3.3, the formula there for T shows that

$$\sigma(s) = \left(-\frac{a}{c} \sin \frac{s}{c}, \frac{a}{c} \cos \frac{s}{c}, \frac{b}{c} \right).$$

Thus the spherical image of a helix lies on the circle cut from the unit sphere by the plane $z = b/c$.

There is no loss of generality in assuming that the original curve β has unit speed, but we *cannot* also expect σ to have unit speed. In fact, since $\sigma = T$, we have $\sigma' = T' = \kappa N$. Thus σ moves always in the principal normal direction of β, with speed $\| \sigma' \|$ equal to the curvature κ of β.

Next we assume $\kappa > 0$, and use the Frenet formulas for β to compute the curvature of σ. Now

$$\sigma'' = (\kappa N)' = \frac{d\kappa}{ds} N + \kappa N' = -\kappa^2 T + \frac{d\kappa}{ds} N + \kappa\tau B.$$

Thus

$$\sigma' \times \sigma'' = -\kappa^3 N \times T + \kappa^2 \tau N \times B = \kappa^2 (\kappa B + \tau T).$$

By Theorem 4.3 the curvature of the spherical image σ is

$$\kappa_\sigma = \frac{\|\sigma' \times \sigma''\|}{v^3} = \frac{\sqrt{\kappa^2 + \tau^2}}{\kappa} = \left(1 + \left(\frac{\tau}{\kappa}\right)^2\right)^{1/2} > 1$$

and thus depends only on the ratio of torsion to curvature for the original curve β.

Here is a closely related application in which this ratio τ/κ turns out to be decisive.

4.5 Definition   A regular curve α in $\mathbf{E}^3$ is a *cylindrical helix* provided the unit tangent T of α has constant angle ϑ with some fixed unit vector $\mathbf{u}$; that is, $T(t) \cdot \mathbf{u} = \cos \vartheta$ for all t.

This condition is not altered by reparametrization, so for theoretical purposes we need only deal with a cylindrical helix β which has unit speed. So suppose β is a unit-speed curve with $T \cdot \mathbf{u} = \cos \vartheta$. If we pick a reference point, say $\beta(0)$, on β, then the real-valued function

$$h(s) = (\beta(s) - \beta(0)) \cdot \mathbf{u}$$

tells how far $\beta(s)$ has "risen" in the $\mathbf{u}$ direction since leaving $\beta(0)$ (Fig. 2.18). But

$$\frac{dh}{ds} = \beta' \cdot \mathbf{u} = T \cdot \mathbf{u} = \cos \vartheta$$

so β is rising at a constant rate *relative to arc length*, and $h(s) = s \cos \vartheta$. (If we shift to an arbitrary parametrization, this formula becomes

$$h(t) = s(t) \cos \vartheta,$$

where s is the arc-length function.)

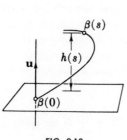

FIG. 2.18

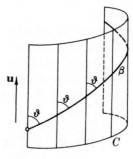

FIG. 2.19

By drawing a line through each point of β in the $\mathbf{u}$ direction, we construct a generalized cylinder C on which β moves in such a way as to cut each ruling (or "element") at constant angle ϑ, as in Fig. 2.19. In the special case when this cylinder is circular, β is evidently a helix of the type defined in Example 3.3.

It turns out to be quite easy to identify cylindrical helices.

4.6 Theorem   A regular curve α with $\kappa > 0$ is a cylindrical helix if and only if the ratio τ/κ is constant.

Proof. It suffices to consider the case where α has unit speed. If α is a cylindrical helix with $T \cdot \mathbf{u} = \cos \vartheta$, then

$$0 = (T \cdot \mathbf{u})' = T' \cdot \mathbf{u} = \kappa N \cdot \mathbf{u}.$$

Since $\kappa > 0$, we conclude that $N \cdot \mathbf{u} = 0$. Thus for each s, $\mathbf{u}$ lies in the plane determined by $T(s)$ and $B(s)$. Orthonormal expansion yields

$$\mathbf{u} = \cos \vartheta \, T + \sin \vartheta \, B.$$

As usual we differentiate and apply Frenet formulas to obtain

$$0 = (\kappa \cos \vartheta - \tau \sin \vartheta)N.$$

Hence $\tau \sin \vartheta = \kappa \cos \vartheta$, so that τ/κ has constant value $\cot \vartheta$.

Conversely, suppose that τ/κ is constant. Choose an angle ϑ such that $\cot \vartheta = \tau/\kappa$. If

$$U = \cos \vartheta \, T + \sin \vartheta \, B,$$

we find

$$U' = (\kappa \cos \vartheta - \tau \sin \vartheta)N = 0.$$

This parallel vector field U then determines (as in Remark 3.4) a unit vector $\mathbf{u}$ such that $T \cdot \mathbf{u} = \cos \vartheta$, so α is a cylindrical helix.   ∎

This proof also shows how to compute the unit vector $\mathbf{u}$ and angle ϑ. For example, the curve α in Example 4.4 is a cylindrical helix, since $\kappa = \tau$. The angle ϑ satisfies the equation $\cot \vartheta = \tau/\kappa = 1$; we take $\vartheta = \pi/4$. Then $\cos \vartheta = \sin \vartheta = 1/\sqrt{2}$, so by the proof above, $\mathbf{u} = (1/\sqrt{2})(T + B)$. The data in Example 4.4 then yield $\mathbf{u} = (0, 0, 1)$. (There is no need to convert α to unit speed—that would merely reparametrize κ, τ, T, and B, with no effect on ϑ and $\mathbf{u}$.)

In Exercise 10 this information about cylindrical helices is used to show that *circular* helices are characterized by constancy of curvature and torsion (see also Corollary 5.5, of Chapter III).

Simple hypotheses on a regular curve in $\mathbf{E}^3$ thus have the following effects ($\Leftrightarrow$ means "if and only if")

$\kappa = 0$	$\Leftrightarrow$ straight line
$\tau = 0$	$\Leftrightarrow$ planar
κ constant > 0 and $\tau = 0$	$\Leftrightarrow$ circle
κ constant > 0 and τ constant $\neq 0$	$\Leftrightarrow$ circular helix
τ/κ constant	$\Leftrightarrow$ cylindrical helix

EXERCISES

1. Consider the curve $\alpha \colon \mathbf{R} \to \mathbf{E}^3$ such that $\alpha(t) = (2t, t^2, t^3/3)$.
   (a) Compute the Frenet apparatus of α: κ, τ, T, N, B.
   (b) Make a careful sketch of this curve for $-4 \le t \le 4$ showing T, N, and B at $t = 0, 2, 4$. (*Hint:* Begin with its projection $(2t, t^2, 0)$ in the xy plane.)
   (c) Find the limiting position of the Frenet frame T, N, B of α as $t \to +\infty$ and $t \to -\infty$.

2. Compute the Frenet apparatus of the curve $\alpha(t) = (\cosh t, \sinh t, t)$. Express the curvature and torsion of α as functions $\kappa(s)$ and $\tau(s)$ of arc length s measured from $t = 0$.

3. For the curve $\alpha(t) = (t \cos t, t \sin t, t)$,
   (a) compute the Frenet apparatus at $t = 0$. (Evaluate $\alpha', \alpha'', \alpha'''$ at $t = 0$ before using Theorem 4.3.)
   (b) sketch this curve for $-2\pi \le t \le 2\pi$, showing T, N, B at $t = 0$. (*Hint:* Ex. 2 of II.2.)

4. For the curve α in Example 4.4, check Lemma 4.2 by direct substitution. Make a sketch, in scale, showing the vectors $T(0), N(0), \alpha'(0)$, and $\alpha''(0)$.

5. Prove that the curvature of a regular curve in $\mathbf{E}^3$ is given by
$$\kappa^2 v^4 = \| \alpha'' \|^2 - (dv/dt)^2.$$

6. If α is a curve with *constant speed* $c > 0$, show that

$$T = \alpha'/c \qquad\qquad \kappa = \| \alpha'' \|/c^2$$
$$N = \alpha''/\| \alpha'' \|$$
$$B = \alpha' \times \alpha''/c \| \alpha'' \| \qquad \tau = \frac{\alpha' \times \alpha'' \cdot \alpha'''}{c^2 \| \alpha'' \|^2}$$

where for N, B, τ, we assume α'' never zero, that is, $\kappa > 0$.

7. Use the formulas in the preceding exercise to compute the Frenet apparatus of the helix α in Example 4.2 in Chapter I.

8. Let α be a cylindrical helix with unit vector $\mathbf{u}$, angle ϑ, and arc-length function s (measured from, say, $t = 0$.) The unique curve γ such that

$\alpha(t) = \gamma(t) + s(t) \cos \vartheta \, \mathbf{u}$ is called the *cross-section curve* of the cylinder on which α lies. Prove that

(a) γ lies in the plane through $\alpha(0)$ orthogonal to $\mathbf{u}$.

(b) The curvature of γ is $\kappa/\sin^2 \vartheta$, where κ is the curvature of α.

(*Hint:* For (b) it suffices to assume α has unit speed.)

9. (*Continuation*). The following curves are cylindrical helices; for each find the unit vector $\mathbf{u}$, angle ϑ, and cross-section curve γ; verify condition (a) above.

(a) The curve in Exercise 1.

(b) The curve in Example 4.4.

(c) The curve in Exercise 2.

10. If β is a unit-speed curve with $\kappa > 0$ and $\tau \neq 0$ both constant, prove that β is a (circular) helix.

11. Let σ be the spherical image (Section 4) of a unit-speed curve β. Prove that the curvature and torsion of σ are

$$\kappa_\sigma = \sqrt{1 + (\tau/\kappa)^2} \qquad \tau_\sigma = \frac{(d/ds)(\tau/\kappa)}{\kappa[1 + (\tau/\kappa)^2]}$$

where κ and τ are the curvature and torsion of β.

12. (a) Prove that a curve is a cylindrical helix if and only if its spherical image is part of a circle. (No computations needed.)

(b) Sketch the spherical image of the cylindrical helix in Exercise 1. Is it a complete circle? Find its center.

13. If α is a curve with $\kappa > 0$, then the *central curve* $\alpha^* = \alpha + (1/\kappa)N$ consists of all centers of curvature of α (Ex. 6 of II.3.) For any two nonzero numbers a and b, let β_{ab} be the helix in Example 3.3. Show that the central curve of β_{ab} is $\beta_{\bar{a}b}$, where $\bar{a} = -b^2/a$. Deduce that the central curve of $\beta_{\bar{a}b}$ is the original helix β_{ab}.

14. If $\alpha(t) = (x(t), y(t))$ is a regular curve in $\mathbf{E}^2$, show that its curvature (Ex. 8, II.3) is

$$\kappa = \frac{\alpha'' \cdot J(\alpha')}{v^3} = \frac{x'y'' - x''y'}{(x'^2 + y'^2)^{3/2}}.$$

Here J is the operator such that

$$J(t_1, t_2) = (-t_2, t_1).$$

15. For a regular curve α in $\mathbf{E}^2$, the central curve $\alpha^* = \alpha + (1/\kappa)N$ is called the *evolute* of α. (It is, of course, not defined at points where $\kappa = 0$.)

(a) Show that α^* is uniquely determined by the condition that its tangent line at each point $\alpha^*(t)$ is the normal line to α at $\alpha(t)$.

(b) Prove that

$$\alpha^* = \alpha + \frac{\alpha' \cdot \alpha'}{\alpha'' \cdot J(\alpha')} J(\alpha')$$

(notation as in Exercise 14)

(c) Find the evolute of the cycloid

$$\alpha(t) = (t + \sin t, 1 + \cos t), \qquad -\pi < t < \pi.$$

Sketch both curves.

16. The *total curvature* of a unit-speed curve α defined on I, is $\int_I \kappa(s)\, ds$ (an improper integral when the interval is infinite). If α is merely regular, the definition becomes $\int_I \kappa(t) v(t)\, dt$; this makes total curvature independent of the parametrization of α. Find the total curvature of the following curves—the first three defined on the whole real line.

(a) The curve in Example 4.4.

(b) The helix in Example 3.3.

(c) The curve in Exercise 2.

(d) The ellipse $\alpha(t) = (a \cos t, b \sin t)$. Since this is a closed curve, consider only a single period $0 \leq t \leq 2\pi$.

17. Let $f > 0$ and g be arbitrary differentiable real-valued functions on some interval in $\mathbf{R}$. Consider the curve

$$\alpha(t) = \left(\int f(t) \sin t, \int f(t) \cos t, \int f(t)g(t) \right)$$

where $\int h$ denotes any function whose derivative is h. Show that the curvature and torsion of α are given by

$$\kappa = \frac{1}{f} \sqrt{\frac{1 + g^2 + g'^2}{(1 + g^2)^3}} \qquad \tau = \frac{-1}{f} \frac{g + g''}{(1 + g^2 + g'^2)}.$$

18. Consider the general cubic curve $\gamma(t) = (at, bt^2, ct^3)$, where $abc \neq 0$.

(a) Compute

$$\tau/\kappa = \frac{3ac}{2b^2} \left[\frac{9c^2 t^4 + 4b^2 t^2 + a^2}{9c^2 t^4 + 9(a^2 c^2/b^2)t^2 + a^2} \right]^{3/2}$$

and deduce that the cubic curve γ is a cylindrical helix if and only if $3ac = \pm 2b^2$.

(b) In the case where $3ac = 2b^2$, find the unit vector $\mathbf{u}$ and angle ϑ.

19. One of the standard tricks of advanced calculus is the construction of an (infinitely) differentiable function f on the real line such that $f(t) = 0$ for $t \leq 0$, and $f(t) > 0$ for $t > 0$. (Also $f''(t) > 0$ for $t > 0$.) If $g(t) = f(-t)$, consider the curve

$$\alpha(t) = (t, f(t), g(t)).$$

(a) Prove that the curvature of α is zero only when $t = 0$.

(b) Sketch this curve for $|t|$ small, and show some principal normals for $t > 0$ and $t < 0$.

This example shows that the condition $\kappa > 0$ cannot be avoided in a detailed study of the geometry of curves in $\mathbf{E}^3$, for if κ is zero even at a single point, the geometric character of the curve can change radically at that point. (Note that this difficulty is not a serious one for curves in $\mathbf{E}^2$; see Ex. 8 of II.3.)

5   Covariant Derivatives

In Chapter I, each time we defined a new object (curve, differential form, mapping, ...) we usually followed by defining an appropriate notion of derivative of the object. Vector fields were an exception; we have delayed defining their derivatives since (as later results will show) this notion belongs properly to the *geometry* of Euclidean space.

The definition generalizes that of the derivative $\mathbf{v}[f]$ of a function f with respect to a tangent vector $\mathbf{v}$ at a point $\mathbf{p}$ (Definition 3.1 of Chapter I). In fact, replacing f by a vector field W, we observe that the function $t \rightarrow W(\mathbf{p} + t\mathbf{v})$ is a vector field on the curve $t \rightarrow \mathbf{p} + t\mathbf{v}$. (The derivative of such a vector field was defined in Section 2.) Now the derivative of W with respect to $\mathbf{v}$ will be the derivative of $t \rightarrow W(\mathbf{p} + t\mathbf{v})$ at $t = 0$.

5.1 Definition   Let W be a vector field on $\mathbf{E}^3$ and let $\mathbf{v}$ be a tangent vector to $\mathbf{E}^3$ at the point $\mathbf{p}$. Then the *covariant derivative* of W with respect to $\mathbf{v}$

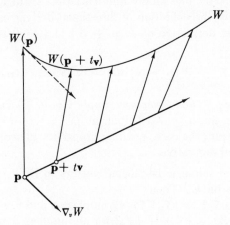

FIG. 2.20

is the tangent vector

$$\nabla_v W = W(\mathbf{p} + t\mathbf{v})'(0)$$

at the point $\mathbf{p}$.

Evidently $\nabla_v W$ *measures the initial rate of change of* $W(\mathbf{p})$ *as* $\mathbf{p}$ *moves in the* $\mathbf{v}$ *direction* (Fig. 2.20). (The term "covariant" derives from the generalization of this notion discussed in Chapter VII.)

For example, suppose $W = x^2 U_1 + yz U_3$, and $\mathbf{v} = (-1, 0, 2)$ at

$$\mathbf{p} = (2, 1, 0).$$

Then

$$\mathbf{p} + t\mathbf{v} = (2 - t, 1, 2t),$$

so

$$W(\mathbf{p} + t\mathbf{v}) = (2 - t)^2 U_1 + 2t U_3,$$

where strictly speaking U_1 and U_2 are also evaluated at $\mathbf{p} + t\mathbf{v}$. Thus,

$$\nabla_v W = W(\mathbf{p} + t\mathbf{v})'(0) = -4 U_1(\mathbf{p}) + 2 U_3(\mathbf{p}).$$

5.2 Lemma   If $W = \sum w_i U_i$ is a vector field on $\mathbf{E}^3$, and $\mathbf{v}$ is a tangent vector at $\mathbf{p}$, then

$$\nabla_v W = \sum \mathbf{v}[w_i] \, U_i(\mathbf{p}).$$

Proof. We have

$$W(\mathbf{p} + t\mathbf{v}) = \sum w_i(\mathbf{p} + t\mathbf{v}) \, U_i(\mathbf{p} + t\mathbf{v})$$

for the restriction of W to the curve $t \to \mathbf{p} + t\mathbf{v}$. To differentiate such a vector field (at $t = 0$), one simply differentiates its Euclidean coordinates (at $t = 0$). But by the definition of directional derivative (Definition 3.1 of Chapter I), the derivative of $w_i(\mathbf{p} + t\mathbf{v})$ at $t = 0$ is precisely $\mathbf{v}[w_i]$. Thus

$$\nabla_v W = W(\mathbf{p} + t\mathbf{v})'(0) = \sum \mathbf{v}[w_i] \, U_i(\mathbf{p}). \qquad \blacksquare$$

In short, *to apply* ∇_v *to a vector field, apply* $\mathbf{v}$ *to its Euclidean coordinates.* Thus the following linearity and Leibnizian properties of covariant derivative follow easily from the corresponding properties (Theorem 3.3 of Chapter I) of directional derivatives.

5.3 Theorem   Let $\mathbf{v}$ and $\mathbf{w}$ be tangent vectors to $\mathbf{E}^3$ at $\mathbf{p}$, and let Y and Z be vector fields on $\mathbf{E}^3$. Then

(1) $\nabla_{a\mathbf{v} + b\mathbf{w}} Y = a\nabla_v Y + b\nabla_w Y$ for all numbers a and b.

(2) $\nabla_v (aY + bZ) = a\nabla_v Y + b\nabla_v Z$ for all numbers a and b.

(3) $\nabla_v(fY) = \mathbf{v}[f]Y(\mathbf{p}) + f(\mathbf{p})\,\nabla_vY$ for all (differentiable) functions f.

(4) $\mathbf{v}[Y\cdot Z] = \nabla_vY\cdot Z(\mathbf{p}) + Y(\mathbf{p})\cdot\nabla_vZ$.

Proof. For example, let us prove (4). If

$$Y = \sum y_iU_i \quad\text{and}\quad Z = \sum z_iU_i,$$

then

$$Y\cdot Z = \sum y_iz_i.$$

Hence by Theorem 3.3 of Chapter I,

$$\mathbf{v}[Y\cdot Z] = \mathbf{v}\Big[\sum y_iz_i\Big] = \sum \mathbf{v}[y_i]\,z_i(\mathbf{p}) + \sum y_i(\mathbf{p})\,\mathbf{v}[z_i]$$

But by the preceding lemma,

$$\nabla_vY = \sum \mathbf{v}[y_i]U_i(\mathbf{p}) \quad\text{and}\quad \nabla_vZ = \sum \mathbf{v}[z_i]U_i(\mathbf{p}).$$

Thus the two sums displayed above are precisely $\nabla_vY\cdot Z(\mathbf{p})$ and $Y(\mathbf{p})\cdot\nabla_vZ$. ∎

Using the pointwise principle (Chapter I, Section 2) once again, we can take the covariant derivative of a vector field W with respect to a *vector field* V, rather than a single tangent vector $\mathbf{v}$. The result is the *vector field* ∇_VW whose value at each point $\mathbf{p}$ is $\nabla_{V(p)}W$. Thus ∇_VW consists of all the covariant derivatives of W with respect to the vectors of V. It follows immediately from the lemma above that if $W = \sum w_iU_i$, then

$$\nabla_VW = \sum V[w_i]U_i.$$

Coordinate computations are easy using the basic identity $U_i[f] = \partial f/\partial x_i$. For example, suppose $V = (y - x)U_1 + xyU_3$ and (as in the example above) $W = x^2U_1 + yzU_3$. Then

$$V[x^2] = (y - x)U_1[x^2] = 2x(y - x)$$

$$V[yz] = xyU_3[yz] = xy^2.$$

Hence

$$\nabla_VW = 2x(y - x)U_1 + xy^2U_3.$$

Now the vector field V was also selected with the earlier example in mind. In fact, the value of V at $\mathbf{p} = (2, 1, 0)$ is

$$V(\mathbf{p}) = (1 - 2)U_1(\mathbf{p}) + 2U_3(\mathbf{p}) = (-1, 0, 2)_p = \mathbf{v}_p,$$

as before. Thus the value of the vector field ∇_VW at this point $\mathbf{p}$ must agree with the earlier computation of ∇_vW. And, in fact, for $\mathbf{p} = (2,1,0)$,

$$\nabla_V(W)(\mathbf{p}) = 2\cdot 2(1 - 2)U_1(\mathbf{p}) + 2U_3(\mathbf{p}) = -4U_1(\mathbf{p}) + 2U_3(\mathbf{p}).$$

For the covariant derivative $\nabla_V W$ as expressed entirely in terms of vector fields, the properties in the preceding theorem take the following form.

5.4 Corollary Let V, W, Y, and Z be vector fields on $\mathbf{E}^3$. Then
(1) $\nabla_V(aY + bZ) = a\nabla_V Y + b\nabla_V Z$, for all numbers a and b.
(2) $\nabla_{fV+gW}Y = f\nabla_V Y + g\nabla_W Y$, for all functions f and g.
(3) $\nabla_V(fY) = V[f]Y + f\nabla_V Y$, for all functions f.
(4) $V[Y \cdot Z] = \nabla_V Y \cdot Z + Y \cdot \nabla_V Z$.

We shall omit the proof, which is an exercise in the use of parentheses based on the (pointwise principle) definition $(\nabla_V Y)(\mathbf{p}) = \nabla_{V(p)} Y$.

Note that $\nabla_V Y$ does not behave symmetrically with respect to V and Y. This is to be expected, since it is Y that is being differentiated, while the role of V is merely algebraic. In particular, $\nabla_{fV} Y$ is $f\nabla_V Y$, but $\nabla_V(fY)$ is not $f\nabla_V Y$: There is an extra term arising from the differentiation of f by V.

EXERCISES

1. Consider the tangent vector $\mathbf{v} = (1, -1, 2)$ at the point $\mathbf{p} = (1, 3, -1)$. Compute $\nabla_V W$ directly from the definition, where
   (a) $W = x^2 U_1 + yU_2$.   (b) $W = xU_1 + x^2 U_2 - z^2 U_3$.

2. Let $V = -yU_1 + xU_3$ and $W = \cos x U_1 + \sin x U_2$.
   Express the following covariant derivatives in terms of U_1, U_2, U_3:
   (a) $\nabla_V W$.   (c) $\nabla_V(z^2 W)$.   (e) $\nabla_V(\nabla_V W)$.
   (b) $\nabla_V V$.   (d) $\nabla_W(V)$.   (f) $\nabla_V(xV - zW)$.

3. If W is a vector field with constant length $\| W \|$, prove that for any vector field V, the covariant derivative $\nabla_V W$ is everywhere orthogonal to W.

4. Let X be the special vector field $\sum x_i U_i$, where x_1, x_2, x_3 are the natural coordinate functions of $\mathbf{E}^3$. Prove that $\nabla_V X = V$ for every vector field V.

5. If $W = \sum w_i U_i$ is a vector field on $\mathbf{E}^3$, the *covariant differential* of W is defined to be $\nabla W = \sum dw_i U_i$. Here ∇W is the function on all tangent vectors whose value on $\mathbf{v}$ is

$$\sum dw_i(\mathbf{v}) U_i(\mathbf{p}) = \nabla_v W.$$

Compute the covariant differential of

$$W = xy^3 U_1 - x^2 z^2 U_3,$$

and use it to find $\nabla_v W$, where

(a) $\mathbf{v} = (1, 0, -3)$ at $\mathbf{p} = (-1, 2, -1)$.

(b) $\mathbf{v} = (-1, 2, -1)$ at $\mathbf{p} = (1, 3, 2)$.

6. Let W be a vector field defined on a region containing a curve α. Then $t \to W(\alpha(t))$ is a vector field on α called the *restriction* of W to α and denoted by W_α.

(a) Prove that $\nabla_{\alpha'(t)} W = (W_\alpha)'(t)$.

(b) Deduce that the straight line in Definition 5.1 may be replaced by *any* curve with initial velocity $\mathbf{v}$. Thus the derivative Y' of a vector field Y on a curve α is (almost) $\nabla_{\alpha'} Y$.

7. The *bracket* of two vector fields is the vector field $[V, W] = \nabla_V W - \nabla_W V$. Establish the following properties of the bracket:

(a) $[V, W][f] = VW[f] - WV[f]$ (here $VW[f]$ denotes the "second derivative" $V[W[f]]$).

(b) $[W, V] = -[V, W]$.

(c) $[U, [V, W]] + [V, [W, U]] + [W, [U, V]] = 0$.

(d) $[fV, gW] = fV[g]W - gW[f]V + fg[V, W]$.

(*Hint:* $Z[f] = 0$ for all f implies $Z = 0$.)

6   Frame Fields

When the Frenet formulas were discovered (by Frenet in 1847, and independently by Serret in 1851), the theory of *surfaces* in $\mathbf{E}^3$ was already a richly developed branch of geometry. The success of the Frenet approach to curves led Darboux (around 1880) to adapt this "method of moving frames" to the study of surfaces. Then, as we mentioned earlier, it was Cartan who brought the method to full generality. His essential idea was very simple: To each point of the object under study (a curve, a surface, Euclidean space itself, ...) assign a frame; then using orthonormal expansion express the rate of change of the frame in terms of the frame itself. This, of course, is just what the Frenet formulas do in the case of a curve.

In the next three sections we shall carry out this scheme in detail for the Euclidean space $\mathbf{E}^3$. We shall see that geometry of curves and surfaces in $\mathbf{E}^3$ is not merely an analogue, but actually a *corollary*, of these basic results. Since the main application (to surface theory) comes only in Chapter VI, these sections may be postponed, and read as a preliminary to that chapter.

By means of the pointwise principle (Chapter I, Section 2) we can automatically extend operations on individual tangent vectors to operations on vector fields. For example, if V and W are vector fields on $\mathbf{E}^3$, then the *dot product* $V \cdot W$ of V and W is the (differentiable) real-valued function on $\mathbf{E}$ whose value at each point $\mathbf{p}$ is $V(\mathbf{p}) \cdot W(\mathbf{p})$. The *norm* $\| V \|$ of V is the real-valued function on $\mathbf{E}^3$ whose value at $\mathbf{p}$ is $\| V(\mathbf{p}) \|$. Thus $\| V \| = (V \cdot V)^{1/2}$. By contrast with $V \cdot W$, the norm function $\| V \|$ need not be differentiable at points for which $V(\mathbf{p}) = 0$, since the square-root function is badly behaved at 0.

At each point $\mathbf{p}$ of $\mathbf{E}^3$, the three tangent vectors $U_1(\mathbf{p})$, $U_2(\mathbf{p})$, $U_3(\mathbf{p})$ constitute a frame at $\mathbf{p}$. This remark is concisely expressed in terms of dot products of vector fields by writing $U_i \cdot U_j = \delta_{ij}$ $(1 \leq i,j \leq 3)$. We used U_1, U_2, U_3 throughout Chapter I. Now, with the dot product at our disposal, we can introduce a simple but crucial generalization.

6.1 Definition Vector fields E_1, E_2, E_3 on $\mathbf{E}^3$ constitute a *frame field* on $\mathbf{E}^3$ provided

$$E_i \cdot E_j = \delta_{ij} \qquad (1 \leq i,j \leq 3)$$

where δ_{ij} is the Kronecker delta.

The term *frame field* is justified by the fact that at each point $\mathbf{p}$ the three vectors $E_1(\mathbf{p})$, $E_2(\mathbf{p})$, $E_3(\mathbf{p})$ form a frame at $\mathbf{p}$. We anticipated this in Chapter I by calling U_1, U_2, U_3 the natural frame field on $\mathbf{E}^3$.

6.2 Example (1) *The cylindrical frame field* (Fig. 2.21). Let r, ϑ, z be the usual cylindrical coordinate functions on $\mathbf{E}^3$. We shall pick a unit vector field in the direction in which each coordinate increases (when the other two are held constant). For r, this is evidently

$$E_1 = \cos \vartheta \, U_1 + \sin \vartheta \, U_2,$$

pointing straight out from the z axis. Then

$$E_2 = - \sin \vartheta \, U_1 + \cos \vartheta \, U_2$$

points in the direction of increasing ϑ as in Fig. 2.21. Finally the direction

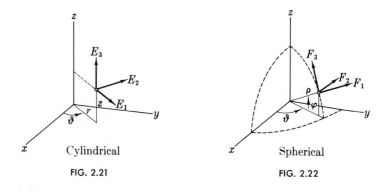

Cylindrical   Spherical

FIG. 2.21   FIG. 2.22

of increase of z is, of course, straight up, so

$$E_3 = U_3.$$

It is easy to check that $E_i \cdot E_j = \delta_{ij}$, so this is a frame field (defined on all of $\mathbf{E}^3$ except the z axis). We call it the *cylindrical frame field* on $\mathbf{E}^3$.

(2) *The spherical frame field on* $\mathbf{E}^3$ (Fig. 2.22). In a similar way, a frame field F_1, F_2, F_3 can be derived from the spherical coordinate functions ρ, ϑ, φ on $\mathbf{E}^3$. As indicated in the figure, we shall measure φ up from the xy plane rather than (as is usually done) down from the z axis.

Let E_1, E_2, E_3 be the cylindrical frame field. For spherical coordinates, the unit vector field F_2 in the direction of increasing ϑ is the same as above, so $F_2 = E_2$. The unit vector field F_1, in the direction of increasing ρ, points straight out from the origin; hence it may be expressed as

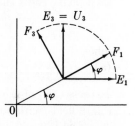

$$F_1 = \cos \varphi \, E_1 + \sin \varphi \, E_3,$$

(Fig. 2.23). Similarly the vector field for increasing φ is

FIG. 2.23

$$F_3 = -\sin \varphi \, E_1 + \cos \varphi \, E_3.$$

Thus the formulas for E_1, E_2, E_3 in (1) yield

$$F_1 = \cos \varphi (\cos \vartheta \, U_1 + \sin \vartheta \, U_2) + \sin \varphi \, U_3$$
$$F_2 = -\sin \vartheta \, U_1 + \cos \vartheta \, U_2$$
$$F_3 = -\sin \varphi (\cos \vartheta \, U_1 + \sin \vartheta \, U_2) + \cos \varphi \, U_3.$$

By repeated use of the identity $\sin^2 + \cos^2 = 1$, we check that F_1, F_2, F_3 is a frame field—the *spherical frame field* on $\mathbf{E}^3$. (Its actual domain of definition is $\mathbf{E}^3$ minus the z axis, as in the cylindrical case.)

The following useful result is an immediate consequence of orthonormal expansion.

6.3 Lemma   Let E_1, E_2, E_3 be a frame field on $\mathbf{E}^3$.

(1) If V is a vector field on $\mathbf{E}^3$, then $V = \sum f_i E_i$, where the functions $f_i = V \cdot E_i$ are called the *coordinate functions* of V with respect to E_1, E_2, E_3.

(2) If $V = \sum f_i E_i$ and $W = \sum g_i E_i$, then $V \cdot W = \sum f_i g_i$. In particular, $\| V \| = (\sum f_i^2)^{1/2}$.

Thus a given vector field V has a different set of coordinate functions with respect to each choice of a frame field E_1, E_2, E_3. The *Euclidean* coordinate functions (Lemma 2.5 of Chapter I), of course, come from the natural frame field U_1, U_2, U_3. In Chapter I, we used this natural frame field exclusively, but now we shall gradually shift to arbitrary frame fields.

The reason is clear: In studying curves and surfaces in $\mathbf{E}^3$, we shall then be able to choose a frame field *specifically adapted to the problem at hand*. Not only does this simplify computations, but it gives a clearer understanding of geometry than if we had insisted on using the same frame field in every situation.

EXERCISES

1. If V and W are vector fields on $\mathbf{E}^3$ which are linearly independent at each point, show that

$$E_1 = \frac{V}{\| V \|}, \qquad E_2 = \frac{\widetilde{W}}{\| \widetilde{W} \|}, \qquad E_3 = E_1 \times E_2$$

   is a frame field, where $\widetilde{W} = W - W \cdot E_1 \, E_1$.

2. Express each of the following vector fields (i) in terms of the cylindrical frame field (with coefficients in terms of r, ϑ, z) and (ii) in terms of the spherical frame field (with coefficients in terms of ρ, ϑ, φ):

   (a) U_1.   (b) $\cos \vartheta \; U_1 + \sin \vartheta \; U_2 + U_3$.

   (c) $x U_1 + y U_2 + z U_3$.

3. Find a frame field E_1, E_2, E_3 such that

$$E_1 = \cos x \; U_1 + \sin x \cos z \; U_2 + \sin x \sin z \; U_3.$$

4. *Toroidal frame field.* Let $\mathcal{O}$ be all of $\mathbf{E}^3$ except the z axis and the circle C of radius R in the xy plane. The toroidal coordinate functions ρ, ϑ, φ are defined on $\mathcal{O}$ as suggested in Fig. 2.24, so that

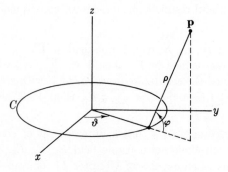

FIG. 2.24

$$x = (R + \rho \cos \varphi) \cos \vartheta$$
$$y = (R + \rho \cos \varphi) \sin \vartheta$$
$$z = \rho \sin \varphi.$$

If E_1, E_2, and E_3 are unit vector fields in the direction of increasing ρ, ϑ, and φ, respectively, express E_1, E_2, E_3 in terms of U_1, U_2, U_3 and prove that it is a frame field.

7 Connection Forms

Once more we state the essential point: The power of the Frenet formulas stems not from the fact that they tell what the derivatives T', N', B' are, but from the fact that *they express these derivatives in terms of T, N, B—* and thereby define curvature and torsion. We shall now do the same thing with an arbitrary frame field E_1, E_2, E_3 on $\mathbf{E}^3$; namely, *express the covariant derivatives of these vector fields in terms of the vector fields themselves.* We begin with the covariant derivative with respect to an arbitrary tangent vector $\mathbf{v}$ at the point $\mathbf{p}$. Then

$$\nabla_v E_1 = c_{11} E_1(\mathbf{p}) + c_{12} E_2(\mathbf{p}) + c_{13} E_3(\mathbf{p})$$
$$\nabla_v E_2 = c_{21} E_1(\mathbf{p}) + c_{22} E_2(\mathbf{p}) + c_{23} E_3(\mathbf{p})$$
$$\nabla_v E_3 = c_{31} E_1(\mathbf{p}) + c_{32} E_2(\mathbf{p}) + c_{33} E_3(\mathbf{p})$$

where by orthonormal expansion the coefficients of these equations are

$$c_{ij} = \nabla_v E_i \cdot E_j(\mathbf{p}) \qquad \text{for} \qquad 1 \leq i,j \leq 3.$$

These coefficients c_{ij} depend on the particular tangent vector $\mathbf{v}$, so a better notation for them is

$$\omega_{ij}(\mathbf{v}) = \nabla_v E_i \cdot E_j(\mathbf{p}), \qquad (1 \leq i,j \leq 3).$$

Thus for each choice of i and j, ω_{ij} is a real-valued function defined on all tangent vectors. But we have met that kind of function before.

7.1 Lemma Let E_1, E_2, E_3 be a frame field on $\mathbf{E}^3$. For each tangent vector $\mathbf{v}$ to $\mathbf{E}^3$ at the point $\mathbf{p}$, let

$$\omega_{ij}(\mathbf{v}) = \nabla_v E_i \cdot E_j(\mathbf{p}), \qquad (1 \leq i,j \leq 3).$$

Then each ω_{ij} is a 1-form, and $\omega_{ij} = -\omega_{ji}$. These 1-forms are called the *connection forms* of the frame field E_1, E_2, E_3.

Proof. By definition, ω_{ij} is a real-valued function on tangent vectors, so to verify that ω_{ij} is a 1-form (Definition 5.1 of Chapter I), it suffices to check the linearity condition. But using Theorem 5.3, we get

$$\omega_{ij}(a\mathbf{v} + b\mathbf{w}) = \nabla_{av+bw}E_i \cdot E_j(\mathbf{p})$$
$$= (a\nabla_v E_i + b\nabla_w E_i) \cdot E_j(\mathbf{p})$$
$$= a\nabla_v E_i \cdot E_j(\mathbf{p}) + b\nabla_w E_i \cdot E_j(\mathbf{p})$$
$$= a\omega_{ij}(\mathbf{v}) + b\omega_{ij}(\mathbf{w}).$$

To prove that $\omega_{ij} = -\omega_{ji}$ we must show that $\omega_{ij}(\mathbf{v}) = -\omega_{ji}(\mathbf{v})$ for every tangent vector $\mathbf{v}$. By definition of frame field, $E_i \cdot E_j = \delta_{ij}$, and since each Kronecker delta has constant value 0 or 1, we have $\mathbf{v}[\delta_{ij}] = 0$. Thus by the Leibnizian formula (4) of Theorem 5.3,

$$0 = \mathbf{v}[E_i \cdot E_j] = \nabla_{\mathbf{v}}E_i \cdot E_j(\mathbf{p}) + E_i(\mathbf{p}) \cdot \nabla_{\mathbf{v}}E_j.$$

By the symmetry of the dot product, the two vectors in this last term may be reversed, so we have found that $0 = \omega_{ij}(\mathbf{v}) + \omega_{ji}(\mathbf{v})$. ∎

The geometric significance of the connection forms is no mystery. The definition $\omega_{ij}(\mathbf{v}) = \nabla_{\mathbf{v}}E_i \cdot E_j(\mathbf{p})$ shows that $\omega_{ij}(\mathbf{v})$ *is the initial rate at which E_i rotates toward E_j as $\mathbf{p}$ moves in the $\mathbf{v}$ direction.* Thus the 1-forms ω_{ij} contain this information for *all* tangent vectors to $\mathbf{E}^3$!

The following basic result is little more than a rephrasing of the definition of connection forms.

7.2 Theorem Let ω_{ij} $(1 \leq i,j \leq 3)$ be the connection forms of a frame field E_1, E_2, E_3 on $\mathbf{E}^3$. Then for any vector field V on $\mathbf{E}^3$,

$$\nabla_V E_i = \sum_j \omega_{ij}(V)E_j, \qquad (1 \leq i \leq 3).$$

We call these the *connection equations* of the frame field E_1, E_2, E_3.

Proof. For fixed i, both sides of this equation are vector fields. Thus we must show that at each point $\mathbf{p}$,

$$\nabla_{V(\mathbf{p})}E_i = \sum \omega_{ij}(V(\mathbf{p}))E_j(\mathbf{p}).$$

But as we have already seen, the very definition of connection form makes this equation a consequence of orthonormal expansion. ∎

When $i = j$, the skew-symmetry condition $\omega_{ij} = -\omega_{ji}$ becomes

$$\omega_{ii} = -\omega_{ii};$$

thus

$$\omega_{11} = \omega_{22} = \omega_{33} = 0.$$

Hence this condition has the effect of reducing the nine 1-forms ω_{ij} for

$1 \leq i,j \leq 3$ to essentially only three, say ω_{12}, ω_{13}, ω_{23}. It is perhaps best to regard the connection forms ω_{ij} as the entries of a skew-symmetric matrix of 1-forms,

$$\omega = \begin{pmatrix} \omega_{11} & \omega_{12} & \omega_{13} \\ \omega_{21} & \omega_{22} & \omega_{23} \\ \omega_{31} & \omega_{32} & \omega_{33} \end{pmatrix} = \begin{pmatrix} 0 & \omega_{12} & \omega_{13} \\ -\omega_{12} & 0 & \omega_{23} \\ -\omega_{13} & -\omega_{23} & 0 \end{pmatrix}.$$

Thus in expanded form, the connection equations (Theorem 7.2) become

$$\begin{aligned} \nabla_V E_1 &= & \omega_{12}(V)E_2 &+ \omega_{13}(V)E_3 \\ \nabla_V E_2 &= -\omega_{12}(V)E_1 & &+ \omega_{23}(V)E_3 \\ \nabla_V E_3 &= -\omega_{13}(V)E_1 &- \omega_{23}(V)E_2. \end{aligned}$$

showing an obvious relation to the Frenet formulas

$$\begin{aligned} T' &= & \kappa N \\ N' &= -\kappa T & &+ \tau B \\ B' &= & -\tau N. \end{aligned}$$

The absence from the Frenet formulas of terms corresponding to $\omega_{13}(V)E_3$ and $-\omega_{13}(V)E_1$ is a consequence of the special way the Frenet frame field is fitted to its curve. Having gotten $T(\sim E_1)$, we chose $N(\sim E_2)$ so that the derivative T' would be a scalar multiple of N alone and not involve $B(\sim E_3)$.

Another difference between the Frenet formulas and the equations above stems from the fact that $\mathbf{E}^3$ has three dimensions, while a curve has but one. The coefficients—curvature κ and torsion τ—in the Frenet formulas measure the rate of change of the frame field T, N, B only along its curve, that is, in the direction of T alone. But the coefficients in the connection equations must be able to make this measurement for E_1, E_2, E_3 with respect to *arbitrary* vector fields in $\mathbf{E}^3$. This is why the connection forms are 1-forms and not just functions.

These formal differences aside, a more fundamental distinction stands out. It is because a Frenet frame field is specially fitted to its curve that the Frenet formulas give information about that curve. Since the frame field E_1, E_2, E_3 used above is completely arbitrary, the connection equations give no direct information about $\mathbf{E}^3$, but only information about the "rate of rotation" of that particular frame field. This is not a weakness, but a strength, since as indicated earlier, if we can fit a frame field to a geometric problem arising in $\mathbf{E}^3$, then the connection equations will give direct information about that problem. Thus these equations play a fundamental role in all the differential geometry of $\mathbf{E}^3$. In particular, the *Frenet formulas can be deduced from them* (Exercise 8). For the sake of motivation, however, we have preferred to deal with the simpler Frenet case first.

Given an arbitrary frame field E_1, E_2, E_3 on $\mathbf{E}^3$, it is fairly easy to find an explicit formula for its connection forms. First use orthonormal expansion to express the vector fields E_1, E_2, E_3 in terms of the natural frame field U_1, U_2, U_3 on $\mathbf{E}^3$:

$$E_1 = a_{11}U_1 + a_{12}U_2 + a_{13}U_3$$
$$E_2 = a_{21}U_1 + a_{22}U_2 + a_{23}U_3$$
$$E_3 = a_{31}U_1 + a_{32}U_2 + a_{33}U_3.$$

Here each $a_{ij} = E_i \cdot U_j$ is a real-valued function on $\mathbf{E}^3$. The matrix

$$A = (a_{ij}) = \begin{pmatrix} a_{11} & a_{12} & a_{13} \\ a_{21} & a_{22} & a_{23} \\ a_{31} & a_{32} & a_{33} \end{pmatrix}$$

with these functions as entries is called the *attitude matrix* of the frame field E_1, E_2, E_3. In fact, at each point $\mathbf{p}$, the numerical matrix

$$A(\mathbf{p}) = (a_{ij}(\mathbf{p}))$$

is exactly the attitude matrix of the frame $E_1(\mathbf{p})$, $E_2(\mathbf{p})$, $E_3(\mathbf{p})$ as in Definition 1.6. Since attitude matrices are orthogonal, the transpose ${}^tA$ of A is equal to its inverse A^{-1}.

Define the differential of $A = (a_{ij})$ to be $dA = (da_{ij})$, so dA is a matrix whose entries are 1-forms. We can now give a simple expression for the connection forms in terms of the attitude matrix.

7.3 Theorem   If $A = (a_{ij})$ is the attitude matrix and $\omega = (\omega_{ij})$ the matrix of connection forms of a frame field E_1, E_2, E_3, then

$$\omega = dA \,{}^tA \qquad \text{(matrix multiplication)},$$

or, equivalently

$$\omega_{ij} = \sum_k a_{jk}\, da_{ik} \qquad \text{for} \qquad 1 \leqq i,j \leqq 3.$$

Proof. If $\mathbf{v}$ is a tangent vector at $\mathbf{p}$, then by definition,

$$\omega_{ij}(\mathbf{v}) = \nabla_v E_i \cdot E_j(\mathbf{p}).$$

Since A is the attitude matrix,

$$E_i = \sum a_{ik} U_k,$$

and thus, by Lemma 5.2,

$$\nabla_v E_i = \sum \mathbf{v}[a_{ik}] \, U_k(\mathbf{p}).$$

The dot product of this vector with

$$E_j(\mathbf{p}) = \sum a_{jk}(\mathbf{p}) \, U_k(\mathbf{p})$$

is then

$$\omega_{ij}(\mathbf{v}) = \sum_k \mathbf{v}[a_{ik}] \, a_{jk}(\mathbf{p}).$$

But by the definition of differential,

$$\mathbf{v}[a_{ik}] = da_{ik}(\mathbf{v});$$

hence

$$\omega_{ij}(\mathbf{v}) = \sum_k a_{jk}(\mathbf{p}) \, da_{ik}(\mathbf{v}) = \left(\sum_k a_{jk} \, da_{ik}\right)(\mathbf{v}).$$

Since this equation holds for all tangent vectors, the two 1-forms ω_{ij} and $\sum a_{jk} \, da_{ik}$ are equal. It is easy to get the neater matrix formula. In fact, the transpose ${}^t A$ has entries ${}^t a_{kj} = a_{jk}$, so

$$\omega_{ij} = \sum_k da_{ik} \, {}^t a_{kj} \qquad \text{for} \qquad 1 \leq i,j \leq 3,$$

which in terms of matrix multiplication is just $\omega = dA \, {}^t A$. ∎

Using this result, let us compute the connection forms of the cylindrical frame field in Example 6.2. From the definition, we read off the attitude matrix

$$A = \begin{pmatrix} \cos\vartheta & \sin\vartheta & 0 \\ -\sin\vartheta & \cos\vartheta & 0 \\ 0 & 0 & 1 \end{pmatrix}.$$

Thus

$$dA = \begin{pmatrix} -\sin\vartheta \, d\vartheta & \cos\vartheta \, d\vartheta & 0 \\ -\cos\vartheta \, d\vartheta & -\sin\vartheta \, d\vartheta & 0 \\ 0 & 0 & 0 \end{pmatrix}.$$

Since

$${}^t A = \begin{pmatrix} \cos\vartheta & -\sin\vartheta & 0 \\ \sin\vartheta & \cos\vartheta & 0 \\ 0 & 0 & 1 \end{pmatrix},$$

we easily compute

$$\omega = dA \, {}^t\!A = \begin{pmatrix} 0 & d\vartheta & 0 \\ -d\vartheta & 0 & 0 \\ 0 & 0 & 0 \end{pmatrix}.$$

Thus $\omega_{12} = d\vartheta$ and all the other connection forms (except, of course, $\omega_{21} = -\omega_{12}$) are zero. Then the connection equations (Theorem 7.2) of the cylindrical frame field become

$$\nabla_V E_1 = d\vartheta\,(V)E_2 = V[\vartheta]E_2$$
$$\nabla_V E_2 = -d\vartheta\,(V)E_1 = -V[\vartheta]E_1$$
$$\nabla_V E_3 = 0$$

for all vector fields V.

These equations have obvious geometric significance. The third equation says that the vector field E_3 is parallel. We knew this already, since in the cylindrical frame field, E_3 is just U_3. The first two equations say that the covariant derivatives of E_1 and E_2 with respect to an arbitrary vector field V depend only on the rate of change of the angle ϑ in the direction of V. From the way the function ϑ is defined, it is clear that $V[\vartheta] = 0$ whenever V is, at each point, tangent to the plane through the z axis. Thus for a vector field of this type, the connection equations above predict that $\nabla_V E_1 = \nabla_V E_2 = 0$. From Fig. 2.21, it is clear that E_1 and E_2 do stay parallel on any plane through the z axis.

EXERCISES

1. For any function f, show that the vector fields
$$E_1 = (\sin f\, U_1 + U_2 - \cos f\, U_3)/\sqrt{2}$$
$$E_2 = (\sin f\, U_1 - U_2 - \cos f\, U_3)/\sqrt{2}$$
$$E_3 = \cos f\, U_1 + \sin f\, U_3$$
form a frame field, and find its connection forms.

2. Find the connection forms of the natural frame field U_1, U_2, U_3.

3. For any function f, show that
$$A = \begin{pmatrix} \cos^2 f & \cos f \sin f & \sin f \\ \sin f \cos f & \sin^2 f & -\cos f \\ -\sin f & \cos f & 0 \end{pmatrix}$$
is the attitude matrix of a frame field, and compute its connection forms.

4. Prove that the connection forms of the spherical frame field are

$$\omega_{12} = \cos \varphi \, d\vartheta, \qquad \omega_{13} = d\varphi, \qquad \omega_{23} = \sin \varphi \, d\vartheta.$$

5. If E_1, E_2, E_3 is a frame field and $W = \sum f_i E_i$, prove the *covariant derivative formula:*

$$\nabla_V W = \sum_j \{ V[f_j] + \sum_i f_i \omega_{ij}(V) \} E_j.$$

6. Let E_1, E_2, E_3 be the cylindrical frame field. If V is a vector field such that $V[\vartheta] = 1$, compute $\nabla_V (r \cos \vartheta \, E_1 + r \sin \vartheta \, E_3)$.

7. If F_1, F_2, F_3 is the spherical frame field,
  (a) Prove that $F_1[\rho] = 1$ and $F_1[\vartheta] = F_1[\varphi] = 0$.
  (b) Compute $\nabla_{F_1} (\cos \rho \, F_2 + \sin \rho \, F_3)$.

8. Let β be a unit-speed curve in $\mathbf{E}^3$ with $\kappa > 0$, and suppose that E_1, E_2, E_3 is a frame field on $\mathbf{E}^3$ such that the restriction of these vector fields to β gives the Frenet-frame field T, N, B of β. Prove that

$$\omega_{12}(T) = \kappa, \qquad \omega_{13}(T) = 0, \qquad \omega_{23}(T) = \tau.$$

Then deduce the Frenet formulas from the connection equations. (*Hint*: Ex. 6 of II.5.)

8 The Structural Equations

We have seen that 1-forms—the connection forms—give the simplest description of the rate of rotation of a frame field. Furthermore, the frame field itself can be described in terms of 1-forms.

8.1 Definition If E_1, E_2, E_3 is a frame field on $\mathbf{E}^3$, then the *dual 1-forms* θ_1, θ_2, θ_3 of the frame field are the 1-forms such that

$$\theta_i(\mathbf{v}) = \mathbf{v} \cdot E_i(\mathbf{p})$$

for each tangent vector $\mathbf{v}$ to $\mathbf{E}^3$ at $\mathbf{p}$.

Note that θ_i is linear on the tangent vectors at each point; hence it *is* a 1-form. (Readers familiar with the notion of dual vector spaces will recognize that at each point, θ_1, θ_2, θ_3 gives the dual basis of E_1, E_2, E_3.)

In the case of the natural frame field U_1, U_2, U_3, the dual forms are just dx_1, dx_2, dx_3. In fact, from Example 5.3 of Chapter I we get

$$dx_i(\mathbf{v}) = v_i = \mathbf{v} \cdot U_i(\mathbf{p})$$

for each tangent vector $\mathbf{v}$; hence $dx_i = \theta_i$.

Using dual forms, the orthonormal expansion formula in Lemma 6.3

may be written $V = \sum \theta_i(V)E_i$. In the characteristic fashion of duality, this formula becomes the following lemma.

8.2 Lemma   Let θ_1, θ_2, θ_3 be the dual 1-forms of a frame field E_1, E_2, E_3. Then any 1-form ϕ on $\mathbf{E}^3$ has a unique expression

$$\phi = \sum \phi(E_i)\theta_i.$$

Proof. Two 1-forms are the same if they have the same value on any vector field V. But

$$\left(\sum \phi(E_i)\theta_i\right)(V) = \sum \phi(E_i)\theta_i(V)$$
$$= \phi\left(\sum \theta_i(V)E_i\right) = \phi(V).$$

These functions $\phi(E_i)$ are the only possible coordinate functions for ϕ in terms of θ_1, θ_2, θ_3, since if $\phi = \sum f_i\theta_i$, then

$$\phi(E_j) = \sum f_i\theta_i(E_j) = \sum f_i\delta_{ij} = f_j. \qquad \blacksquare$$

Thus ϕ is expressed in terms of dual forms of E_1, E_2, E_3 by evaluating it on E_1, E_2, E_3. This useful fact is the generalization to arbitrary-frame fields of Lemma 5.4 of Chapter I.

We compared a frame field E_1, E_2, E_3 to the natural frame field by means of its attitude matrix $A = (a_{ij})$, for which

$$E_i = \sum a_{ij}U_j \qquad (1 \leqq i \leqq 3).$$

The dual formulation is just

$$\theta_i = \sum a_{ij}\,dx_j$$

with *the same coefficients*. In fact, by Lemma 8.2 (or rather its special case, Lemma 5.4 of Chapter I), we have

$$\theta_i = \sum \theta_i(U_j)\,dx_j.$$

But

$$\theta_i(U_j) = E_i \cdot U_j = \left(\sum a_{ik}U_k\right) \cdot U_j = a_{ij}.$$

These formulas for E_i and θ_i show plainly that θ_1, θ_2, θ_3 is merely the dual description of the frame field E_1, E_2, E_3.

In calculus, when a new function appears on the scene, it is natural to ask what its derivative is. Similarly with 1-forms—having associated with each frame field its dual forms and connection forms—it is reasonable to ask what their exterior derivatives are. The answer is given by two neat sets of equations discovered by Cartan.

8.3 Theorem (Cartan structural equations)   Let E_1, E_2, E_3 be a frame field on $\mathbf{E}^3$ with dual forms θ_1, θ_2, θ_3 and connection forms ω_{ij} $(1 \leqq i,j \leqq 3)$. The exterior derivatives of these forms satisfy

(1) the *first structural equations:*

$$d\theta_i = \sum_j \omega_{ij} \wedge \theta_j \qquad (1 \leqq i \leqq 3);$$

(2) the *second structural equations:*

$$d\omega_{ij} = \sum_k \omega_{ik} \wedge \omega_{kj} \qquad (1 \leqq i,j \leqq 3).$$

Because θ_i is the dual of E_i, the first structural equations may be easily recognized as the dual of the connection equations. Only on the basis of later experience will we discover that the second structural equation shows that $\mathbf{E}^3$ is flat—roughly speaking, in the same sense that the plane $\mathbf{E}^2$ is flat.

Proof. We have seen that

$$\theta_i = \sum a_{ij}\, dx_j,$$

hence

$$d\theta_i = \sum da_{ij} \wedge dx_j.$$

Since the attitude matrix $A = (a_{ij})$ is orthogonal, the expression in Theorem 7.3 for ω_{ij} in terms of da_{ij} may be solved for da_{ij} by the usual formalism of linear algebra to give

$$da_{ij} = \sum_k \omega_{ik} a_{kj}.$$

Thus

$$\begin{aligned} d\theta_i &= \sum_j \{ (\sum_k \omega_{ik} a_{kj}) \wedge dx_j\} \\ &= \sum_k \{\omega_{ik} \wedge \sum_j a_{kj}\, dx_j\} \\ &= \sum_k \omega_{ik} \wedge \theta_k \end{aligned}$$

which is the first structural equation.

We could give a similar index proof for the second structural equation, but a liberal use of matrix notation will give a better idea of what is really going on. To apply the exterior derivative d to a matrix of functions or 1-forms, we apply it to each entry. The matrix formula $\omega = dA\, {}^t A$ in Theorem 7.3, for example, means

$$\omega_{ij} = \sum da_{ik} a_{jk}.$$

But

$$d\omega_{ij} = -\sum da_{ik} \wedge da_{jk}.$$

(Note the minus sign!) Hence in matrix notation, where we shall suppress the wedge,

$$d\omega = -dA \ ^t(dA).$$

Multiplying both sides of $\omega = dA \ ^tA$ by A yields

$$dA = \omega A,$$

since $^tA = A^{-1}$. Then the rule for transpose of a matrix product gives

$$^t(dA) = \ ^t(\omega A) = \ ^tA \ ^t\omega.$$

By substituting in the equation above for $d\omega$ we find

$$d\omega = -\omega A \ ^tA \ ^t\omega = -\omega \ ^t\omega = \omega \omega$$

since ω is skew-symmetric. But this is just the second structural equation $d\omega_{ij} = \sum \omega_{ik} \wedge \omega_{kj}$ in matrix notation. ∎

8.4 Example Structural equations for the spherical frame field (Example 6.2). The dual forms and connection forms are

$$\theta_1 = d\rho \qquad\qquad \omega_{12} = \cos \varphi \ d\vartheta$$
$$\theta_2 = \rho \cos \varphi \ d\vartheta \qquad \omega_{13} = d\varphi$$
$$\theta_3 = \rho \ d\varphi \qquad\qquad \omega_{23} = \sin \varphi \ d\vartheta.$$

Let us check, say, the first structural equation

$$d\theta_3 = \sum \omega_{3j} \wedge \theta_j = \omega_{31} \wedge \theta_1 + \omega_{32} \wedge \theta_2.$$

Using the skew-symmetry $\omega_{ij} = -\omega_{ji}$ and the general properties of forms developed in Chapter I, we get

$$\omega_{31} \wedge \theta_1 = -d\varphi \wedge d\rho = d\rho \wedge d\varphi$$
$$\omega_{32} \wedge \theta_2 = (-\sin \varphi \ d\vartheta) \wedge (\rho \cos \varphi \ d\vartheta) = 0$$

(the latter since $d\vartheta \wedge d\vartheta = 0$). The sum of these terms is, correctly,

$$d\theta_3 = d(\rho \ d\varphi) = d\rho \wedge d\varphi.$$

Second structural equations involve only one wedge product. For example, since $\omega_{11} = \omega_{22} = 0$,

$$d\omega_{12} = \sum \omega_{1k} \wedge \omega_{k2} = \omega_{13} \wedge \omega_{32}.$$

In this case,

$$\omega_{13} \wedge \omega_{32} = d\varphi \wedge (-\sin \varphi \ d\vartheta) = -\sin \varphi \ d\varphi \wedge d\vartheta$$

which is the same as

$$d\omega_{12} = d(\cos \varphi \ d\vartheta) = d(\cos \varphi) \wedge d\vartheta = -\sin \varphi \ d\varphi \wedge d\vartheta.$$

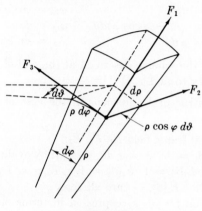

FIG. 2.25

To derive the expressions given above for the dual 1-forms, first compute dx_1, dx_2, dx_3 by differentiating the well-known equations

$$x_1 = \rho \cos \varphi \cos \vartheta$$
$$x_2 = \rho \cos \varphi \sin \vartheta$$
$$x_3 = \rho \sin \varphi.$$

Then substitute in the formula $\theta_i = \sum a_{ij} \, dx_j$, where $A = (a_{ij})$ is the attitude matrix to be found in Example 6.2.

We shall later find a more efficient computational technique related to the following reliable nonsense from elementary calculus: If at each point the spherical coordinates ρ, ϑ, φ are altered by $d\rho$, $d\vartheta$, $d\varphi$, then the sides of the resulting infinitesimal box are $d\rho$, $\rho \cos \varphi \, d\vartheta$, $\rho \, d\varphi$ (Fig. 2.25). But these are precisely the formulas for the dual forms θ_1, θ_2, θ_3.

As we mentioned earlier, our main use of the Cartan structural equations is in studying the geometry of surfaces (Chapters VI and VII). A wider variety of applications is given by Flanders [1].

EXERCISES

1. For a 1-form $\phi = \sum f_i \theta_i$, prove that

$$d\phi = \sum_j \{ df_j + \sum_i f_i \omega_{ij} \} \wedge \theta_j.$$

(Compare Ex. 5 of II.7.)

2. For the toroidal frame field in Ex. 4 of II.6, show that

$$\theta_1 = d\rho \qquad\qquad \omega_{12} = \cos\varphi \, d\vartheta$$
$$\theta_2 = (R + \rho\cos\varphi)d\vartheta \qquad \omega_{13} = d\varphi$$
$$\theta_3 = \rho \, d\varphi \qquad\qquad \omega_{23} = \sin\varphi \, d\vartheta.$$

(*Hint:* Find θ_i by the scheme described at the end of II.8. No computations are necessary to find ω_{ij}!)

3. Check the first structural equations in the case of the toroidal frame field.

4. For the cylindrical frame field E_1, E_2, E_3:
   (a) Prove that $\theta_1 = dr$, $\theta_2 = r \, d\vartheta$, $\theta_3 = dz$ by evaluating on U_1, U_2, U_3.
   (b) Deduce that $E_1[r] = 1$, $E_2[\vartheta] = 1/r$, $E_3[z] = 1$ and that the other six possibilities $E_1[\vartheta]$, $\cdots$ are all zero.
   (c) For a function $f(r, \vartheta, z)$ expressed in terms of cylindrical coordinates, show that

$$E_1[f] = \frac{\partial f}{\partial r} \qquad E_2[f] = \frac{1}{r}\frac{\partial f}{\partial \vartheta} \qquad E_3[f] = \frac{\partial f}{\partial z}.$$

5. Frame fields on $\mathbf{E}^2$. For a frame field E_1, E_2 on the plane $\mathbf{E}^2$:
   (a) Find the connection equations.
   (b) If

$$E_1 = \cos\varphi \, U_1 + \sin\varphi \, U_2$$
$$E_2 = -\sin\varphi \, U_1 + \cos\varphi \, U_2$$

where φ is an arbitrary function, express the dual 1-forms θ_1, θ_2 and the connection form ω_{12} in terms of φ.
   (c) Prove the structural equations for this case.

9 Summary

We have accomplished the aims set at the beginning of this chapter. The idea of a moving frame has been expressed rigorously as a *frame field*—either on a curve in $\mathbf{E}^3$, or on an open set of $\mathbf{E}^3$ itself. In the case of a curve, we used only the Frenet frame field T, N, B of the curve. Expressing the derivatives of these vector fields in terms of the vector fields themselves, we discovered the *curvature* and *torsion* of the curve. It is already clear that curvature and torsion tell a lot about the geometry of a curve; we shall find in Chapter III that they tell everything. In the case of an open set of $\mathbf{E}^3$, we dealt with an arbitrary frame field E_1, E_2, E_3. Cartan's generalization (Theorem 7.2) of the Frenet formulas followed the same pattern of expressing the (covariant) derivatives of these vector fields in terms of the vector fields themselves. Omitting the vector field V from the notation in Theorem 7.2 we have

$$
\begin{array}{ll}
\qquad \textit{Cartan} & \qquad\qquad \textit{Frenet} \\[4pt]
\nabla E_1 = \qquad\quad \omega_{12}E_2 + \omega_{13}E_3 & T' = \qquad\qquad\quad \kappa N \\
\nabla E_2 = -\omega_{12}E_1 \qquad + \omega_{23}E_3 & N' = -\kappa T \qquad\qquad + \tau B \\
\nabla E_3 = -\omega_{13}E_1 - \omega_{23}E_2 & B' = \qquad\qquad -\tau N
\end{array}
$$

Cartan's equations are not conspicuously more complicated than Frenet's because the notion of 1-form is available for the coefficients ω_{ij}, the connection forms.

CHAPTER III

Euclidean Geometry

We recall some familiar features of plane geometry. First of all, two triangles are *congruent* if there is a rigid motion of the plane which carries one triangle exactly onto the other. Corresponding angles of congruent triangles are equal, corresponding sides have the same length, the areas enclosed are equal, and so on. Indeed, any geometric property of a given triangle is automatically shared by every congruent triangle. Conversely, there are a number of simple ways in which one can decide if two given triangles are congruent—for example, if for each the same three numbers occur as lengths of sides.

In this chapter we shall investigate the rigid motions (isometries) of Euclidean space, and see how these remarks about triangles can be extended to other geometric objects.

1 Isometries of $\mathbf{E}^3$

An isometry, or rigid motion, of Euclidean space is a special type of mapping which preserves the Euclidean distance between points (Definition 1.2, Chapter II).

1.1 Definition   An *isometry* of $\mathbf{E}^3$ is a mapping $F: \mathbf{E}^3 \to \mathbf{E}^3$ such that

$$d(F(\mathbf{p}), F(\mathbf{q})) = d(\mathbf{p}, \mathbf{q})$$

for all points $\mathbf{p}, \mathbf{q}$ in $\mathbf{E}^3$.

1.2 Example

(1) *Translations.*   Fix a point $\mathbf{a}$ in $\mathbf{E}^3$ and let T be the mapping that adds $\mathbf{a}$ to every point of $\mathbf{E}^3$. Thus $T(\mathbf{p}) = \mathbf{p} + \mathbf{a}$ for all points $\mathbf{p}$. T is called *translation* by $\mathbf{a}$. It is easy to see that T is an isometry, since

$$d(T(\mathbf{p}), T(\mathbf{q})) = d(\mathbf{p} + \mathbf{a}, \mathbf{q} + \mathbf{a})$$
$$= \|(\mathbf{p} + \mathbf{a}) - (\mathbf{q} + \mathbf{a})\|$$
$$= \|\mathbf{p} - \mathbf{q}\| = d(\mathbf{p}, \mathbf{q}).$$

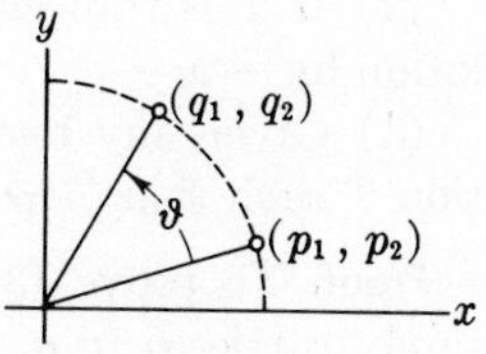

(2) **Rotation around a coordinate axis.** A rotation of the xy plane through an angle ϑ carries the point (p_1, p_2) to the point (q_1, q_2) with coordinates (Fig. 3.1)

FIG. 3.1

$$q_1 = p_1 \cos \vartheta - p_2 \sin \vartheta$$
$$q_2 = p_1 \sin \vartheta + p_2 \cos \vartheta.$$

Thus a *rotation* C of $\mathbf{E}^3$ around the z axis (through an angle ϑ) has the formula

$$C(\mathbf{p}) = C(p_1, p_2, p_3) = (p_1 \cos \vartheta - p_2 \sin \vartheta, \; p_1 \sin \vartheta + p_2 \cos \vartheta, \; p_3)$$

for all points $\mathbf{p}$. Evidently, C is a linear transformation, hence, in particular, a mapping. A straightforward computation shows that C preserves Euclidean distance, so C is an isometry.

Recall that if F and G are mappings of $\mathbf{E}^3$, the composite function GF is a mapping of $\mathbf{E}^3$ obtained by applying first F, then G.

1.3 Lemma If F and G are isometries of $\mathbf{E}^3$, then the composite mapping GF is also an isometry of $\mathbf{E}^3$.

Proof. Since G is an isometry, the distance from $G(F(\mathbf{p}))$ to $G(F(\mathbf{q}))$ is $d(F(\mathbf{p}), F(\mathbf{q}))$. But since F is an isometry, this distance equals $d(\mathbf{p}, \mathbf{q})$. Thus GF preserves distance, hence is an isometry. ∎

In short, a composition of isometries is again an isometry.

We also recall that if $F: \mathbf{E}^3 \to \mathbf{E}^3$ is both one-to-one and onto, then F has a unique inverse function $F^{-1}: \mathbf{E}^3 \to \mathbf{E}^3$, which sends each point $F(\mathbf{p})$ back to $\mathbf{p}$. The relationship between F and F^{-1} is best described by the formulas

$$FF^{-1} = I, \qquad F^{-1}F = I$$

where I is the *identity mapping* of $\mathbf{E}^3$, that is, the mapping such that $I(\mathbf{p}) = \mathbf{p}$ for all $\mathbf{p}$.

The translations of $\mathbf{E}^3$ (as defined in Example 1.2) are the simplest type of isometry.

1.4 Lemma (1) If S and T are translations, then $ST = TS$ is also a translation.

(2) If T is translation by $\mathbf{a}$, then T has an inverse T^{-1}, which is translation by $-\mathbf{a}$.

(3) Given any two points $\mathbf{p}$ and $\mathbf{q}$ of $\mathbf{E}^3$, there exists a unique translation T such that $T(\mathbf{p}) = \mathbf{q}$.

Proof. To prove (3), for example, note that translation by $\mathbf{q} - \mathbf{p}$ certainly carries $\mathbf{p}$ to $\mathbf{q}$. This is the only possibility, since if T is translation by $\mathbf{a}$ and $T(\mathbf{p}) = \mathbf{q}$, then $\mathbf{p} + \mathbf{a} = \mathbf{q}$; hence $\mathbf{a} = \mathbf{q} - \mathbf{p}$. ∎

A useful special case of (3) is that if T is a translation such that for some one point $T(\mathbf{p}) = \mathbf{p}$, then $T = I$.

The rotation in Example 1.2 is an example of an *orthogonal transformation* of $\mathbf{E}^3$, that is, a linear transformation $C: \mathbf{E}^3 \rightarrow \mathbf{E}^3$ which preserves dot products in the sense that

$$C(\mathbf{p}) \cdot C(\mathbf{q}) = \mathbf{p} \cdot \mathbf{q} \qquad \text{for all } \mathbf{p}, \mathbf{q}.$$

1.5 Lemma   If $C: \mathbf{E}^3 \rightarrow \mathbf{E}^3$ is an orthogonal transformation, then C is an isometry of $\mathbf{E}^3$.

Proof. First we show that C preserves norms. By definition $\| \mathbf{p} \|^2 = \mathbf{p} \cdot \mathbf{p}$; hence

$$\| C(\mathbf{p}) \|^2 = C(\mathbf{p}) \cdot C(\mathbf{p}) = \mathbf{p} \cdot \mathbf{p} = \| \mathbf{p} \|^2.$$

Thus $\| C(\mathbf{p}) \| = \| \mathbf{p} \|$ for all points $\mathbf{p}$. Since C is linear, it follows easily that C is an isometry:

$$d(C(\mathbf{p}), C(\mathbf{q})) = \| C(\mathbf{p}) - C(\mathbf{q}) \| = \| C(\mathbf{p} - \mathbf{q}) \| = \| \mathbf{p} - \mathbf{q} \|$$

$$= d(\mathbf{p}, \mathbf{q}) \qquad \text{for all } \mathbf{p}, \mathbf{q}. \qquad ∎$$

Our goal now is Theorem 1.7, which asserts that every isometry can be expressed as an orthogonal transformation followed by a translation. The main part of the proof is the following converse of Lemma 1.5.

1.6 Lemma   If F is an isometry of $\mathbf{E}^3$ such that $F(\mathbf{0}) = \mathbf{0}$, then F is an orthogonal transformation.

Proof. First we show that F preserves dot products; then we show that F is a linear transformation. Note that by definition of Euclidean distance, the norm $\| \mathbf{p} \|$ of a point $\mathbf{p}$ is just the Euclidean distance $d(\mathbf{0}, \mathbf{p})$ from the origin to $\mathbf{p}$. By hypothesis, F preserves Euclidean distance, and $F(\mathbf{0}) = \mathbf{0}$; hence

$$\| F(\mathbf{p}) \| = d(\mathbf{0}, F(\mathbf{p})) = d(F(\mathbf{0}), F(\mathbf{p})) = d(\mathbf{0}, \mathbf{p}) = \| \mathbf{p} \|.$$

Thus F preserves norms. Now by a standard trick ("polarization") we shall deduce that it also preserves dot products. Since F is an isometry,

$$d(F(\mathbf{p}), F(\mathbf{q})) = d(\mathbf{p}, \mathbf{q})$$

for any pair of points. Hence

$$\| F(\mathbf{p}) - F(\mathbf{q}) \| = \| \mathbf{p} - \mathbf{q} \|,$$

which, by definition of norm, implies

$$(F(\mathbf{p}) - F(\mathbf{q})) \cdot (F(\mathbf{p}) - F(\mathbf{q})) = (\mathbf{p} - \mathbf{q}) \cdot (\mathbf{p} - \mathbf{q}).$$

Hence

$$\| F(\mathbf{p}) \|^2 - 2F(\mathbf{p}) \cdot F(\mathbf{q}) + \| F(\mathbf{q}) \|^2 = \| \mathbf{p} \|^2 - 2\mathbf{p} \cdot \mathbf{q} + \| \mathbf{q} \|^2.$$

The norm terms here cancel, since F preserves norms, and we find

$$F(\mathbf{p}) \cdot F(\mathbf{q}) = \mathbf{p} \cdot \mathbf{q},$$

as required.

It remains to prove that F is linear. Let $\mathbf{u}_1, \mathbf{u}_2, \mathbf{u}_3$ be the unit points $(1, 0, 0)$, $(0, 1, 0)$, $(0, 0, 1)$, respectively. Then we have the identity

$$\mathbf{p} = (p_1, p_2, p_3) = \sum p_i \mathbf{u}_i.$$

Also, the points $\mathbf{u}_1, \mathbf{u}_2, \mathbf{u}_3$ are orthonormal; that is, $\mathbf{u}_i \cdot \mathbf{u}_j = \delta_{ij}$.

We know that F preserves dot products, so $F(\mathbf{u}_1)$, $F(\mathbf{u}_2)$, $F(\mathbf{u}_3)$ must also be orthonormal. Thus orthonormal expansion gives

$$F(\mathbf{p}) = \sum F(\mathbf{p}) \cdot F(\mathbf{u}_i) F(\mathbf{u}_i).$$

But

$$F(\mathbf{p}) \cdot F(\mathbf{u}_i) = \mathbf{p} \cdot \mathbf{u}_i = p_i,$$

so

$$F(\mathbf{p}) = \sum p_i F(\mathbf{u}_i).$$

Using this identity, it is a simple matter to check the linearity condition

$$F(a\mathbf{p} + b\mathbf{q}) = aF(\mathbf{p}) + bF(\mathbf{q}). \qquad \blacksquare$$

We now give a concrete description of what an arbitrary isometry is like.

1.7 Theorem   If F is an isometry of $\mathbf{E}^3$, then there exist a unique translation T and a unique orthogonal transformation C such that

$$F = TC.$$

Proof. Let T be translation by $F(\mathbf{0})$. We saw in Lemma 1.4 that T^{-1} is translation by $-F(\mathbf{0})$. But $T^{-1}F$ is an isometry, by Lemma 1.3, and furthermore,

$$(T^{-1}F)(\mathbf{0}) = T^{-1}(F(\mathbf{0})) = F(\mathbf{0}) - F(\mathbf{0}) = \mathbf{0}.$$

Thus by Lemma 1.6, $T^{-1}F$ is an orthogonal transformation, say $T^{-1}F = C$. Applying T on the left, we get $F = TC$.

To prove the required uniqueness, we suppose that F can also be expressed as $\bar{T}\bar{C}$, where $\bar{T}$ is a translation and $\bar{C}$ an orthogonal transformation. We must prove $\bar{T} = T$ and $\bar{C} = C$. Now $TC = \bar{T}\bar{C}$; hence $C = T^{-1}\bar{T}\bar{C}$. Since C and $\bar{C}$ are linear transformation, they of course send the origin to itself. It follows that $(T^{-1}\bar{T})(\mathbf{0}) = \mathbf{0}$. But since $T^{-1}\bar{T}$ is a translation, we conclude that $T^{-1}\bar{T} = I$; hence $\bar{T} = T$. Then the equation $TC = \bar{T}\bar{C}$ becomes $TC = T\bar{C}$. Applying T^{-1}, we get $C = \bar{C}$. ∎

Thus every isometry of $\mathbf{E}^3$ *can be* uniquely *described as an orthogonal transformation followed by a translation.* When $F = TC$ as in Theorem 1.7, we call C the *orthogonal part* of F, and T the *translation part* of F. Note that CT is generally not the same as TC (Exercise 1).

The decomposition theorem above is the decisive fact about isometries of $\mathbf{E}^3$ (and its proof holds for $\mathbf{E}^n$ as well). For example, we shall now find explicit formulas for an arbitrary isometry $F = TC$. If (c_{ij}) is the *matrix* of the linear transformation C, we have the explicit formula

$$C(p_1, p_2, p_3) = \left(\sum c_{1j}p_j, \sum c_{2j}p_j, \sum c_{3j}p_j \right)$$

for all points $\mathbf{p} = (p_1, p_2, p_3)$. We are using the *column-vector* conventions, under which $\mathbf{q} = C(\mathbf{p})$ means

$$\begin{pmatrix} q_1 \\ q_2 \\ q_3 \end{pmatrix} = \begin{pmatrix} c_{11} & c_{12} & c_{13} \\ c_{21} & c_{22} & c_{23} \\ c_{31} & c_{32} & c_{33} \end{pmatrix} \begin{pmatrix} p_1 \\ p_2 \\ p_3 \end{pmatrix}.$$

Since C is an *orthogonal* linear transformation, it is easy to show that its matrix (c_{ij}) is orthogonal in the sense: inverse equals transpose.

Returning to the decomposition $F = TC$, suppose that T is translation by $\mathbf{a} = (a_1, a_2, a_3)$. Then

$$F(\mathbf{p}) = TC(\mathbf{p}) = \mathbf{a} + C(\mathbf{p}).$$

Using the above formula for $C(\mathbf{p})$, we get

$$F(\mathbf{p}) = F(p_1, p_2, p_3) = \left(a_1 + \sum c_{1j}p_j, a_2 + \sum c_{2j}p_j, a_3 + \sum c_{3j}p_j \right).$$

Alternatively, using the column-vector conventions, $\mathbf{q} = F(\mathbf{q})$ means

$$\begin{pmatrix} q_1 \\ q_2 \\ q_3 \end{pmatrix} = \begin{pmatrix} a_1 \\ a_2 \\ a_3 \end{pmatrix} + \begin{pmatrix} c_{11} & c_{12} & c_{13} \\ c_{21} & c_{22} & c_{23} \\ c_{31} & c_{32} & c_{33} \end{pmatrix} \begin{pmatrix} p_1 \\ p_2 \\ p_3 \end{pmatrix}.$$

EXERCISES

Throughout these exercises, A, B, and C denote orthogonal transformations (or their matrices), and T_a is translation by $\mathbf{a}$.

1. Prove that $CT_a = T_{C(a)}C$.

2. Given isometries $F = T_aA$ and $G = T_bB$, find the translation and orthogonal part of FG and GF.

3. Show that an isometry $F = T_aC$ has an inverse mapping F^{-1}, which is also an isometry. Find the translation and orthogonal parts of F^{-1}.

4. If

$$C = \begin{pmatrix} -\frac{2}{3} & \frac{2}{3} & -\frac{1}{3} \\ \frac{2}{3} & \frac{1}{3} & -\frac{2}{3} \\ \frac{1}{3} & \frac{2}{3} & \frac{2}{3} \end{pmatrix} \quad \text{and} \quad \begin{cases} \mathbf{p} = (3, 1, -6), \\ \mathbf{q} = (1, 0, 3) \end{cases}$$

show that C is orthogonal; then compute $C(\mathbf{p})$ and $C(\mathbf{q})$, and check that $C(\mathbf{p}) \cdot C(\mathbf{q}) = \mathbf{p} \cdot \mathbf{q}$.

5. Let $F = T_aC$, where $\mathbf{a} = (1, 3, -1)$ and

$$C = \begin{pmatrix} 1/\sqrt{2} & 0 & -1/\sqrt{2} \\ 0 & 1 & 0 \\ 1/\sqrt{2} & 0 & 1/\sqrt{2} \end{pmatrix}.$$

If $\mathbf{p} = (2, -2, 8)$, find the coordinates of the point $\mathbf{q}$ for which

(a) $\mathbf{q} = F(\mathbf{p})$.   (b) $\mathbf{q} = F^{-1}(\mathbf{p})$.   (c) $\mathbf{q} = (CT_a)(\mathbf{p})$.

6. In each case decide whether F is an isometry of $\mathbf{E}^3$. If so, find its translation and orthogonal parts.
   (a) $F(\mathbf{p}) = -\mathbf{p}$.
   (b) $F(\mathbf{p}) = \mathbf{p} \cdot \mathbf{a} \, \mathbf{a}$, where $\| \mathbf{a} \| = 1$.
   (c) $F(\mathbf{p}) = (p_3 - 1, p_2 - 2, p_1 - 3)$.
   (d) $F(\mathbf{p}) = (p_1, p_2, 1)$.

A *group* G is a set furnished with an *operation* that assigns to each pair g_1, g_2 of elements of G an element g_1g_2, subject to these rules: (1) associative law: $(g_1g_2)g_3 = g_1(g_2g_3)$, (2) there is a unique *identity element* e such that $eg = ge = g$ for all g in G, and (3) inverses: For each g in G there is an element g^{-1} in G such that $gg^{-1} = g^{-1}g = e$.

Groups occur naturally in many parts of geometry, and we shall mention a few in subsequent exercises. Basic properties of groups may be found, for example, in Birkhoff and MacLane [2].

7. Prove that the set $\mathcal{E}$ of all isometries of $\mathbf{E}^3$ forms a group—with composition of functions as the operation. $\mathcal{E}$ is called the *Euclidean group* (of order 3), or the group of Euclidean motions of $\mathbf{E}^3$.

A subset H of a group G is a *subgroup* of G provided (1) if g_1 and g_2 are in H, then so is $g_1 g_2$, (2) if g is in H, so is g^{-1}, and (3) the identity element e of G is in H. A subgroup H of G is automatically a group.

8. Prove that the set $\mathfrak{I}$ of all translations of $\mathbf{E}^3$ and the set $O(3)$ of all orthogonal transformations of $\mathbf{E}^3$ are each subgroups of the Euclidean group $\mathcal{E}$. $O(3)$ is called the *orthogonal group* of order 3. Which isometries of $\mathbf{E}^3$ are in both these subgroups?

2 The Derivative Map of an Isometry

In Chapter I we showed that an arbitrary mapping $F\colon \mathbf{E}^3 \to \mathbf{E}^3$ has a derivative map F_* which carries each tangent vector $\mathbf{v}$ at $\mathbf{p}$ to a tangent vector $F_*(\mathbf{v})$ at $F(\mathbf{p})$. If F is an isometry, its derivative map is remarkably simple. (Since the distinction between tangent vector and point is crucial here, we temporarily restore the point of application to the notation.)

2.1 Theorem   Let F be an isometry of $\mathbf{E}^3$ with orthogonal part C. Then

$$F_*(\mathbf{v}_p) \;=\; (C\mathbf{v})_{F(p)}$$

for all tangent vectors $\mathbf{v}_p$ to $\mathbf{E}^3$.

Verbally: To get $F_*(\mathbf{v}_p)$, first shift the tangent vector $\mathbf{v}_p$ to the canonically corresponding point $\mathbf{v}$ of $\mathbf{E}^3$, then apply the orthogonal part C of F, and finally shift this point $C(\mathbf{v})$ to the canonically corresponding tangent vector at $F(\mathbf{p})$ (Fig. 3.2). *Thus all tangent vectors at all points $\mathbf{p}$ of $\mathbf{E}^3$ are "rotated" in exactly the same way by F_*—only the new point of application $F(\mathbf{p})$ depends on $\mathbf{p}$.*

Proof. Write $F = TC$ as in Theorem 1.7. Let T be translation by $\mathbf{a}$, so $F(\mathbf{p}) = \mathbf{a} + C(\mathbf{p})$. If $\mathbf{v}_p$ is a tangent vector to $\mathbf{E}^3$, then by Definition 7.4 of Chapter I, $F_*(\mathbf{v}_p)$ is the initial velocity of the curve $t \to F(\mathbf{p} + t\mathbf{v})$. But using the linearity of C, we obtain

$$F(\mathbf{p} + t\mathbf{v}) = TC(\mathbf{p} + t\mathbf{v}) = T(C(\mathbf{p}) + tC(\mathbf{v})) = \mathbf{a} + C(\mathbf{p}) + tC(\mathbf{v})$$
$$= F(\mathbf{p}) + tC(\mathbf{v}).$$

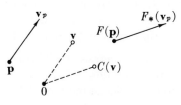

FIG. 3.2

Thus $F_*(v_p)$ is the initial velocity of the curve $t \to F(\mathbf{p}) + tC(\mathbf{v})$, which is precisely the tangent vector $(C\mathbf{v})_{F(p)}$. ∎

Expressed in terms of Euclidean coordinates, this result becomes

$$F_*(\sum_j v_j U_j) = \sum_{i,j} c_{ij} v_j \bar{U}_i$$

where $C = (c_{ij})$ is the orthogonal part of the isometry F, and if U_i is evaluated at $\mathbf{p}$, then $\bar{U}_i$ is evaluated at $F(\mathbf{p})$.

2.2 Corollary Isometries preserve dot products of tangent vectors. That is, if v_p and w_p are tangent vectors to $\mathbf{E}^3$ at the same point, and F is an isometry, then

$$F_*(v_p) \cdot F_*(w_p) = v_p \cdot w_p.$$

Proof. Let C be the orthogonal part of F, and recall that C, being an orthogonal transformation, preserves dot products in $\mathbf{E}^3$. By Theorem 2.1,

$$F_*(v_p) \cdot F_*(w_p) = (C\mathbf{v})_{F(p)} \cdot (C\mathbf{w})_{F(p)} = C\mathbf{v} \cdot C\mathbf{w}$$

$$= \mathbf{v} \cdot \mathbf{w} = v_p \cdot w_p$$

where we have twice used Definition 1.3 of Chapter II (dot products of tangent vectors). ∎

By proving this fundamental corollary and the following theorem, the initial result (Theorem 2.1) has largely accomplished its mission. Thus it should be safe to drop the point of application from the notation once again, and write simply $F_*(\mathbf{v}) \cdot F_*(\mathbf{w}) = \mathbf{v} \cdot \mathbf{w}$. In fancier language, the corollary asserts that for each point $\mathbf{p}$, the derivative map F_{*p} at $\mathbf{p}$ is an orthogonal transformation of tangent spaces (differing from C only by the canonical isomorphisms of $\mathbf{E}^3$).

Since dot products are preserved, it follows automatically that derived concepts such as norm and orthogonality are preserved. Explicitly, if F is an isometry, then $\| F_*(\mathbf{v}) \| = \| \mathbf{v} \|$, and if $\mathbf{v}$ and $\mathbf{w}$ are orthogonal, so are $F_*(\mathbf{v})$ and $F_*(\mathbf{w})$. Thus frames are also preserved: if e_1, e_2, e_3 is a frame at some point $\mathbf{p}$ of $\mathbf{E}^3$ and F is an isometry, then $F_*(e_1), F_*(e_2), F_*(e_3)$ is a frame at $F(\mathbf{p})$. (A direct proof is easy: $e_i \cdot e_j = \delta_{ij}$, so by Corollary 2.2, $F_*(e_i) \cdot F_*(e_j) = e_i \cdot e_j = \delta_{ij}$.)

Assertion (3) of Lemma 1.4 shows how two *points* uniquely determine a translation. We now show that two *frames* uniquely determine an isometry.

2.3 Theorem Given any two frames on $\mathbf{E}^3$, say e_1, e_2, e_3 at the point $\mathbf{p}$ and f_1, f_2, f_3 at the point $\mathbf{q}$, there exists a unique isometry F of $\mathbf{E}^3$ such that $F_*(e_i) = f_i$ for $1 \leq i \leq 3$.

Proof. First we show that there is such an isometry. Let e_1, e_2, e_3, and f_1, f_2, f_3 be the points of $\mathbf{E}^3$ canonically corresponding to the vectors in the two frames. Let C be the unique linear transformation of $\mathbf{E}^3$ such that $C(e_i) = f_i$ for $1 \leq i \leq 3$. It is easy to check that C is orthogonal. Then let T be a translation by the point $\mathbf{q} - C(\mathbf{p})$. Now we assert that the isometry $F = TC$ carries the $\mathbf{e}$ frame to the $\mathbf{f}$ frame. First note that

$$F(\mathbf{p}) = T(C\mathbf{p}) = \mathbf{q} - C(\mathbf{p}) + C(\mathbf{p}) = \mathbf{q}.$$

Then using Theorem 2.1 we get

$$F_*(\mathbf{e}_i) = (Ce_i)_{F(p)} = (f_i)_{F(p)} = (f_i)_q = \mathbf{f}_i$$

for $1 \leq i \leq 3$.

To prove uniqueness, we observe that by Theorem 2.1 this choice of C is the *only* possibility for the orthogonal part of the required isometry. The translation part is then completely determined also, since it must carry $C(\mathbf{p})$ to $\mathbf{q}$. Thus the isometry $F = TC$ is uniquely determined. ∎

Explicit computation of the isometry in the theorem is not difficult. Let $e_i = (a_{i1}, a_{i2}, a_{i3})$ and $f_i = (b_{i1}, b_{i2}, b_{i3})$ for $1 \leq i \leq 3$. Thus the (orthogonal) matrices $A = (a_{ij})$ and $B = (b_{ij})$ are the attitude matrices of the frames $\mathbf{e}_1, \mathbf{e}_2, \mathbf{e}_3$ and $\mathbf{f}_1, \mathbf{f}_2, \mathbf{f}_3$, respectively. We claim that C in the theorem (or, strictly speaking, its matrix) is ${}^t B \cdot A$. It suffices to show that ${}^t BA(e_i) = f_i$, since this uniquely characterizes C. But using the column-vector conventions, we have

$$
{}^t BA \begin{pmatrix} a_{11} \\ a_{12} \\ a_{13} \end{pmatrix} = {}^t B \begin{pmatrix} 1 \\ 0 \\ 0 \end{pmatrix} = \begin{pmatrix} b_{11} \\ b_{12} \\ b_{13} \end{pmatrix};
$$

that is, ${}^t BA(e_1) = f_1$ (the cases $i = 2, 3$ are similar). Thus $C = {}^t BA$. As noted above, T is then necessarily translation by $\mathbf{q} - C(\mathbf{p})$.

EXERCISES

1. If T is a translation, then for every tangent vector $\mathbf{v}$ show that $T_*(\mathbf{v})$ is parallel to $\mathbf{v}$ (same Euclidean coordinates).

2. Prove the general formulas $(GF)_* = G_* F_*$ and $(F^{-1})_* = (F_*)^{-1}$ in the special case where F and G are isometries of $\mathbf{E}^3$.

3. (a) Let $\mathbf{e}_1, \mathbf{e}_2, \mathbf{e}_3$ be a frame at $\mathbf{p}$ with attitude matrix A. If F is the isometry that carries the natural frame at $\mathbf{0}$ to this frame, show that $F = T_p A^{-1}$ $(A^{-1} = {}^t A)$.

   (b) Now let $\mathbf{f}_1, \mathbf{f}_2, \mathbf{f}_3$ be a frame at $\mathbf{q}$ with attitude matrix B. Use Exercise 2 to prove the result in the text that the isometry which carries the $\mathbf{e}$ frame to the $\mathbf{f}$ frame has orthogonal part $B^{-1}A$.

4. (a) Prove that an isometry $F = TC$ carries the plane through **p** orthogonal to **q** to the plane through $F(\mathbf{p})$ orthogonal to $C(\mathbf{q})$.

   (b) If P is the plane through $(\frac{1}{2}, -1, 0)$ orthogonal to $(0, 1, 0)$ find an isometry $F = TC$ such that $F(P)$ is the plane through $(1, -2, 1)$ orthogonal to $(1, 0, -1)$.

5. Given the frame $\mathbf{e}_1 = (2, 2, 1)/3$, $\mathbf{e}_2 = (-2, 1, 2)/3$, $\mathbf{e}_3 = (1, -2, 2)/3$ at $\mathbf{p} = (0, 1, 0)$ and the frame

$$\mathbf{f}_1 = (1, 0, 1)/\sqrt{2}, \qquad \mathbf{f}_2 = (0, 1, 0), \qquad \mathbf{f}_3 = (1, 0, -1)/\sqrt{2}$$

at $\mathbf{q} = (3, -1, 1)$, find the isometry $F = TC$ which carries the **e** frame to the **f** frame.

3 Orientation

We now come to one of the most interesting and elusive ideas in geometry. Intuitively, it is *orientation* that distinguishes between a right-handed glove and a left-handed glove in ordinary space. To handle this concept mathematically, we replace gloves by frames, and separate all the frames on $\mathbf{E}^3$ into two classes as follows. Recall that associated with each frame $\mathbf{e}_1$, $\mathbf{e}_2$, $\mathbf{e}_3$ at a point of $\mathbf{E}^3$ is its attitude matrix A. According to the exercises for Section 1 of Chapter II,

$$\mathbf{e}_1 \cdot \mathbf{e}_2 \times \mathbf{e}_3 = \det A = \pm 1.$$

When this number is $+1$, we shall say that the frame $\mathbf{e}_1$, $\mathbf{e}_2$, $\mathbf{e}_3$ is *positively oriented* (or, right-handed); when it is -1, the frame is *negatively oriented* (or, left-handed).

We omit the easy proof of the following facts.

3.1 Remark   (1) At each point of $\mathbf{E}^3$ the frame assigned by the natural frame field U_1, U_2, U_3 is positively oriented.

(2) A frame $\mathbf{e}_1$, $\mathbf{e}_2$, $\mathbf{e}_3$ is positively oriented if and only if $\mathbf{e}_1 \times \mathbf{e}_2 = \mathbf{e}_3$. Thus the orientation of a frame can be determined, for practical purposes, by the "right-hand rule" given at the end of Section 1 of Chapter II. Pictorially, the frame (P) in Fig. 3.3 is positively oriented, whereas the frame (N) is negatively oriented. In particular, *Frenet frames are always positively oriented*, since by definition $B = T \times N$.

(3) For a positively oriented frame $\mathbf{e}_1$, $\mathbf{e}_2$, $\mathbf{e}_3$, the cross products are

$$\mathbf{e}_1 = \mathbf{e}_2 \times \mathbf{e}_3 = -\mathbf{e}_3 \times \mathbf{e}_2$$
$$\mathbf{e}_2 = \mathbf{e}_3 \times \mathbf{e}_1 = -\mathbf{e}_1 \times \mathbf{e}_3$$
$$\mathbf{e}_3 = \mathbf{e}_1 \times \mathbf{e}_2 = -\mathbf{e}_2 \times \mathbf{e}_1.$$

For a negatively oriented frame, reverse the vectors in each cross product.

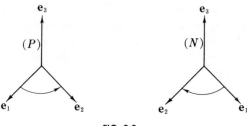

FIG. 3.3

(One need not memorize these formulas—the right-hand rule will give them all correctly.)

Having attached a sign to each frame on $\mathbf{E}^3$, we next attach a sign to each isometry F of $\mathbf{E}^3$. In Chapter II we proved the well-known fact that the determinant of an orthogonal matrix is either $+1$ or -1. Thus if C is the orthogonal part of the isometry F, we define the *sign* of F to be the determinant of C, with notation

$$\operatorname{sgn} F = \det C.$$

We know that the derivative map of an isometry carries frames to frames. The following result tells what happens to their orientations.

3.2 Lemma   If $\mathbf{e}_1$, $\mathbf{e}_2$, $\mathbf{e}_3$ is a frame at some point of $\mathbf{E}^3$ and F is an isometry, then

$$F_*(\mathbf{e}_1)\cdot F_*(\mathbf{e}_2) \times F_*(\mathbf{e}_3) = \operatorname{sgn} F\; \mathbf{e}_1\cdot\mathbf{e}_2 \times \mathbf{e}_3.$$

Proof. If $\mathbf{e}_j = \sum a_{jk}U_k$, then by the coordinate form of Theorem 2.1 we have

$$F_*(\mathbf{e}_j) = \sum_{i,k} c_{ik}a_{jk}\bar{U}_i$$

where $C = (c_{ij})$ is the orthogonal part of F. Thus the attitude matrix of the frame $F_*(\mathbf{e}_1)$, $F_*(\mathbf{e}_2)$, $F_*(\mathbf{e}_3)$ is the matrix

$$\left(\sum_k c_{ik}a_{jk}\right) = \left(\sum_k c_{ik}\,{}^t a_{kj}\right) = C\,{}^t A.$$

But the triple scalar product of a frame is the determinant of its attitude matrix, and by definition $\operatorname{sgn} F = \det C$. Consequently,

$$F_*(\mathbf{e}_1)\cdot F_*(\mathbf{e}_2) \times F_*(\mathbf{e}_3) = \det (C\,{}^t A)$$

$$= \det C\cdot\det{}^t A = \det C\cdot\det A$$

$$= \operatorname{sgn} F\; \mathbf{e}_1\cdot\mathbf{e}_2 \times \mathbf{e}_3. \qquad\blacksquare$$

This lemma shows that if $\operatorname{sgn} F = +1$, then F_* carries positively oriented frames to positively oriented frames, and carries negatively oriented to negatively oriented frames. On the other hand, if sgn $F = -1$, positive goes to negative, and negative to positive.

3.3 Definition   An isometry F of $\mathbf{E}^3$ is said to be

$$orientation\text{-}preserving \text{ if sgn } F = \det C = +1$$

$$orientation\text{-}reversing \text{ if sgn } F = \det C = -1$$

where C is the orthogonal part of F.

3.4 Example

(1) *Translations.* All translations are orientation-preserving. Geometrically this is clear, and in fact the orthogonal part of a translation T is just the identity mapping I, so sgn $T = \det I = +1$.

(2) *Rotations.* Consider the orthogonal transformation C given in Example 1.2, which rotates $\mathbf{E}^3$ through angle θ around the z axis. Its matrix is

$$\begin{pmatrix} \cos\theta & -\sin\theta & 0 \\ \sin\theta & \cos\theta & 0 \\ 0 & 0 & 1 \end{pmatrix}$$

Hence sgn $C = \det C = +1$, so C is orientation-preserving (see Exercise 4).

(3) *Reflections.* One can (literally) see reversal of orientation by using a mirror. Suppose the yz plane of $\mathbf{E}^3$ is the mirror. If one looks toward that plane, the point $\mathbf{p} = (p_1, p_2, p_3)$ appears to be located at the point

$$R(\mathbf{p}) = (-p_1, p_2, p_3)$$

(Fig. 3.4). The mapping R so defined is called *reflection* in the yz plane. Evidently it is an orthogonal transformation with matrix

$$\begin{pmatrix} -1 & 0 & 0 \\ 0 & 1 & 0 \\ 0 & 0 & 1 \end{pmatrix}$$

Thus R is an orientation-reversing isometry, as confirmed by the experimental fact that the mirror image of a right hand is a left hand.

Both dot and cross product were originally defined in terms of *Euclidean* coordinates. We have seen that the dot product is given by the same formula,

$$\mathbf{v} \cdot \mathbf{w} = \left(\sum v_i \mathbf{e}_i\right) \cdot \left(\sum w_i \mathbf{e}_i\right) = \sum v_i w_i,$$

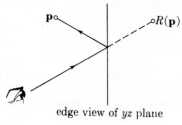

edge view of yz plane

FIG. 3.4

no matter what frame e_1, e_2, e_3 is used to get coordinates for $\mathbf{v}$ and $\mathbf{w}$. We have almost the same result for cross products, but orientation is now involved.

3.5 Lemma Let e_1, e_2, e_3 be a frame at a point of $\mathbf{E}^3$. If $\mathbf{v} = \sum v_i e_i$ and $\mathbf{w} = \sum w_i e_i$, then

$$\mathbf{v} \times \mathbf{w} = \epsilon \begin{vmatrix} e_1 & e_2 & e_3 \\ v_1 & v_2 & v_3 \\ w_1 & w_2 & w_3 \end{vmatrix}$$

where $\epsilon = e_1 \cdot e_2 \times e_3 = \pm 1$.

Proof. It suffices merely to expand the cross product

$$\mathbf{v} \times \mathbf{w} = (v_1 e_1 + v_2 e_2 + v_3 e_3) \times (w_1 e_1 + w_2 e_2 + w_3 e_3)$$

using the formulas (3) of Remark 3.1. For example, if the frame is positively oriented, for the e_1 component of $\mathbf{v} \times \mathbf{w}$, we get

$$v_2 e_2 \times w_3 e_3 + v_3 e_3 \times w_2 e_2 = (v_2 w_3 - v_3 w_2) e_1.$$

Since $\epsilon = 1$ in this case, we get the same result from the right side of the equation to be proved. ∎

It follows immediately that the effect of an isometry on cross products also involves orientation.

3.6 Theorem Let $\mathbf{v}$ and $\mathbf{w}$ be tangent vectors to $\mathbf{E}^3$ at $\mathbf{p}$. If F is an isometry of $\mathbf{E}^3$, then

$$F_*(\mathbf{v} \times \mathbf{w}) = \operatorname{sgn} F\, F_*(\mathbf{v}) \times F_*(\mathbf{w}).$$

Proof. Write $\mathbf{v} = \sum v_i U_i(\mathbf{p})$ and $\mathbf{w} = \sum w_i U_i(\mathbf{p})$. Now let

$$e_i = F_*(U_i(\mathbf{p})).$$

Since F_* is linear,

$$F_*(\mathbf{v}) = \sum v_i e_i \quad \text{and} \quad F_*(\mathbf{w}) = \sum w_i e_i.$$

A straightforward computation using Lemma 3.5 shows that

$$F_*(\mathbf{v}) \times F_*(\mathbf{w}) = \epsilon F_*(\mathbf{v} \times \mathbf{w}),$$

where

$$\epsilon = e_1 \cdot e_2 \times e_3 = F_*(U_1(\mathbf{p})) \cdot F_*(U_2(\mathbf{p})) \times F_*(U_3(\mathbf{p})).$$

But U_1, U_2, U_3 is positively oriented, so by Lemma 3.2, $\epsilon = \operatorname{sgn} F$. ∎

EXERCISES

1. Prove

$$\text{sgn } (FG) = \text{sgn } F \cdot \text{sgn } G = \text{sgn } (GF).$$

Deduce that sgn $F = \text{sgn } (F^{-1})$.

2. If H_0 is an orientation-reversing isometry of $\mathbf{E}^3$, show that *every* orientation-reversing isometry has a unique expression $H_0 F$, where F is orientation-preserving.

3. Let $\mathbf{v} = (3, 1, -1)$ and $\mathbf{w} = (-3, -3, 1)$ be tangent vectors at some point. If C is the orthogonal transformation given in Exercise 4 of Section 1, check the formula

$$C_*(\mathbf{v} \times \mathbf{w}) = \text{sgn } C \, C_*(\mathbf{v}) \times C_*(\mathbf{w}).$$

4. A *rotation* is an orthogonal transformation C such that det $C = +1$. Prove that C does, in fact, rotate $\mathbf{E}^3$ around an axis. Explicitly, given a rotation C, show that there exists a number ϑ and points $\mathbf{e}_1$, $\mathbf{e}_2$, $\mathbf{e}_3$ with $\mathbf{e}_i \cdot \mathbf{e}_j = \delta_{ij}$ such that (Fig. 3.5)

$$C(\mathbf{e}_1) = \cos \vartheta \, \mathbf{e}_1 + \sin \vartheta \, \mathbf{e}_2$$

$$C(\mathbf{e}_2) = -\sin \vartheta \, \mathbf{e}_1 + \cos \vartheta \, \mathbf{e}_2$$

$$C(\mathbf{e}_3) = \mathbf{e}_3.$$

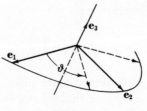

FIG. 3.5

(*Hint:* The fact that the dimension of $\mathbf{E}^3$ is odd means that C has a characteristic root $+1$, so there is a point $\mathbf{p} \neq \mathbf{0}$ such that $C(\mathbf{p}) = \mathbf{p}$.)

5. Let $\mathbf{a}$ be a point of $\mathbf{E}^3$ such that $\| \mathbf{a} \| = 1$. Prove that the formula

$$C(\mathbf{p}) = \mathbf{a} \times \mathbf{p} + \mathbf{p} \cdot \mathbf{a} \, \mathbf{a}$$

defines an orthogonal transformation. Describe its general effect on $\mathbf{E}^3$.

6. Prove
   (a) The set $O^+(3)$ of all rotations of $\mathbf{E}^3$ is a subgroup of the orthogonal group $O(3)$ (see Ex. 8 of III.1).
   (b) The set $\mathcal{E}^+$ of all orientation-preserving isometries of $\mathbf{E}^3$ is a subgroup of the Euclidean group $\mathcal{E}$.

7. Find a single formula for all isometries of the real line $\mathbf{E}^1$. Do the same for the plane $\mathbf{E}^2$ (use $\epsilon = \pm 1$). Which of these isometries are orientation-preserving?

4 Euclidean Geometry

In the discussion at the beginning of this chapter, we recalled a fundamental feature of plane geometry: If there is an isometry carrying one triangle onto another, then the two (congruent) triangles have exactly the same geometric properties. A close examination of this statement will show that it does not admit a proof—it is, in fact, just the definition of "geometric property of a triangle." More generally, *Euclidean geometry* can be defined as the totality of concepts that are preserved by isometries of Euclidean space. For example, Corollary 2.2 shows that the notion of dot product on tangent vectors belongs to Euclidean geometry. Similarly Theorem 3.6 shows that the cross product is preserved by isometries (except possibly for sign).

This famous definition of Euclidean geometry is somewhat generous, however. In practice, the label "Euclidean geometry" is usually attached only to those concepts that are preserved by isometries, but *not* by arbitrary mappings, or even the more restrictive class of mappings (diffeomorphisms) that possess inverse mappings. An example should make this distinction clearer. If $\alpha = (\alpha_1, \alpha_2, \alpha_3)$ is a curve in $\mathbf{E}^3$, then the various derivatives

$$\alpha' = \left(\frac{d\alpha_1}{dt}, \frac{d\alpha_2}{dt}, \frac{d\alpha_3}{dt}\right), \quad \alpha'' = \left(\frac{d^2\alpha_1}{dt^2}, \frac{d^2\alpha_2}{dt^2}, \frac{d^2\alpha_3}{dt^2}\right), \cdots$$

look pretty much alike. Now, we interpreted Theorem 7.8 of Chapter I as saying that *velocity is preserved by arbitrary mappings* $F\colon \mathbf{E}^3 \to \mathbf{E}^3$. That is, if $\beta = F(\alpha)$, then $\beta' = F_*(\alpha')$. But it is easy to see that *acceleration is not preserved by arbitrary mappings*. For example, if $\alpha(t) = (t, 0, 0)$ and $F = (x^2, y, z)$, then $\alpha'' = 0$; hence $F_*(\alpha'') = 0$. But $\beta = F(\alpha)$ has the formula $\beta(t) = (t^2, 0, 0)$, so $\beta'' = 2U_1$. Thus in this case, $\beta = F(\alpha)$, but $\beta'' \neq F_*(\alpha'')$. We shall see in a moment, however, that acceleration is preserved by *isometries*.

For this reason, the notion of velocity belongs to the *calculus* of Euclidean space, while the notion of acceleration belongs to Euclidean *geometry*. In this section we examine some of the concepts introduced in Chapter II and prove that they are, in fact, preserved by isometries. (We leave largely to the reader the easier task of showing that they are not preserved by diffeomorphisms.)

Recall the notion of vector field on a curve (Definition 2.2 of Chapter II). Now if Y is a vector field on $\alpha\colon I \to \mathbf{E}^3$, and $F\colon \mathbf{E}^3 \to \mathbf{E}^3$ is any mapping, then $\bar{Y} = F_*(Y)$ is a vector field on the image curve $\bar{\alpha} = F(\alpha)$. In fact, for each t in I, $Y(t)$ is a tangent vector to $\mathbf{E}^3$ at the point $\alpha(t)$. But then $\bar{Y}(t) = F_*(Y(t))$ is a tangent vector to $\mathbf{E}^3$ at the point $F(\alpha(t)) = \bar{\alpha}(t)$.

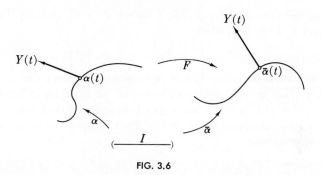

FIG. 3.6

(These relationships are illustrated in Fig. 3.6.) Isometries preserve the *derivatives* of such vector fields.

4.1 Corollary Let Y be a vector field on a curve α in $\mathbf{E}^3$, and let F be an isometry of $\mathbf{E}^3$. Then $\bar{Y} = F_*(Y)$ is a vector field on $\bar{\alpha} = F(\alpha)$, and

$$\bar{Y}' = F_*(Y').$$

Proof. We compute $F_*(Y')$ and $\bar{Y}'$ starting from the expression

$$Y = \sum y_j U_j$$

for Y in terms of its Euclidean coordinate functions. To differentiate such a vector field Y, one simply differentiates its Euclidean coordinate functions, so

$$Y' = \sum \frac{dy_j}{dt} U_j.$$

Thus by the coordinate version of Theorem 2.1, we get

$$F_*(Y') = \sum c_{ij} \frac{dy_j}{dt} \bar{U}_i.$$

On the other hand,

$$\bar{Y} = F_*(Y) = \sum c_{ij} y_j \bar{U}_i.$$

But each c_{ij} is constant, being by definition an entry in the matrix of the orthogonal part of the isometry F. Hence

$$\bar{Y}' = \sum \frac{d}{dt} (c_{ij} y_j) \bar{U}_i = \sum c_{ij} \frac{dy_j}{dt} \bar{U}_i.$$

Thus the vector fields $F_*(Y')$ and $\bar{Y}'$ are the same. ∎

We claimed earlier that isometries preserve acceleration: If $\bar{\alpha} = F(\alpha)$, where F is an isometry, then $\bar{\alpha}'' = F_*(\alpha'')$. This is an immediate consequence of the preceding result, for if we set $Y = \alpha'$, then by Theorem 7.8 of Chapter I, $\bar{Y} = \bar{\alpha}'$; hence

$$\bar{\alpha}'' = \bar{Y}' = F_*(Y') = F_*(\alpha'').$$

Now we show that the Frenet apparatus of a curve is preserved by isometries. This is certainly to be expected on intuitive grounds, since a rigid motion ought to carry one curve into another that turns and twists in exactly the same way. And this is what happens *when the isometry is orientation-preserving*.

4.2 Theorem Let β be a unit-speed curve in $\mathbf{E}^3$ with positive curvature, and let $\bar{\beta} = F(\beta)$ be the image curve of β under an isometry F of $\mathbf{E}^3$. Then

$$\bar{\kappa} = \kappa \qquad\qquad \bar{T} = F_*(T)$$

$$\bar{\tau} = \operatorname{sgn} F\, \tau \qquad \bar{N} = F_*(N)$$

$$\bar{B} = \operatorname{sgn} F\, F_*(B)$$

where $\operatorname{sgn} F = \pm 1$ is the sign of the isometry F.

Proof. Note that $\bar{\beta}$ is also a unit-speed curve, since

$$\| \bar{\beta}' \| = \| F_*(\beta') \| = \| \beta' \| = 1.$$

Thus the definitions in Section 3 of Chapter II apply to both β and $\bar{\beta}$, so

$$\bar{T} = \bar{\beta}' = F_*(\beta') = F_*(T).$$

Since F_* preserves both acceleration and norms, it follows from the definition of curvature that

$$\bar{\kappa} = \| \bar{\beta}'' \| = \| F_*(\beta'') \| = \| \beta'' \| = \kappa.$$

To get the full Frenet frame, we now use the hypothesis $\kappa > 0$ (which implies $\bar{\kappa} > 0$, since $\bar{\kappa} = \kappa$). By definition, $N = \beta''/\kappa$; hence using preceding facts, we find

$$\bar{N} = \frac{\bar{\beta}''}{\bar{\kappa}} = \frac{F_*(\beta'')}{\kappa} = F_*\left(\frac{\beta''}{\kappa} \right) = F_*(N).$$

It remains only to prove the interesting cases B and τ. Since the definition $B = T \times N$ involves a cross product, we use Theorem 3.6 to get

$$\bar{B} = \bar{T} \times \bar{N} = F_*(T) \times F_*(N) = \operatorname{sgn} F\, F_*(T \times N) = \operatorname{sgn} F\, F_*(B).$$

The definition of torsion is essentially $\tau = -B' \cdot N = B \cdot N'$. Thus using the results above for B and N, we get

$$\bar{\tau} = \bar{B} \cdot \bar{N}' = \operatorname{sgn} F\, F_*(B) \cdot F_*(N') = \operatorname{sgn} F\, B \cdot N' = \operatorname{sgn} F\, \tau. \quad\blacksquare$$

The presence of $\operatorname{sgn} F$ in the formula for the torsion of $F(\beta)$ shows that the torsion of a curve gives a more subtle description of the curve than has been apparent so far. *The sign of τ measures the orientation of the twisting of the curve.* If F is orientation-reversing, the formula $\bar\tau = -\tau$ proves that the twisting of the image of curve $F(\beta)$ is exactly opposite to that of β 'self.

A simple example will illustrate this reversal.

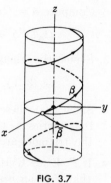

FIG. 3.7

4.3 Example Let β be the unit-speed helix

$$\beta(s) = \left(\cos\frac{s}{c}, \sin\frac{s}{c}, \frac{s}{c}\right),$$

gotten from Example 3.3 of Chapter II by setting $a = b = 1$; hence $c = \sqrt{2}$. We know from the general formulas for helices that $\kappa = \tau = \frac{1}{2}$. Now let R be reflection in the xy plane, so R is the isometry $R(x,y,z) = (x, y, -z)$. Thus the image curve $\bar\beta = R(\beta)$ is the mirror image

$$\bar\beta(s) = \left(\cos\frac{s}{c}, \sin\frac{s}{c}, -\frac{s}{c}\right)$$

of the original curve. One can see in Fig. 3.7 that the mirror has its usual effect: β and $\bar\beta$ twist in opposite ways—if β is "right-handed," then $\bar\beta$ is "left-handed." (The fact that β is going up and $\bar\beta$ down is, in itself, irrelevant.) Formally: The reflection R is orientation-reversing; hence the theorem predicts $\bar\kappa = \kappa = \frac{1}{2}$ and $\bar\tau = -\tau = -\frac{1}{2}$. Since $\bar\beta$ is just the helix gotten in Example 3.3 of Chapter II by taking $a = 1$ and $b = -1$, this may be checked by the general formulas there.

EXERCISES

1. Let $F = TC$ be an isometry of $\mathbf{E}^3$, β a unit speed curve in $\mathbf{E}^3$. Prove
   (a) If β is a cylindrical helix, then $F(\beta)$ is a cylindrical helix.
   (b) If β has spherical image $\bar\beta$, then $F(\beta)$ has spherical image $C(\bar\beta)$.

2. Let $Y = (t, 1 - t^2, 1 + t^2)$ be a vector field on the helix

$$\alpha(t) = (\cos t, \sin t, 2t),$$

and let C be the orthogonal transformation

$$C = \begin{pmatrix} -1 & 0 & 0 \\ 0 & 1/\sqrt{2} & -1/\sqrt{2} \\ 0 & 1/\sqrt{2} & 1/\sqrt{2} \end{pmatrix}$$

Compute $\bar{\alpha} = C(\alpha)$ and $\bar{Y} = C_*(Y)$, and check that

$$C_*(Y') = \bar{Y}', \qquad C_*(\alpha'') = \bar{\alpha}'', \qquad Y' \cdot \alpha'' = \bar{Y}' \cdot \bar{\alpha}''.$$

3. Sketch the triangles in $\mathbf{E}^2$ that have vertices

$$\Delta_1: (3, 1), (7, 1), (7, 4) \qquad \Delta_2: (2, 0), (2, 5), (-\tfrac{2}{5}, \tfrac{16}{5})$$

Show that these triangles are congruent by exhibiting an isometry F that carries Δ_1 to Δ_2. (*Hint:* the orthogonal part of F is not altered if the triangles are translated.)

4. If $F: \mathbf{E}^3 \to \mathbf{E}^3$ is a mapping such that F_* preserves dot products, show that F is an isometry. (*Hint:* Use Ex. 11 of II.2.)

5. Let F be an isometry of $\mathbf{E}^3$. For each vector field V let $\bar{V}$ be the vector field such that $F_*(V(\mathbf{p}) = \bar{V}(F(\mathbf{p}))$ for all $\mathbf{p}$. Prove that isometries preserve covariant derivatives; that is, show $\overline{\nabla_V W} = \nabla_{\bar{V}} \bar{W}$.

5 Congruence of Curves

In the case of curves in $\mathbf{E}^3$, the general notion of congruence takes the following form.

5.1 Definition Two curves $\alpha, \beta: I \to \mathbf{E}^3$ are *congruent* provided there exists an isometry F of $\mathbf{E}^3$ such that $\beta = F(\alpha)$; that is, $\beta(t) = F(\alpha(t))$ for all t in I.

Intuitively speaking, congruent curves are the same except for position in space. They represent *trips at the same speed along routes of the same shape*. For example, the helix $\alpha(t) = (\cos t, \sin t, t)$ spirals around the z axis in exactly the same way the helix $\beta(t) = (t, \cos t, \sin t)$ spirals around the x axis. Evidently these two curves are congruent, since if F is the isometry such that

$$F(p_1, p_2, p_3) = (p_3, p_1, p_2),$$

then $F(\alpha) = \beta$.

To decide whether given curves α and β are congruent, it is hardly practical to try all the isometries of $\mathbf{E}^3$ to see if there is one that carries α to β. What we want is a description of the shape of a unit-speed curve so accurate that if α and β have the same description, then they must be congruent. The proper description, as the reader will doubtless suspect, is given by curvature and torsion. To prove this we need one preliminary result.

Curves whose congruence is established by a translation are said to be *parallel*. Thus curves $\alpha, \beta: I \to \mathbf{E}^3$ are parallel if and only if there is a point

p in $\mathbf{E}^3$ such that $\beta(s) = \alpha(s) + \mathbf{p}$ for all s in I, or, in functional notation, $\beta = \alpha + \mathbf{p}$.

5.2 Lemma Two curves $\alpha, \beta \colon I \to \mathbf{E}^3$ are parallel if their velocity vectors $\alpha'(s)$ and $\beta'(s)$ are parallel for each s in I. In this case, if $\alpha(s_0) = \beta(s_0)$ for some one s_0 in I, then $\alpha = \beta$.

Proof. By definition, if $\alpha'(s)$ and $\beta'(s)$ are parallel, they have the same Euclidean coordinates. Thus

$$\frac{d\alpha_i}{ds}(s) = \frac{d\beta_i}{ds}(s) \qquad \text{for } 1 \leqq i \leqq 3$$

where α_i and β_i are the Euclidean coordinate functions of α and β. But by elementary calculus, the equation $d\alpha_i/ds = d\beta_i/ds$ implies that there is a constant p_i such that $\beta_i = \alpha_i + p_i$. Hence $\beta = \alpha + \mathbf{p}$. Furthermore, if $\alpha(s_0) = \beta(s_0)$, we deduce that $\mathbf{p} = \mathbf{0}$; hence $\alpha = \beta$. ∎

5.3 Theorem If $\alpha, \beta \colon I \to E^3$ are unit-speed curves such that $\kappa_\alpha = \kappa_\beta$ and $\tau_\alpha = \pm\tau_\beta$, then α and β are congruent.

Proof. There are two main steps:

(1) Replace α by a suitably chosen congruent curve $F(\alpha)$.

(2) Show that $F(\alpha) = \beta$ (Fig. 3.8).

Our guide for the choice in (1) is Theorem 4.2. Fix a number, say 0, in the interval I. If $\tau_\alpha = \tau_\beta$, then let F be the (orientation-preserving) isometry that carries the Frenet frame $T_\alpha(0)$, $N_\alpha(0)$, $B_\alpha(0)$ of α at $\alpha(0)$ to the Frenet frame $T(0)$, $N(0)$, $B(0)$, of β at $\beta(0)$. (The existence of this isometry is guaranteed by Theorem 2.3.) Denote the Frenet apparatus of $\bar{\alpha} = F(\alpha)$ by $\bar{\kappa}, \bar{\tau}, \bar{T}, \bar{N}, \bar{B}$; then it follows immediately from Theorem 4.2 and the information above that

$$
\begin{array}{ll}
\bar{\alpha}(0) = \beta(0) & \bar{T}(0) = T(0) \\
\bar{\kappa} = \kappa_\beta & \bar{N}(0) = N(0) \qquad\qquad (\ddagger) \\
\bar{\tau} = \tau_\beta & \bar{B}(0) = B(0).
\end{array}
$$

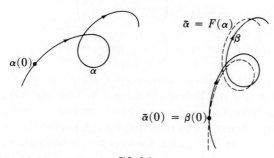

FIG. 3.8

On the other hand, if $\tau_\alpha = -\tau_\beta$, we choose F to be the (orientation-reversing) isometry that carries $T_\alpha(0)$, $N_\alpha(0)$, $B_\alpha(0)$ at $\alpha(0)$ to the frame $T(0)$, $N(0)$, $-B(0)$ at $\beta(0)$. (Frenet frames are positively oriented; hence this last frame is negatively oriented: This is why F is orientation-reversing.) Then it follows from Theorem 4.2 that the equations ($\ddagger$) hold also for $\bar{\alpha} = F(\alpha)$ and β. For example,

$$\bar{B}(0) = -F_*(B_\alpha(0)) = B(0).$$

For step (2) of the proof, we shall show $\bar{T} = T$; that is, the unit tangents of $\bar{\alpha} = F(\alpha)$ and β are parallel at each point. Since $\bar{\alpha}(0) = \beta(0)$, it will follow from Lemma 5.2 that $F(\alpha) = \beta$. On the interval I, consider the real-valued function $f = \bar{T} \cdot T + \bar{N} \cdot N + \bar{B} \cdot B$. Since these are *unit* vector fields, the Schwarz inequality (Section 1, Chapter II) shows that

$$\bar{T} \cdot T \leqq 1;$$

furthermore $\bar{T} \cdot T = 1$ if and only if $\bar{T} = T$. Similar remarks hold for the other two terms in f. Thus *it suffices to show that f has constant value 3.* By ($\ddagger$), $f(0) = 3$. Now consider

$$f' = \bar{T}' \cdot T + \bar{T} \cdot T' + \bar{N}' \cdot N + \bar{N} \cdot N' + \bar{B}' \cdot B + \bar{B} \cdot B'.$$

A simple computation completes the proof. Substitute the Frenet formulas in this expression and use the equations $\bar{\kappa} = \kappa$, $\bar{\tau} = \tau$ from ($\ddagger$). The resulting eight terms cancel in pairs, so $f' = 0$, and f has, indeed, constant value 3. ∎

Thus *a unit-speed curve is determined but for position in $\mathbf{E}^3$ by its curvature and torsion.*

Actually the proof of Theorem 5.3 does more than establish that α and β are congruent; it shows how to compute *explicitly* an isometry carrying α to β. We illustrate this in a special case.

5.4 Example   Consider the unit-speed curves $\alpha, \beta: \mathbf{R} \to \mathbf{E}^3$ such that

$$\alpha(s) = \left(\cos \frac{s}{c}, \sin \frac{s}{c}, \frac{s}{c} \right)$$

$$\beta(s) = \left(\cos \frac{s}{c}, \sin \frac{s}{c}, -\frac{s}{c} \right)$$

where $c = \sqrt{2}$. Obviously these curves are congruent by means of a reflection—they are the helices considered in Example 4.3—but we shall ignore this in order to describe a general method for computing the required isometry. According to Example 3.3 of Chapter II, α and β have the same curvature, $\kappa_\alpha = \frac{1}{2} = \kappa_\beta$; but torsions of opposite sign, $\tau_\alpha = \frac{1}{2} = -\tau_\beta$. Thus the theorem predicts congruence by means of an orientation-reversing isometry F. From its proof we see that F must carry the Frenet frame

$$T_\alpha(0) = (0, a, a)$$
$$N_\alpha(0) = (-1, 0, 0)$$
$$B_\alpha(0) = (0, -a, a)$$

to the frame

$$T_\beta(0) = (0, a, -a)$$
$$N_\beta(0) = (-1, 0, 0)$$
$$-B_\beta(0) = (0, -a, -a)$$

where $a = 1/\sqrt{2}$. (These explicit formulas also come from Example 3.3 of Chapter II.) By the remark following Theorem 2.3, the isometry F has orthogonal part $C = {}^tBA$, where A and B are the attitude matrices of the two frames above. Thus

$$C = \begin{pmatrix} 0 & -1 & 0 \\ a & 0 & -a \\ -a & 0 & -a \end{pmatrix} \begin{pmatrix} 0 & a & a \\ -1 & 0 & 0 \\ 0 & -a & a \end{pmatrix} = \begin{pmatrix} 1 & 0 & 0 \\ 0 & 1 & 0 \\ 0 & 0 & -1 \end{pmatrix}$$

since $a = 1/\sqrt{2}$. These two frames have the same point of application $\alpha(0) = \beta(0) = (1, 0, 0)$. But C does not move this point, so the translation part of F is just the identity map. Thus we have (correctly) found that the reflection $F = C$ carries α to β.

From the viewpoint of Euclidean geometry, two curves in $\mathbf{E}^3$ are "the same" if they differ only by an isometry of $\mathbf{E}^3$. What, for example, is a helix? Not just a curve that spirals around the z axis as in Example 3.3 of Chapter II, but any curve congruent to one of these special helixes. One can give general formulas, but the best characterization follows.

5.5 Corollary Let α be a unit speed curve in $\mathbf{E}^3$. Then α is a helix if and only if both its curvature and torsion are nonzero constants.

Proof. For any numbers $a > 0$ and $b \neq 0$, let $\beta_{a,b}$ be the special helix given in Example 3.3 of Chapter II. If α is congruent to $\beta_{a,b}$, then (changing the sign of b if necessary) we can assume the isometry is orientation-preserving. Thus, α has curvature and torsion

$$\kappa = \frac{a}{a^2 + b^2} \qquad \tau = \frac{b}{a^2 + b^2}.$$

Conversely, suppose α has constant nonzero κ and τ. Solving the preceding equations, we get

$$a = \frac{\kappa}{\kappa^2 + \tau^2} \qquad b = \frac{\tau}{\kappa^2 + \tau^2}.$$

Thus α and $\beta_{a,b}$ have the same curvature and torsion, hence they are congruent. ∎

Our results so far demand unit speed, but it is easy to weaken this restriction.

5.6 Corollary   Let $\alpha, \beta \colon I \to \mathbf{E}^3$ be arbitrary-speed curves. If

$$v_\alpha = v_\beta > 0, \qquad \kappa_\alpha = \kappa_\beta > 0, \qquad \text{and} \qquad \tau_\alpha = \pm\tau_\beta,$$

then the curves α and β are congruent.

Proof. Let $\bar{\alpha}$ and $\bar{\beta}$ be unit-speed reparametrizations of α and β, both based at, say, $t = 0$. Since α and β have the same speed function, it follows immediately that they also have the same arc length function $s = s(t)$ and hence the same inverse function $t = t(s)$. But since

$$\kappa_\alpha = \kappa_\beta \quad \text{and} \quad \tau_\alpha = \pm\tau_\beta,$$

we deduce from the general definitions of curvature and torsion in that Section 4 of Chapter II

$$\kappa_{\bar{\alpha}}(s) = \kappa_\alpha(t(s)) = \kappa_\beta(t(s)) = \kappa_{\bar{\beta}}(s)$$
$$\tau_{\bar{\alpha}}(s) = \tau_\alpha(t(s)) = \pm\tau_\beta(t(s)) = \pm\tau_{\bar{\beta}}(s).$$

Hence the congruence theorem (5.3) shows that $\bar{\alpha}$ and $\bar{\beta}$ are congruent —say, $F(\bar{\alpha}) = \bar{\beta}$. But then the same isometry carries α to β, since

$$F(\alpha(t)) = F(\bar{\alpha}(s(t))) = F(\bar{\beta}(s(t))) = F(\beta(t)). \qquad ∎$$

The theory of curves we have presented applies only to regular curves with positive curvature, because only for such curves is it generally possible to define the Frenet frame field. However, a completely arbitrary curve α in $\mathbf{E}^3$ can be studied by means of an *arbitrary* frame field on α, that is, any three unit-vector fields E_1, E_2, E_3 on α that are orthogonal at each point (Fig. 3.9).

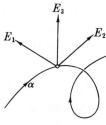

E_3

E_1 E_2

α

FIG. 3.9

For example, the congruence theorem 5.3 can easily be extended to arbitrary curves.

5.7 Theorem Let $\alpha, \beta: I \rightarrow \mathbf{E}^3$ be arbitrary curves, and let E_1, E_2, E_3 be a frame field on α; F_1, F_2, F_3 a frame field on β. If

(1) $\alpha' \cdot E_i = \beta' \cdot F_i$ $(1 \leq i \leq 3)$

(2) $E_i' \cdot E_j = F_i' \cdot F_j$ $(1 \leq i,j \leq 3)$,

then α and β are congruent.

Proof. We need only generalize the argument in Theorem 5.3. Fix a number, say 0, in I. Then let F be the isometry that carries

$$E_1(0),\ E_2(0),\ E_3(0) \qquad \text{to} \qquad F_1(0),\ F_2(0),\ F_3(0).$$

Since F_* preserves dot products, it follows that $\bar{E}_i = F_*(E_i)$ $(1 \leq i \leq 3)$ is a frame field on $\bar\alpha = F(\alpha)$. Since F_* preserves velocities and derivatives of vector fields as well, we deduce

$$
\begin{aligned}
\bar\alpha(0) &= \beta(0) & \bar\alpha' \cdot \bar{E}_i &= \beta' \cdot F_i \\
\bar{E}_i(0) &= F_i(0) & \bar{E}_i' \cdot \bar{E}_j &= F_i' \cdot F_j \qquad \text{for} \quad 1 \leq i,j \leq 3.
\end{aligned}
\qquad (\ddagger)
$$

This last equation means we can write $\bar{E}_i' = \sum a_{ij}\bar{E}_j$ and $F_i' = \sum a_{ij}F_j$ with the same coefficient functions a_{ij}. Note that $a_{ij} + a_{ji} = 0$. (Differentiate $\bar{E}_i \cdot \bar{E}_j = \delta_{ij}$.) If $f = \sum \bar{E}_i \cdot F_i$, we then prove $f = 3$ just as before, since

$$f' = \sum (\bar{E}_i' \cdot F_i + \bar{E}_i \cdot F_i') = \sum (a_{ij} + a_{ji})\bar{E}_j \cdot F_i = 0.$$

Thus $\bar{E}_i = F_i$ (parallelism!) and it follows from $(\ddagger)$ that

$$\bar\alpha' = \sum \bar\alpha' \cdot \bar{E}_i\, \bar{E}_i \qquad \text{and} \qquad \beta' = \sum \beta' \cdot F_i\, F_i$$

are parallel at each point. But $\bar\alpha(0) = \beta(0)$; hence by Lemma 5.2,

$$F(\alpha) = \bar\alpha = \beta. \qquad \blacksquare$$

We shall need this degree of generalization in Section 8 of Chapter VI. A more elegant (but slightly less general) version of this theorem is given in Exercise 3.

EXERCISES

1. Given a curve $\alpha = (\alpha_1, \alpha_2, \alpha_3): I \rightarrow \mathbf{E}^3$, prove that $\beta: I \rightarrow \mathbf{E}^3$ is congruent to α if and only if β can be written

$$\beta(t) = \mathbf{p} + \alpha_1(t)\mathbf{e}_1 + \alpha_2(t)\mathbf{e}_2 + \alpha_3(t)\mathbf{e}_3,$$

where $\mathbf{e}_i \cdot \mathbf{e}_j = \delta_{ij}$.

2. Let α be the curve in Example 4.4 of Chapter II. Find a (congruent) curve of the form $\gamma(t) = (at, bt^2, ct^3)$ and an isometry F such that $F(\alpha) = \gamma$.

3. Let E_1, E_2, E_3 be a frame field on $\mathbf{E}^3$ with dual forms θ_i and connection forms ω_{ij}. Prove that two curves $\alpha, \beta: I \to \mathbf{E}^3$ are congruent if we have $\theta_i(\alpha') = \theta_i(\beta')$ and $\omega_{ij}(\alpha') = \omega_{ij}(\beta')$ for $1 \leqq i,j \leqq 3$.

4. Show that the curve

$$\beta(t) = (t + \sqrt{3} \sin t, 2 \cos t, \sqrt{3}\, t - \sin t)$$

is a helix by computing its curvature and torsion. Find a helix α of the form $(a \cos t, a \sin t, bt)$ and an isometry F such that $F(\alpha) = \beta$.

5. Let $\alpha, \beta: I \to \mathbf{E}^3$ be congruent curves with $\kappa > 0$. Show that there is only *one* isometry F such that $F(\alpha) = \beta$—unless $\tau = 0$, in which case there are exactly *two*.

6. (*Continuation*). Find the two isometries carrying the parabola $\alpha(t) = (\sqrt{2}t, t^2, 0)$ to the parabola $\beta(t) = (-t, t, t^2)$.

7. If β is a unit-speed curve in $\mathbf{E}^3$, then every unit-speed reparametrization $\bar{\beta}$ of β has the form $\bar{\beta}(s) = \beta(\pm s + s_0)$. If β and $\bar{\beta}$ are congruent, this represents a *symmetry* in the common route of β and $\bar{\beta}$. Prove that helical routes are completely symmetric. Explicitly, show that the helix β in Example 3.3 of Chapter II is congruent to *every* unit-speed reparametrization $\bar{\beta}$ by explicitly finding the isometry $F = TC$ such that $F(\beta) = \bar{\beta}$.

8. Two curves $\alpha: I \to \mathbf{E}^3$ and $\beta: I \to \mathbf{E}^3$ have *congruent routes* provided there is an isometry F such that $F(\alpha)$ is a reparametrization of β.
   (a) Show that unit-speed curves α and β have congruent routes if and only if there is a number s_0 such that $\kappa_\alpha(s) = \kappa_\beta(\epsilon s + s_0)$ and $\tau_\alpha(s) = \pm \tau_\beta(\epsilon s + s_0)$, where ϵ is either $+1$ or -1.
   (b) If α is the curve in Ex. 2 of II.4 show α and $\beta = (e^t, e^{-t}/2, t)$ have congruent routes. Exhibit the isometry $F = TC$ and the reparametrization needed to fulfill the definition.
   The following three exercises deal with curves in $\mathbf{E}^2$.

9. Given any differentiable† function κ on an interval I, prove that there is a unit-speed curve α in $\mathbf{E}^2$ such that κ is the curvature function of α. (*Hint:* Find an integral formula for α by reversing the order of results in Ex. 8 of II.3.)

10. Find plane curves—in any convenient parametrization—for which

---

† Even if κ is merely continuous, we obtain a twice-differentiable curve. Similar results can be proved for curves in $\mathbf{E}^3$ using systems of ordinary differential equations. See Willmore [3].

(a) $\kappa(s) = 1/(1 + s^2)$,   (b) $\kappa(s) = 1/s$ $(s > 0)$, where s is the arc length.

11. Prove that two unit-speed curves α and β in $\mathbf{E}^2$ are congruent if and only if $\kappa_\alpha = \pm \kappa_\beta$.

6 Summary

The basic result of this chapter is that an arbitrary isometry of Euclidean space can be uniquely expressed as an orthogonal transformation followed by a translation. Its main consequences are that the derivative map of an isometry F is at every point essentially just the orthogonal part of F, and that there is a unique isometry which carries any one given frame to another. Then it is a routine matter to test the concepts introduced earlier and discover which belong to *Euclidean geometry*, that is, which are preserved by isometries of Euclidean space. Finally, we proved an analogue for curves of the well-known "side-angle-side," "side-side-side" theorems on triangles from elementary plane geometry. Namely, we showed that curvature and torsion (and speed) provide a necessary and sufficient condition for two given curves to be congruent. Furthermore, the required isometry can be explicitly computed.

CHAPTER IV

Calculus on a Surface

This chapter begins with the definition of a surface in $\mathbf{E}^3$ and with some standard ways to construct surfaces. Although this concept is a more-or-less familiar one, it is not as widely known as it should be that each surface has a differential and integral calculus strictly comparable with the usual calculus on the Euclidean plane $\mathbf{E}^2$. The elements of this calculus—functions, vector fields, differential forms, mappings—belong strictly to the surface and not to the Euclidean space $\mathbf{E}^3$ in which the surface is located. Indeed, we shall see in the final section that this calculus survives undamaged when $\mathbf{E}^3$ is removed, leaving just the surface and nothing more.

1 Surfaces in $\mathbf{E}^3$

A surface in $\mathbf{E}^3$ is, to begin with, a *subset* of $\mathbf{E}^3$, that is, a certain collection of points of $\mathbf{E}^3$. Of course, not all subsets are surfaces: We must certainly require that a surface be smooth and two-dimensional. We shall express this requirement in mathematical terms by the next two definitions.

1.1 Definition A *coordinate patch* $\mathbf{x}\colon D \rightarrow \mathbf{E}^3$ is a one-to-one regular mapping of an open set D of $\mathbf{E}^2$ into $\mathbf{E}^3$.

The image $\mathbf{x}(D)$ of a coordinate patch $\mathbf{x}$—that is, the set of all values of $\mathbf{x}$—is a smooth two-dimensional subset of $\mathbf{E}^3$ (Fig. 4.1). Regularity (Definition 7.9 of Chapter I), for a patch as for a curve, is a basic smoothness condition; the one-to-one requirement is included to prevent $\mathbf{x}(D)$ from cutting across itself. Furthermore, in order to avoid certain technical difficulties (Example 1.7), we shall sometimes use *proper* patches, those for which the inverse function $\mathbf{x}^{-1}\colon \mathbf{x}(D) \rightarrow D$ is continuous (that is, has

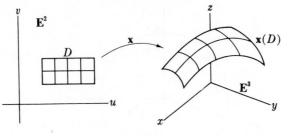

FIG. 4.1

continuous coordinate functions). If we think of D as a thin sheet of rubber, we can get $\mathbf{x}(D)$ by bending and stretching D in a not too violent fashion.

To construct a suitable definition of surface we start from the rough idea that *any small enough region in a surface M resembles a region in the plane* $\mathbf{E}^2$. The discussion above shows that this can be stated somewhat more precisely as: *Near each of its points, M can be expressed as the image of a proper patch.* (When the image of a patch $\mathbf{x}$ is contained in M, we say that $\mathbf{x}$ is a patch *in M*.) To get the final form of the definition, it remains only to define a *neighborhood* $\mathfrak{N}$ *of* $\mathbf{p}$ *in M* to consist of all points *of M* whose Euclidean distance from $\mathbf{p}$ is less than some number $\varepsilon > 0$.

1.2 Definition   A *surface in* $\mathbf{E}^3$ is a subset M of $\mathbf{E}^3$ such that for each point $\mathbf{p}$ of M there exists a proper patch in M whose image contains a neighborhood of $\mathbf{p}$ in M (Fig. 4.2).

The familiar surfaces used in elementary calculus satisfy this definition; for example, let us verify that the unit sphere Σ in $\mathbf{E}^3$ is a surface. By definition, Σ consists of all points at unit distance from the origin—that is, all points $\mathbf{p}$ such that

$$\| \mathbf{p} \| = (p_1^2 + p_2^2 + p_3^2)^{1/2} = 1.$$

To check the definition above, we start by finding a proper patch in Σ covering a neighborhood of the north pole $(0, 0, 1)$. Note that by dropping

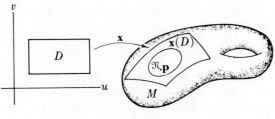

FIG. 4.2

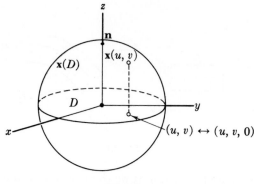

FIG. 4.3

each point (q_1, q_2, q_3) of the northern hemisphere of Σ onto the xy plane at $(q_1, q_2, 0)$ we get a one-to-one correspondence of this hemisphere with a disc D of radius 1 in the xy plane (see Fig. 4.3). If we identify this plane with $\mathbf{E}^2$ by means of the natural association $(q_1, q_2, 0) \leftrightarrow (q_1, q_2)$, then D becomes the disc in $\mathbf{E}^2$ consisting of all points (u, v) such that $u^2 + v^2 < 1$.

Expressing this correspondence as a function on D we find the formula

$$\mathbf{x}(u, v) = (u, v, \sqrt{1 - u^2 - v^2}).$$

Thus $\mathbf{x}$ is a one-to-one function from D onto the northern hemisphere of Σ. We claim that $\mathbf{x}$ is a proper patch. The coordinate functions of $\mathbf{x}$ are differentiable on D, so $\mathbf{x}$ is a mapping. To show that $\mathbf{x}$ is regular, we compute its Jacobian matrix (or transpose)

$$\begin{pmatrix} \dfrac{\partial u}{\partial u} & \dfrac{\partial v}{\partial u} & \dfrac{\partial f}{\partial u} \\[2mm] \dfrac{\partial u}{\partial v} & \dfrac{\partial v}{\partial v} & \dfrac{\partial f}{\partial v} \end{pmatrix} = \begin{pmatrix} 1 & 0 & \dfrac{\partial f}{\partial u} \\[2mm] 0 & 1 & \dfrac{\partial f}{\partial v} \end{pmatrix}$$

where $f = \sqrt{1 - u^2 - v^2}$. Evidently the rows of this matrix are always linearly independent, so its rank at each point is 2. Thus by the criterion following Definition 7.9 of Chapter I, $\mathbf{x}$ is regular, and hence is a patch. Furthermore $\mathbf{x}$ is proper, since its inverse function $\mathbf{x}^{-1} \colon \mathbf{x}(D) \to D$ is given by the formula

$$\mathbf{x}^{-1}(p_1, p_2, p_3) = (p_1, p_2),$$

hence is certainly continuous. Finally we observe that the patch $\mathbf{x}$ covers a neighborhood of $\mathbf{p} = (0, 0, 1)$ in S. Indeed it covers a neighborhood of every point $\mathbf{q}$ in the northern hemisphere (Fig. 4.4).

In a strictly analogous way, we can find a proper patch covering each

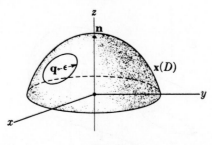

FIG. 4.4

of the other five coordinate hemispheres of Σ, and thus verify, by Definition 1.2, that Σ is a surface. Our real purpose here has been to illustrate Definition 1.2—we shall soon find a much quicker way to prove (in particular) that spheres are surfaces.

The argument above shows that if f is *any* differentiable real-valued function on an open set D in $\mathbf{E}^2$, then the function $\mathbf{x}\colon D \to \mathbf{E}^3$ such that

$$\mathbf{x}(u, v) = (u, v, f(u, v))$$

is a proper patch. We shall call patches of this type *Monge patches*.

We turn now to some standard methods of constructing surfaces. Note that the image $M = \mathbf{x}(D)$ of just one proper patch automatically satisfies 1.2; M is then called a *simple* surface. (Thus Definition 1.2 says that any surface in $\mathbf{E}^3$ can be constructed by gluing together simple surfaces.)

1.3 Example The surface $M\colon z = f(x, y)$. Every differentiable real-valued function f on $\mathbf{E}^2$ determines a surface M in $\mathbf{E}^3$: the graph of f, that is, the set of all points of $\mathbf{E}^3$ whose coordinates satisfy the equation $z = f(x, y)$. Evidently M is the image of the Monge patch

$$\mathbf{x}(u, v) = (u, v, f(u, v));$$

hence by the remarks above, M is a simple surface.

If g is a real-valued function on $\mathbf{E}^3$ and c is a number, denote by $M\colon g = c$ the set of all points $\mathbf{p}$ such that $g(\mathbf{p}) = c$. For example, if g is a temperature distribution in space, then $M\colon g = c$ consists of all points of temperature c. There is a simple condition that tells when such a subset of $\mathbf{E}^3$ is a surface.

1.4 Theorem Let g be a differentiable real-valued function on $\mathbf{E}^3$, and c a number. The subset $M\colon g(x, y, z) = c$ of $\mathbf{E}^3$ is a surface if the differential dg is not zero at any point of M.

(In Definition 1.2 and in this theorem we are tacitly assuming that M has some points in it; thus the equation $x^2 + y^2 + z^2 = -1$, for example, does not define a surface.)

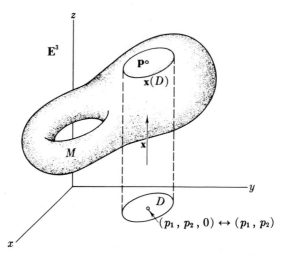

FIG. 4.5

Proof. All we do is give geometric content to a famous result of advanced calculus—the implicit function theorem. If $\mathbf{p}$ is a point of M, we must find a proper patch covering a neighborhood of $\mathbf{p}$ in M (Fig. 4.5). Now

$$dg = \frac{\partial g}{\partial x}\, dx + \frac{\partial g}{\partial y}\, dy + \frac{\partial g}{\partial z}\, dz.$$

Thus the hypothesis on dg is equivalent to assuming that at least one of these partial derivatives is not zero at $\mathbf{p}$, say $(\partial g/\partial z)(\mathbf{p}) \neq 0$. In this case, the implicit function theorem says that near $\mathbf{p}$ the equation $g(x, y, z) = c$ can be solved for z. More precisely, it asserts that there is a differentiable real-valued function h defined on a neighborhood D of (p_1, p_2) such that

(1) For each point (u, v) in D, the point $(u, v, h(u, v))$ lies in M; that is, $g(u, v, h(u, v)) = c$.

(2) Points of the form $(u, v, h(u, v))$, with (u, v) in D, fill a neighborhood of $\mathbf{p}$ in M.

It follows immediately that the Monge patch $\mathbf{x}: D \to \mathbf{E}^3$ such that

$$\mathbf{x}(u, v) = (u, v, h(u, v))$$

satisfies the requirements in Definition 1.2. Since $\mathbf{p}$ was an arbitrary point of M, we conclude that M is a surface. ∎

When $M: g = c$ is a surface, M is said to be defined *implicitly* by the equation $g = c$. It is now very easy to prove that spheres are surfaces. The sphere Σ in $\mathbf{E}^3$ of radius $r > 0$ and center $\mathbf{c} = (c_1, c_2, c_3)$ is the set of all points at distance r from $\mathbf{c}$. If $g = \Sigma(x_i - c_i)^2$, then Σ is defined implicitly by the equation $g = r^2$. Now $dg = 2\Sigma (x_i - c_i) dx_i$, hence dg is zero only at the point $\mathbf{c}$, which is *not* in Σ. Thus Σ is a surface.

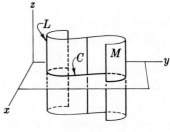

Using this theorem and the notion of curve defined on page 20 we derive two well-known types of surfaces.

1.5 Example Cylinders. As a line L, perpendicular to a plane P, moves along

FIG. 4.6

a curve C in P, it sweeps out a *cylinder*. For definiteness, let P be, say, the xy plane, so that L is always parallel to the z axis as in Fig. 4.6. If the curve C is given by

$$C: f(x, y) = c \qquad \text{in } \mathbf{E}^2,$$

let $\tilde{f}$ be the function on $\mathbf{E}^3$ such that $\tilde{f}(p_1, p_2, p_3) = f(p_1, p_2)$. Then the resulting cylinder is evidently given by

$$M: \tilde{f}(x, y, z) = c \qquad \text{in } \mathbf{E}^3.$$

The definition of curve on page 20 requires that at each point of C either $\partial f/\partial x$ or $\partial f/\partial y$ is nonzero. Since

$$\frac{\partial \tilde{f}}{\partial x} (p_1, p_2, p_3) = \frac{\partial f}{\partial x} (p_1, p_2),$$

and similarly for $\partial/\partial y$, it follows that dg is never zero at a point of M. Thus M is a surface.

When C is a circle, we obtain a *circular cylinder* $M: x^2 + y^2 = r^2$ in $\mathbf{E}^3$.

In Example 1.5 we constructed a surface essentially by *translating* a curve; now we get one by *rotating* a curve.

1.6 Example Surfaces of revolution. Let C be a curve in a plane P, and let A be a line in P which does not meet C. If this *profile curve* C is revolved around the axis A, it sweeps out a *surface of revolution* M in $\mathbf{E}^3$. We now check, using Theorem 1.4, that M really is a surface. For simplicity, assume that P is a coordinate plane and A is a coordinate axis—say the xy plane and x axis, respectively. Since C must not meet A, we assume it is in the upper half $(y > 0)$ of the xy plane. As C is revolved, each point $(q_1, q_2, 0)$ of C gives rise to a whole circle of points

$$(q_1, q_2 \cos v, q_2 \sin v) \text{ in } M, \qquad \text{for} \qquad 0 \leq v \leq 2\pi.$$

Put in reverse, *a point* $\mathbf{p} = (p_1, p_2, p_3)$ *is in M if and only if the point*

$$\bar{\mathbf{p}} = (p_1, \sqrt{p_2{}^2 + p_3{}^2}, 0)$$

is in C (Fig. 4.7).

If the profile curve is $C: f(x, y) = c$, we define a function g on $\mathbf{E}^3$ by $g(x, y, z) = f(x, \sqrt{y^2 + z^2})$. Then the argument above shows that the resulting surface of revolution is exactly $M: g(x, y, z) = c$. Using the chain rule, it is not hard to show that dg is never zero on M, so M is a surface.

The circles in M generated, under revolution, by each point of C are called the *parallels* of M, the different positions of C as it is rotated are called the *meridians* of M. This terminology derives from the geography of the sphere; however, a sphere is *not* a surface of revolution as defined above. Its profile curve must twice meet the axis of revolution, so two "parallels" reduce to single points. To simplify the statements of subsequent theorems we use a slightly different terminology in this case; see Exercise 12.

The necessity of the properness condition on the patches in Definition 1.2 is shown by the following example.

1.7 Example Suppose that a rectangular strip of tin is bent into a figure-8, as in Fig. 4.8. The configuration M which results does not satisfy our intuitive picture of what a surface should be, for along the axis A, M is not like the plane $\mathbf{E}^2$ but is instead like *two* intersecting planes. To express this construction in mathematical terms, let D be the rectangle $-\pi < u < \pi, 0 < v < 1$ in $\mathbf{E}^2$ and define $\mathbf{x}: D \to \mathbf{E}^3$ by $\mathbf{x}(u, v) = (\sin u, \sin 2u, v)$. It is easy to check that $\mathbf{x}$ is a patch, but its image $M = \mathbf{x}(D)$ is

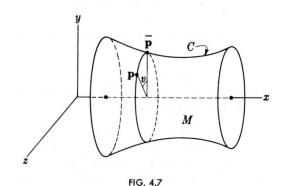

FIG. 4.7

FIG. 4.8

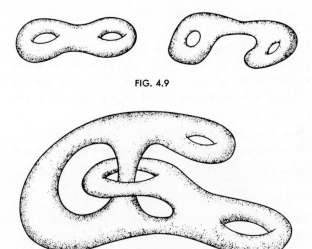

FIG. 4.9

FIG. 4.10

not a surface: **x** is not a *proper* patch. Continuity fails for $\mathbf{x}^{-1}\colon M \to D$ since, roughly speaking, to restore M to D, $\mathbf{x}^{-1}$ must tear M along the axis A (the z axis of $\mathbf{E}^3$).

By Example 1.6, the familiar *torus of revolution* T is a surface (Fig. 4.19). With somewhat more work, one could construct *double toruses* of various shapes, as in Fig. 4.9. By adding "handles" and "tubes" to existing surfaces one can—in principle, at least—construct surfaces of any desired degree of complexity (Fig. 4.10).

EXERCISES

1. None of the following subsets M of $\mathbf{E}^3$ are surfaces. At which points **p** is it impossible to find a proper patch in M that will cover a neighborhood of **p** in M? (Sketch M—formal proofs not required.)
   (a) Cone $M\colon z^2 = x^2 + y^2$.
   (b) Closed disc $M\colon x^2 + y^2 \leqq 1, z = 0$.
   (c) Folded plane $M\colon xy = 0, x \geqq 0, y \geqq 0$.

2. A *plane* in $\mathbf{E}^3$ is a surface $M\colon ax + by + cz = d$, where the numbers a, b, c are necessarily not all zero. Prove that every plane in $\mathbf{E}^3$ may be described by a vector equation as on page 60.

3. Sketch the general shape of the surface $M\colon z = ax^2 + by^2$ in each of the cases:

(a) $a > b > 0$ (c) $a > b = 0$
(b) $a > 0 > b$ (d) $a = b = 0$.

4. In which of the following cases is the mapping $\mathbf{x} : E^2 \to E^3$ a patch?
    (a) $\mathbf{x}(u, v) = (u, uv, v)$.   (c) $\mathbf{x}(u, v) = (u, u^2, v + v^3)$.
    (b) $\mathbf{x}(u, v) = (u^2, u^3, v)$.   (d) $\mathbf{x}(u, v) = (\cos 2\pi u, \sin 2\pi u, v)$.
    (Recall that $\mathbf{x}$ is one-to-one if and only if $\mathbf{x}(u, v) = \mathbf{x}(u_1, v_1)$ implies $(u, v) = (u_1, v_1)$.)

5. (a) Prove that $M : (x^2 + y^2)^2 + 3z^2 = 1$ is a surface.
    (b) For which values of c is $M : z(z - 2) + xy = c$ a surface?

6. Determine the intersection $z = 0$ of the *monkey saddle*

$$M : z = f(x, y), \qquad f = y^3 - 3yx^2,$$

with the xy plane. On which regions of the plane is $f > 0$? $f < 0$? How does this surface get its name?

7. Let $\mathbf{x} : D \to E^3$ be a mapping, with

$$\mathbf{x}(u, v) = (x_1(u, v), x_2(u, v), x_3(u, v)).$$

    (a) Prove that a point $\mathbf{p} = (p_1, p_2, p_3)$ of E^3 is in the image $\mathbf{x}(D)$ if and only if the equations

$$p_1 = x_1(u, v) \qquad p_2 = x_2(u, v) \qquad p_3 = x_3(u, v)$$

    can be solved for u and v, with (u, v) in D.
    (b) If for every point $\mathbf{p}$ in $\mathbf{x}(D)$ these equations have the *unique* solution: $u = f_1(p_1, p_2, p_3)$, $v = f_2(p_1, p_2, p_3)$, with (u, v) in D, prove that $\mathbf{x}$ is one-to-one and that $\mathbf{x}^{-1} : \mathbf{x}(D) \to D$ is given by the formula

$$\mathbf{x}^{-1}(\mathbf{p}) = (f_1(\mathbf{p}), f_2(\mathbf{p})).$$

8. Let $\mathbf{x} : D \to E^3$ be the function given by

$$\mathbf{x}(u, v) = (u^2, uv, v^2)$$

on the first quadrant $D : u > 0$, $v > 0$. Show that $\mathbf{x}$ is one-to-one and find a formula for its inverse function $\mathbf{x}^{-1} : \mathbf{x}(D) \to D$. Then prove that $\mathbf{x}$ is a proper patch.

9. Let $\mathbf{x} : E^2 \to E^3$ be the mapping

$$\mathbf{x}(u, v) = (u + v, u - v, uv).$$

Show that $\mathbf{x}$ is a proper patch, and that the image of $\mathbf{x}$ is the surface $M : z = (x^2 - y^2)/4$.

10. If F is an isometry of E^3 and M is a surface in E^3, prove that the image

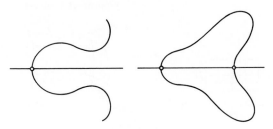

FIG. 4.11

$F(M)$ is also a surface in $\mathbf{E}^3$. (*Hint:* If $\mathbf{x}$ is a patch in M, then the composite function $F(\mathbf{x})$ is regular, since $F(\mathbf{x})_* = F_* \, \mathbf{x}_*$ by Ex. 12 of I.7.)

11. The assertion in Exercise 10 remains true when F is merely a diffeomorphism. Prove this special case: If F is a diffeomorphism of $\mathbf{E}^3$, then the image of the surface $M : g = c$ is $\bar{M} : \bar{g} = c$, where $\bar{g} = g(F^{-1})$— and $\bar{M}$ is a surface. (*Hint:* If $dg(\mathbf{v}) \neq 0$ at $\mathbf{p}$ in M, show by using Ex. 9 of I.7 that $d\bar{g}(F_* \, \mathbf{v}) \neq 0$.

12. If f is a differentiable function and $f(x, y^2) = c$ defines a curve C in the xy plane, then C is symmetric about the x axis and must cross this axis once (if C is an arc) or twice (if C is closed). Prove that revolving C about the x axis gives a surface M in $\mathbf{E}^3$. We call M an *augmented surface of revolution:* If the points on the axis are deleted, it becomes an ordinary surface of revolution (Fig. 4.11).

2   Patch Computations

In Section 1, coordinate patches were used to *define* a surface; now we consider some properties of patches that will be useful in *studying* surfaces.

Let $\mathbf{x} : D \to \mathbf{E}^3$ be a coordinate patch. Holding u or v constant in the function $(u, v) \to \mathbf{x}(u, v)$ produces curves. Explicitly, for each point (u_0, v_0) in D the curve

$$u \to \mathbf{x}(u, v_0)$$

is called the *u-parameter curve,* $v = v_0$, of $\mathbf{x}$; and the curve

$$v \to \mathbf{x}(u_0, v)$$

is the *v-parameter curve,* $u = u_0$ (Fig. 4.12).

Thus the image $\mathbf{x}(D)$ is covered by these two families of curves, which are the images under $\mathbf{x}$ of the horizontal and vertical lines in D, and one curve from each family goes through each point of $\mathbf{x}(D)$.

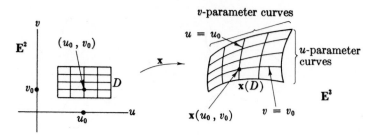

FIG. 4.12

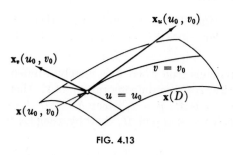

FIG. 4.13

2.1 Definition If $\mathbf{x}: D \to \mathbf{E}^3$ is a patch, for each point (u_0, v_0) in D:

(1) The velocity vector at u_0 of the u-parameter curve, $v = v_0$, is denoted by $\mathbf{x}_u(u_0, v_0)$.

(2) The velocity vector at v_0 of the v-parameter curve, $u = u_0$, is denoted by $\mathbf{x}_v(u_0, v_0)$.

The vectors $\mathbf{x}_u(u_0, v_0)$ and $\mathbf{x}_v(u_0, v_0)$ are called the *partial velocities* of $\mathbf{x}$ at (u_0, v_0) (Fig. 4.13).

Thus $\mathbf{x}_u$ and $\mathbf{x}_v$ are actually functions on D whose values at each point (u_0, v_0) are tangent vectors to $\mathbf{E}^3$ at $\mathbf{x}(u_0, v_0)$. The subscripts u and v are intended to suggest partial differentiation. Indeed if the patch is given in terms of its Euclidean coordinate functions by a formula

$$\mathbf{x}(u, v) = (x_1(u, v), x_2(u, v), x_3(u, v)),$$

then it follows from the definition above that the partial velocity functions are given by

$$\mathbf{x}_u = \left(\frac{\partial x_1}{\partial u}, \frac{\partial x_2}{\partial u}, \frac{\partial x_3}{\partial u} \right)_{\mathbf{x}}$$

$$\mathbf{x}_v = \left(\frac{\partial x_1}{\partial v}, \frac{\partial x_2}{\partial v}, \frac{\partial x_3}{\partial v} \right)_{\mathbf{x}}.$$

The subscript $\mathbf{x}$ (frequently omitted) is a reminder that $\mathbf{x}_u(u, v)$ and $\mathbf{x}_v(u, v)$ have point of application $\mathbf{x}(u, v)$.

2.2 Example The *geographical patch* in the sphere. Let Σ be the sphere of radius $r > 0$ centered at the origin of $\mathbf{E}^3$. Longitude and latitude on the earth suggest a patch in Σ quite different from the Monge patch used on

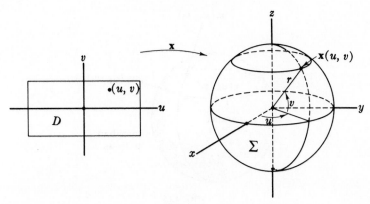

FIG. 4.14

Σ in Section 1. The point $x(u, v)$ of Σ with longitude u $(-\pi < u < \pi)$ and latitude v $(-\pi/2 < v < \pi/2)$ has Euclidean coordinates (Fig. 4.14).

$$x(u, v) = (r \cos v \cos u, r \cos v \sin u, r \sin v).$$

With the domain D of x defined by these inequalities, the image $x(D)$ of x is all of Σ except one semicircle from north pole to south pole. The u-parameter curve, $v = v_0$, is a circle—the parallel of latitude v_0. The v-parameter curve, $u = u_0$, is a semicircle—the meridian of longitude u_0.

We compute the partial velocities of x to be

$$x_u(u, v) = r(-\cos v \sin u, \cos v \cos u, 0)$$

$$x_v(u, v) = r(-\sin v \cos u, -\sin v \sin u, \cos v)$$

where r denotes a scalar multiplication. Evidently x_u always points due east, and x_v due north. In a moment we shall give a formal proof that x is a patch in Σ (Fig. 4.15).

To test whether a given subset M of E^3 is a surface, Definition 1.2 demands proper patches (and Example 1.7 shows why). But once we know that M is a surface, the properness condition need no longer concern us (Exercise 14 of Section 3). Furthermore, in many situations the one-to-one restriction on patches can also be dropped.

2.3 Definition   A regular mapping $x: D \to E^3$ whose image lies in a surface M is called a *parametrization* of the region $x(D)$ in M.

(Thus a patch is merely a one-to-one parametrization.) In favorable cases this image $x(D)$ may be the whole surface M, and we have then the analogue of the more familiar notion of parametrization of a curve (p. 20). Parametrizations will be of first importance in practical computations with

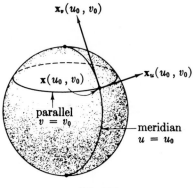

FIG. 4.15

surfaces, so we consider some ways of determining whether a mapping $\mathbf{x}: D \to \mathbf{E}^3$ is a parametrization of (part of) a given surface M.

The image of $\mathbf{x}$ must, of course, lie in M. Note that if the surface is given in the implicit form $M: g = c$, this means that the composite function $g(\mathbf{x})$ must have constant value c.

To test whether $\mathbf{x}$ is regular, note first that parameter curves and partial velocities $\mathbf{x}_u$ and $\mathbf{x}_v$ are well-defined for an arbitrary differentiable mapping $\mathbf{x}: D \to \mathbf{E}^3$. Now the last two rows of the cross product

$$\mathbf{x}_u \times \mathbf{x}_v = \begin{vmatrix} U_1 & U_2 & U_3 \\ \dfrac{\partial x_1}{\partial u} & \dfrac{\partial x_2}{\partial u} & \dfrac{\partial x_3}{\partial u} \\ \dfrac{\partial x_1}{\partial v} & \dfrac{\partial x_2}{\partial v} & \dfrac{\partial x_3}{\partial v} \end{vmatrix}$$

give the (transposed) Jacobian matrix of $\mathbf{x}$ at each point. Thus the regularity of $\mathbf{x}$ is equivalent to the condition that $\mathbf{x}_u \times \mathbf{x}_v$ *is never zero*, or, by properties of the cross product, that at each point (u, v) of D *the partial velocity vectors of* $\mathbf{x}$ *are linearly independent*.

Let us try out these methods on the mapping $\mathbf{x}$ given in Example 2.2. Since the sphere is defined implicitly by $g = x^2 + y^2 + z^2 = r^2$, we must show that $g(\mathbf{x}) = r^2$. Substituting the coordinate functions of $\mathbf{x}$ for x, y, and z, we get

$$r^{-2} g(\mathbf{x}) = (\cos v \cos u)^2 + (\cos v \sin u)^2 + \sin^2 u$$
$$= \cos^2 v + \sin^2 v = 1.$$

A short computation using the formulas for $\mathbf{x}_u$ and $\mathbf{x}_v$ given in Example 2.2 yields

$$r^{-2} \mathbf{x}_u \times \mathbf{x}_v = \cos u \cos^2 v \, U_1 + \sin u \cos^2 v \, U_2 + \cos v \sin v \, U_3.$$

Since $-\pi/2 < v < \pi/2$ for the domain D of $\mathbf{x}$, $\cos v$ is never zero there; but $\sin u$ and $\cos u$ are never simultaneously zero, so $\mathbf{x}_u \times \mathbf{x}_v$ is never zero on D. Thus $\mathbf{x}$ is regular, and hence is a parameterization.

To show that $\mathbf{x}$ is a patch, we prove it is one-to-one, that is, show that $\mathbf{x}(u, v) = \mathbf{x}(u_1, v_1)$ implies $(u, v) = (u_1, v_1)$. If $\mathbf{x}(u, v) = \mathbf{x}(u_1, v_1)$, then the definition of $\mathbf{x}$ gives three coordinate equations:

$$r \cos v \cos u = r \cos v_1 \cos u_1$$

$$r \cos v \sin u = r \cos v_1 \sin u_1$$

$$r \sin v = r \sin v_1.$$

Again, since $-\pi/2 < v < \pi/2$ for all points of D, the last equation implies that $v = v_1$. Thus $r \cos v = r \cos v_1 > 0$ may be cancelled from the first two equations and we conclude that $u = u_1$ also.

For this particular function $\mathbf{x}$ in Σ, these results might almost be considered obvious from the discussion in Example 2.2, but the methods used above will serve in more difficult cases.

We shall now see how to find natural parametrizations in cylinders and surfaces of revolution.

2.4 Example Parametrization of a cylinder M. Suppose, as in Example 1.5, that M is the cylinder over a curve $C\colon f(x, y) = a$ in the xy plane (Fig. 4.16). If $\alpha = (\alpha_1, \alpha_2, 0)$ is a parametrization of C, we assert that

$$\mathbf{x}(u, v) = (\alpha_1(u), \alpha_2(u), v)$$

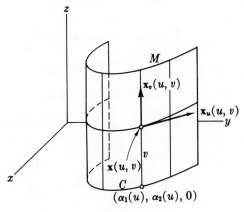

FIG. 4.16

is a parametrization of M. Clearly $\mathbf{x}$ lies in M, covers all of M, and is differentiable. Furthermore $\mathbf{x}$ is regular, for at each point (u, v) the partial velocities

$$\mathbf{x}_u = \left(\frac{d\alpha_1}{du}, \frac{d\alpha_2}{du}, 0 \right)$$

$$\mathbf{x}_v = (0, 0, 1)$$

are linearly independent ($\mathbf{x}_u$ is never zero, since α is by definition regular). If the curve α is defined on an interval I, the domain of $\mathbf{x}$ is the vertical strip D: u in I, v arbitrary. (Thus if I is the whole real line, D is $\mathbf{E}^2$.) The u-parameter curves of $\mathbf{x}$ are merely translates of C and are called the *cross-sectional curves* of the cylinder. The v-parameter curves follow the straight lines called the *rulings* (or "elements") of the cylinder. If C is not a closed curve, then α—and hence also $\mathbf{x}$—is one-to-one, so $\mathbf{x}$ is a patch. But if C is closed, $\mathbf{x}$ wraps D an infinite number of times around C.

2.5 Example Parametrization of a surface of revolution. Suppose that M is obtained, as in Example 1.6, by revolving a curve C in the upper half of the xy plane about the x axis. Now let

$$\alpha(u) = (g(u), h(u), 0)$$

be a parametrization of C (note that $h > 0$). As we observed in Example 1.6, when the point $(g(u), h(u), 0)$ on the profile curve C has been rotated through an angle v, it reaches a point $\mathbf{x}(u, v)$ with the same x coordinate $g(u)$, but new y and z coordinates $h(u) \cos v$ and $h(u) \sin v$, respectively (Fig. 4.17). Thus

$$\mathbf{x}(u, v) = (g(u), h(u) \cos v, h(u) \sin v)$$

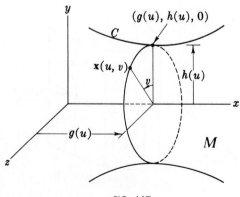

FIG. 4.17

Evidently this formula defines a mapping into M whose image is all of M. A short computation shows that $\mathbf{x}_u$ and $\mathbf{x}_v$ are always linearly independent, so $\mathbf{x}$ is a parametrization of M. As in Example 2.4, the domain D of $\mathbf{x}$ consists of all points (u, v) for which u is in the domain of α. The u-parameter curves of $\mathbf{x}$ parametrize the meridians of M; the v-parameter curves, the parallels. (Thus the parametrization $\mathbf{x}: D \to M$ is never one-to-one.)

Obviously we are not limited to rotating curves in the xy plane about the x axis. But with other choices of coordinates, we maintain the same geometric meaning for the functions g and h: g measures distance *along* the axis of revolution, while h measures distance *from* the axis of revolution.

Actually the geographical patch in the sphere is one instance of Example 2.5 (with u and v reversed); here is another.

2.6 Example *Torus of revolution T.* This is the surface of revolution obtained when the profile curve C is a circle. Suppose that C is the circle in the xz plane with radius $r > 0$ and center $(R, 0, 0)$. We shall rotate about the z axis; hence we must require $R > r$ to keep C from meeting the axis of revolution. A natural parametrization (Fig. 4.18) for C is

$$\alpha(u) = (R + r \cos u, r \sin u).$$

Thus by the remarks above we must have $g(u) = r \sin u$, distance *along* the z axis, and $h(u) = R + r \cos u$, distance *from* the z axis. The general argument in Example 2.5—with coordinate axes permuted—then yields the parametrization

$$\mathbf{x}(u, v) = (h(u) \cos v, h(u) \sin v, g(u))$$

$$= ((R + r \cos u) \cos v, (R + r \cos u) \sin v, r \sin u).$$

The domain of $\mathbf{x}$ is the whole plane $\mathbf{E}^2$, and (as always when the profile curve is closed) $\mathbf{x}$ is periodic in both u and v. Here

$$\mathbf{x}(u + 2\pi, v + 2\pi) = \mathbf{x}(u, v) \qquad \text{for all } (u, v).$$

There are infinitely many different parametrizations (and patches) in

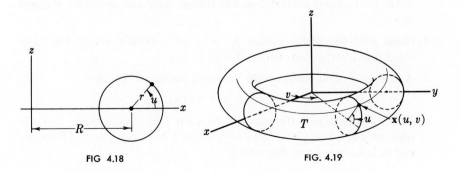

FIG 4.18   FIG. 4.19

any surface. Those we have discussed have been singled out by the natural way they are fitted to their surfaces.

EXERCISES

1. Find a parametrization for the entire surface obtained by revolving:
   (a) $C: y = \cosh x$ around the x axis (catenoid).
   (b) $C: (z - 2)^2 + y^2 = 1$ around the y axis (torus).
   (c) $C: z = x^2$ around the z axis (paraboloid of revolution).

2. Partial velocities $\mathbf{x}_u$ and $\mathbf{x}_v$ are defined for an arbitrary mapping $\mathbf{x}: D \to \mathbf{E}^3$, so we may consider the real-valued functions

$$E = \mathbf{x}_u \cdot \mathbf{x}_u \qquad F = \mathbf{x}_u \cdot \mathbf{x}_v \qquad G = \mathbf{x}_v \cdot \mathbf{x}_v$$

   on D. Prove

$$\| \mathbf{x}_u \times \mathbf{x}_v \|^2 = EG - F^2.$$

   Deduce that $\mathbf{x}$ is a regular mapping if and only if $EG - F^2$ is never zero. (This is often the easiest way to check regularity. The geometric significance of these functions is discussed in V.4).

3. Show that

$$M: (\sqrt{x^2 + y^2} - 4)^2 + z^2 = 4$$

   is a torus of revolution: Find a profile circle and the axis of revolution.

   A *ruled surface* is a surface swept out by a straight line L moving along a curve β. The various positions of the generating line L are called the *rulings* of the surface. Such a surface thus always has a parametrization in *ruled form*

$$\mathbf{x}(u, v) = \beta(u) + v\,\delta(u) \qquad \text{or} \qquad \beta(v) + u\,\delta(v)$$

   where we call β the *base curve*, δ the *director curve*. Alternatively we may vizualize δ as a vector field on β. Frequently it is necessary to restrict v to some interval, so the rulings may not be entire straight lines.

4. Show that the *saddle surface* $M: z = xy$ is doubly ruled: Find two ruled parametrizations with different rulings.

5. A *cone* is a ruled surface with parametrization of the form

$$\mathbf{x}(u, v) = \mathbf{p} + v\,\delta(u).$$

   Thus all rulings pass through the vertex $\mathbf{p}$ (Fig. 4.20). Show that the regularity of $\mathbf{x}$ is equivalent to both v and $\delta \times \delta'$ never zero. (Thus the vertex is never part of the cone.)

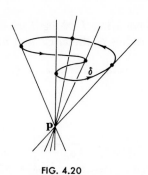

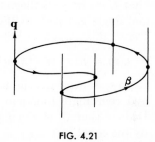

FIG. 4.20   FIG. 4.21

6. A *cylinder* is a ruled surface with parametrization of the form

$$\mathbf{x}(u, v) = \beta(u) + v\mathbf{q}.$$

Thus the rulings are all parallel (Fig. 4.21). Prove that the regularity of $\mathbf{x}$ is equivalent to $\beta' \times \mathbf{q}$ never zero. Show that this definition generalizes Example 2.4.

7. A line L is attached orthogonally to an axis A (Fig. 4.22). If L moves along A and rotates—both at constant speed—then L sweeps out a *helicoid H*.

　　If A is the z axis, then H is the image of the mapping $\mathbf{x}\colon E^2 \to E^3$ such that

$$\mathbf{x}(u, v) = (u \cos v, u \sin v, bv)$$
$$(b \neq 0).$$

(a) Prove that $\mathbf{x}$ is a patch.

(b) Describe the parameter curves of $\mathbf{x}$.

(c) Express the helicoid in implicit form $g = c$.

8. (a) Show that $\mathbf{x}\colon D \to E^3$ is a regular mapping, where

$$\mathbf{x}(u, v) = (u \cos v, u, \sin v)$$
$$\text{on } D\colon u > 0.$$

(b) Find a function $g(x, y, z)$ such that the image of $\mathbf{x}$ is the surface $M\colon g = 0$.

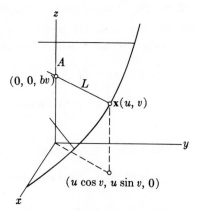

(0, 0, bv)

A

L

$\mathbf{x}(u, v)$

$(u \cos v, u \sin v, 0)$

FIG. 4.22

(c) Show that M is a ruled surface and sketch M. (*Hint:* Begin with the curve sliced from M by the plane $y = 1$.)

9. Let β be a unit-speed parametrization of the unit circle in the xy plane.

Construct a ruled surface as follows: Move a line L along β in such a way that L is always orthogonal to the radius of the circle and makes constant angle $\pi/4$ with β' (Fig. 4.23).

(a) Derive this parametrization of the resulting ruled surface M:

$$\mathbf{x}(u, v) = \beta(u) + v(\beta'(u) + U_3).$$

(b) Express $\mathbf{x}$ explicitly in terms of v and coordinate functions for β.

(c) Deduce that M is given implicitly by the equation

$$x^2 + y^2 - z^2 = 1.$$

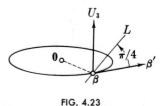

FIG. 4.23

(d) Show that if the angle $\pi/4$ above is changed to $-\pi/4$, the same surface M results. Thus M is doubly ruled.

(e) Sketch this surface M showing the two rulings through each of the points $(1, 0, 0)$ and $(2, 1, 2)$.

A *quadric surface* is a surface $M: g = 0$ for which g involves at most quadratic terms in x_1, x_2, x_3—that is,

$$g = \sum_{i,j} a_{ij} x_i x_j + \sum_i b_i x_i + c.$$

Trivial cases excepted, there are just five types of quadric surfaces, represented by the familiar surfaces in the next three examples (see Theorem 2.2, p. 280, Birkhoff and MacLane [2]).

10. In each case, (i) prove that M is a surface and sketch its general shape, (ii) show that $\mathbf{x}$ is a parametrization and find its image in M.

(a) *Ellipsoid*, $M: \dfrac{x^2}{a^2} + \dfrac{y^2}{b^2} + \dfrac{z^2}{c^2} = 1$

$$\mathbf{x}(u, v) = (a \cos u \cos v, \, b \cos u \sin v, \, c \sin u)$$
$$\text{on } D: \, -\pi/2 < u < \pi/2.$$

(b) *Elliptic hyperboloid*, $M: \dfrac{x^2}{a^2} + \dfrac{y^2}{b^2} - \dfrac{z^2}{c^2} = 1$

$$\mathbf{x}(u, v) = (a \cosh u \cos v, \, b \cosh u \sin v, \, c \sinh u) \qquad \text{on } \mathbf{E}^2.$$

(c) *Elliptic hyperboloid* (two sheets), $M: \dfrac{x^2}{a^2} + \dfrac{y^2}{b^2} - \dfrac{z^2}{c^2} = -1$

$$\mathbf{x}(u, v) = (a \sinh u \cos v, \, b \sinh u \sin v, \, c \cosh u) \qquad \text{on } D: u \neq 0.$$

11. *Elliptic paraboloid*, $M: z = \dfrac{x^2}{a^2} + \dfrac{y^2}{b^2}.$

(a) Show that M is a surface, and that

$$\mathbf{x}(u, v) = (au \cos v, bu \sin v, u^2), \qquad u > 0,$$

is a parametrization that omits only one point of M.

(b) Describe the parameter curves of $\mathbf{x}$ in general, and sketch this surface for $a = 1$, $b = 4$, showing some parameter curves.

12. *Hyperbolic paraboloid*, $M: z = \dfrac{x^2}{a^2} - \dfrac{y^2}{b^2}$.

(a) Show that $\mathbf{x}: \mathbf{E}^2 \to \mathbf{E}^3$ is a proper patch covering all of M, where

$$\mathbf{x}(u, v) = (a(u + v), b(u - v), 4uv).$$

(b) Show that M is a doubly ruled surface by rewriting $\mathbf{x}$ in ruled form in two different ways.

(c) Same as (b) of Exercise 11.

13. Let M be the surface of revolution obtained by revolving the curve $t \to (g(t), h(t), 0)$ about the x axis $(h > 0)$. Show that

(a) If g' is never zero, M has a parametrization of form

$$\mathbf{x}(u, v) = (u, f(u) \cos v, f(u) \sin v).$$

(b) If h' is never zero, M has a parametrization of form

$$\mathbf{x}(u, v) = (f(u), u \cos v, u \sin v).$$

3 Differentiable Functions and Tangent Vectors

We now begin an exposition of the calculus on a surface M in $\mathbf{E}^3$. The space $\mathbf{E}^3$ will gradually fade out of the picture, since our ultimate goal is a calculus for M alone. Generally speaking, we shall follow the order of topics in Chapter I, making such changes as are necessary to adapt the calculus of the plane $\mathbf{E}^2$ to a surface M.

Suppose that f is a real-valued function defined only on a surface M. If $\mathbf{x}: D \to M$ is a coordinate patch in M, then the composite function $f(\mathbf{x})$ is called a *coordinate expression* for f; it is an ordinary real-valued function $(u,v) \to f(\mathbf{x}(u, v))$. We define f to be *differentiable* provided all its coordinate expressions are differentiable in the usual Euclidean sense (Definition 1.3 of Chapter I).

For a function $F: \mathbf{E}^n \to M$, each patch $\mathbf{x}$ in M gives a *coordinate expression* $\mathbf{x}^{-1}(F)$ for F. Evidently this composite function is defined only on the set $\mathcal{O}$ of all points $\mathbf{p}$ of $\mathbf{E}^n$ such that $F(\mathbf{p})$ is in $\mathbf{x}(D)$. Again we define F to be *differentiable* provided all its coordinate expressions are differentiable in the usual Euclidean sense. [We must understand that this includes the requirement that $\mathcal{O}$ be an open set of $\mathbf{E}^n$, so that the differentiability of $\mathbf{x}^{-1}(F)\colon \mathcal{O} \to \mathbf{E}^2$ is well-defined, as in Section 7 of Chapter I.]

In particular, a *curve* $\alpha\colon I \to M$ in a surface M is, as before, a differentiable function from an open interval I into M.

To see how this definition works out in practice, we examine an important special case.

3.1 Lemma   If α is a curve $\alpha\colon I \to M$ whose route lies in the image $\mathbf{x}(D)$ of a single patch $\mathbf{x}$, then there exist unique differentiable functions a_1, a_2 on I such that

$$\alpha(t) \; = \; \mathbf{x}(a_1(t),\, a_2(t)) \qquad \text{for all } t$$

or, in functional notation, $\alpha = \mathbf{x}(a_1, a_2)$. (See Fig. 4.24.)

Proof. By definition, the coordinate expression $\mathbf{x}^{-1}\alpha\colon I \to D$ is differentiable—it is just a curve in $\mathbf{E}^2$ whose route lies in the domain D of $\mathbf{x}$. If a_1, a_2 are the Euclidean coordinate functions of $\mathbf{x}^{-1}\alpha$, then

$$\alpha \; = \; \mathbf{x}\mathbf{x}^{-1}\alpha \; = \; \mathbf{x}(a_1, a_2).$$

These are the only such functions, for if $\alpha = \mathbf{x}(b_1, b_2)$, then

$$(a_1, a_2) \; = \; \mathbf{x}^{-1}\alpha \; = \; \mathbf{x}^{-1}\mathbf{x}(b_1, b_2) \; = \; (b_1, b_2). \qquad \blacksquare$$

These functions a_1, a_2 are called the *coordinate functions* of the curve α with respect to the patch $\mathbf{x}$. For example, the curve α given in (3) of Example 4.2, Chapter I, lies in the part of the sphere Σ of radius 2 that is covered by the patch $\mathbf{x}$ given in Example 2.2. Observe that this curve moves so as to have equal longitude and latitude at each point. In fact its coordinate functions with respect to $\mathbf{x}$ are $a_1(t) = a_2(t) = t$, since by the formula for $\mathbf{x}$,

$$\mathbf{x}(a_1(t),\, a_2(t)) \; = \; \mathbf{x}(t, t) \; = \; 2(\cos^2 t, \, \cos t \sin t, \, \sin t) \; = \; \alpha(t).$$

For an arbitrary patch $\mathbf{x}\colon D \to M$ (as in the case just considered) it is

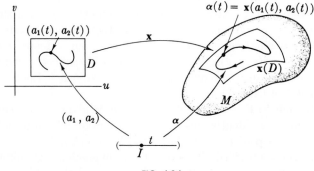

FIG. 4.24

natural to think of the domain D as a *map* of the region $\mathbf{x}(D)$ in M. The functions $\mathbf{x}$ and $\mathbf{x}^{-1}$ establish a one-to-one correspondence between objects in $\mathbf{x}(D)$ and objects in D. If a curve α in $\mathbf{x}(D)$ represents the voyage of a ship, the coordinate curve (a_1, a_2) plots its position on the map D.

A rigorous proof of the following rather technical fact requires the methods of advanced calculus, and we shall not attempt to give a proof here.

3.2 Theorem Let M be a surface in $\mathbf{E}^3$. If $F: \mathbf{E}^n \to \mathbf{E}^3$ is a (differentiable) mapping whose image lies in M, then considered as a function $F: \mathbf{E}^n \to M$ into M, F is differentiable (as on p. 143.)

This theorem links the calculus of M tightly to the calculus of $\mathbf{E}^3$. For example, it implies the "obvious" result that a curve in $\mathbf{E}^3$ which lies in M is a curve of M.

Since a patch is a differentiable function from (an open set of) $\mathbf{E}^2$ into $\mathbf{E}^3$, it follows that a patch is a differentiable function into M. Hence its coordinate expressions are all differentiable, so *patches overlap smoothly.*

3.3 Corollary If $\mathbf{x}$ and $\mathbf{y}$ are patches in a surface M in $\mathbf{E}^3$ whose images overlap, then the composite functions $\mathbf{x}^{-1}\mathbf{y}$ and $\mathbf{y}^{-1}\mathbf{x}$ are (differentiable) mappings defined on open sets of $\mathbf{E}^2$.

(The function $\mathbf{y}^{-1}\mathbf{x}$, for example, is defined only for those points (u, v) in D such that $\mathbf{x}(u, v)$ lies in the image $\mathbf{y}(E)$ of $\mathbf{y}$ (Fig. 4.25).

By an argument like that for Lemma 3.1, Corollary 3.3 can be rewritten.

3.4 Corollary If $\mathbf{x}$ and $\mathbf{y}$ are overlapping patches in M, then there exist unique differentiable functions $\bar{u}$ and $\bar{v}$ such that

$$\mathbf{y}(u,v) = \mathbf{x}(\bar{u}(u,v), \bar{v}(u,v))$$

for all (u, v) in the domain of $\mathbf{x}^{-1}\mathbf{y}$. In functional notation: $\mathbf{y} = \mathbf{x}(\bar{u}, \bar{v})$.

There are, of course, symmetrical equations expressing $\mathbf{x}$ in terms of $\mathbf{y}$.

Corollary 3.3 makes it much easier to prove differentiability. For example, if f is a real-valued function on M, instead of verifying that *all* coordi-

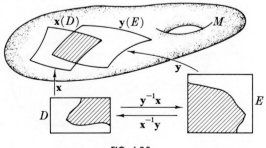

FIG. 4.25

nate expressions $f(\mathbf{x})$ are Euclidean differentiable, we need only do so for enough patches $\mathbf{x}$ to cover all of M (so a single patch will often be enough). The proof is an exercise in checking domains of composite functions: For an *arbitrary* patch $\mathbf{y}$, $f\mathbf{x}$ and $\mathbf{x}^{-1}\mathbf{y}$ differentiable imply $f\mathbf{x}\mathbf{x}^{-1}\mathbf{y}$ differentiable. This function is in general not $f\mathbf{y}$, because its domain is too small. But since there are enough $\mathbf{x}$'s to cover M, such functions constitute all of $f\mathbf{y}$, and thus prove it is differentiable.

It is intuitively clear what it means for a vector to be tangent to a surface M in $\mathbf{E}^3$. A formal definition can be based on the idea that a curve in M must have all its velocity vectors tangent to M.

3.5 Definition Let $\mathbf{p}$ be a point of a surface M in $\mathbf{E}^3$. A tangent vector $\mathbf{v}$ to $\mathbf{E}^3$ at $\mathbf{p}$ is *tangent to M at $\mathbf{p}$* provided $\mathbf{v}$ is a velocity vector of some curve in M (Fig. 4.26).

The set of all tangent vectors to M at $\mathbf{p}$ is called the *tangent plane of M at $\mathbf{p}$*, and is denoted by $T_p(M)$. The following result shows, in particular, that at each point $\mathbf{p}$ of M the tangent plane $T_p(M)$ is actually a two-dimensional vector subspace of the tangent space $T_p(\mathbf{E}^3)$.

3.6 Lemma Let $\mathbf{p}$ be a point of a surface M in $\mathbf{E}^3$, and let $\mathbf{x}$ be a patch in M such that $\mathbf{x}(u_0, v_0) = \mathbf{p}$. A tangent vector $\mathbf{v}$ to $\mathbf{E}^3$ at $\mathbf{p}$ is tangent to M if and only if $\mathbf{v}$ can be written as a linear combination of $\mathbf{x}_u(u_0, v_0)$ and $\mathbf{x}_v(u_0, v_0)$.

Since partial velocities are always linearly independent, we deduce that they provided a *basis* for the tangent plane of M at each point of $\mathbf{x}(D)$.

Proof. Note that the parameter curves of $\mathbf{x}$ are curves in M, so these partial velocities are tangent to M at $\mathbf{p}$.

First suppose that $\mathbf{v}$ is tangent to M at $\mathbf{p}$; thus there is a curve α in M such that $\alpha(0) = \mathbf{p}$ and $\alpha'(0) = \mathbf{v}$. Now by Lemma 3.1, α may be written $\alpha = \mathbf{x}(a_1, a_2)$; hence by the chain rule

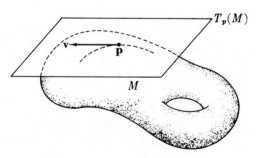

FIG. 4.26

$$\alpha' = \mathbf{x}_u(a_1, a_2)\,\frac{da_1}{dt} + \mathbf{x}_v(a_1, a_2)\,\frac{da_2}{dt}.$$

But since $\alpha(0) = \mathbf{p} = \mathbf{x}(u_0, v_0)$, we have $a_1(0) = u_0$, $a_2(0) = v_0$. Hence evaluation at $t = 0$ yields

$$\mathbf{v} = \alpha'(0) = \frac{da_1}{dt}\,(0)\,\mathbf{x}_u(u_0, v_0) + \frac{da_2}{dt}\,(0)\,\mathbf{x}_v(u_0, v_0).$$

Conversely, suppose that a tangent vector $\mathbf{v}$ to $\mathbf{E}^3$ can be written

$$\mathbf{v} = c_1\mathbf{x}_u(u_0, v_0) + c_2\mathbf{x}_v(u_0, v_0).$$

By computations as above, $\mathbf{v}$ is the velocity vector at $t = 0$ of the curve

$$t \to \mathbf{x}(u_0 + tc_1, v_0 + tc_2).$$

Thus $\mathbf{v}$ is tangent to M at $\mathbf{p}$. ∎

A reasonable deduction, based on the general properties of derivatives, is that the tangent plane $T_p(M)$ is the linear approximation of the surface M near $\mathbf{p}$.

3.7 Definition A Euclidean vector field Z on a surface M in $\mathbf{E}^3$ is a function that assigns to each point $\mathbf{p}$ of M a tangent vector $Z(\mathbf{p})$ to $\mathbf{E}^3$ at $\mathbf{p}$.

A Euclidean vector field V for which each vector $V(\mathbf{p})$ is tangent to M at $\mathbf{p}$ is called a *tangent vector field* on M (Fig. 4.27). Frequently these vector fields are defined, not on all of M, but only on some region in M. As usual, we always assume differentiability (Exercise 12).

A Euclidean vector $\mathbf{z}$ at a point $\mathbf{p}$ of M is *normal* to M if it is orthogonal to the tangent plane $T_p(M)$—that is, to every tangent vector to M at $\mathbf{p}$. And a Euclidean vector field Z on M is a *normal vector field* on M provided each vector $Z(\mathbf{p})$ is normal to M.

Because $T_p(M)$ is a two-dimensional subspace of $T_p(\mathbf{E}^3)$, there is only one direction normal to M at $\mathbf{p}$: All normal vectors $\mathbf{z}$ at $\mathbf{p}$ are collinear.

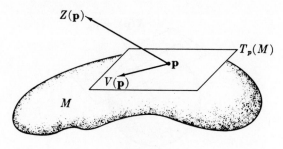

FIG. 4.27

Thus if $\mathbf{z}$ is not zero, it follows that $T_p(M)$ *consists of precisely those vectors in* $T_p(\mathbf{E}^3)$ *that are orthogonal to* $\mathbf{z}$.

It is particularly easy to deal with tangent and normal vector fields on a surface given in implicit form.

3.8 Lemma If $M: g = c$ is a surface in $\mathbf{E}^3$, then the *gradient* vector field $\nabla g = \sum (\partial g/\partial x_i) U_i$ (considered only at points of M) is a nonvanishing normal vector field on the entire surface M.

Proof. The gradient is nonvanishing (that is, never zero) on M since by Theorem 1.4 the partial derivatives $\partial g/\partial x_i$ cannot simultaneously be zero at any point of M.

We must show that $(\nabla g)(\mathbf{p}) \cdot \mathbf{v} = 0$ for every tangent vector $\mathbf{v}$ to M at $\mathbf{p}$. First note that if α is a curve in M, then $g(\alpha) = g(\alpha_1, \alpha_2, \alpha_3)$ has constant value c. Thus by the chain rule

$$\sum \frac{\partial g}{\partial x_i}(\alpha) \frac{d\alpha_i}{dt} = 0.$$

Now choose α to have initial velocity

$$\alpha'(0) = \mathbf{v} = (v_1, v_2, v_3)$$

at $\alpha(0) = \mathbf{p}$. Then

$$0 = \sum \frac{\partial g}{\partial x_i}(\alpha(0)) \frac{d\alpha_i}{dt}(0) = \sum \frac{\partial g}{\partial x_i}(\mathbf{p})v_i = (\nabla g)(\mathbf{p}) \cdot \mathbf{v} = 0. \quad \blacksquare$$

3.9 Example Vector fields on the sphere $\Sigma: g = \sum x_i^2 = r^2$. The lemma shows that

$$X = \tfrac{1}{2}\nabla g = \sum x_i U_i$$

is a normal vector field on Σ (Fig. 4.28). This is geometrically evident, since $X(\mathbf{p}) = \sum p_i U_i(\mathbf{p})$ is the vector $\mathbf{p}$ with point of application $\mathbf{p}$! It follows by a remark above that $\mathbf{v}_p$ is tangent to Σ if and only if the dot product

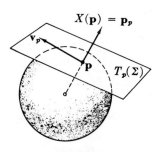

FIG. 4.28

$\mathbf{v}_p \cdot \mathbf{p}_p = \mathbf{v} \cdot \mathbf{p}$ is zero. Similarly a vector field V on Σ is a *tangent* vector field if and only if $V \cdot X = 0$. For example, $V(\mathbf{p}) = (-p_2, p_1, 0)$ defines a tangent vector field on Σ that points "due east" and vanishes at the north and south poles $(0, 0, \pm r)$.

We must emphasize that only the *tangent* vector fields on M belong to the calculus of M itself, since they derive ultimately from curves in M (Definition 3.5). This is certainly not the case with *normal* vector fields. However, as we shall see in the next chapter, normal vector fields are quite useful in studying M from the viewpoint of an observer in $\mathbf{E}^3$.

Finally, we shall adapt the notion of directional derivative to a surface. Definition 3.1 of Chapter I uses straight lines in $\mathbf{E}^3$; thus we must use the less restrictive formulation based on Lemma 4.6 of Chapter I.

3.10 Definition Let $\mathbf{v}$ be a tangent vector to M at $\mathbf{p}$, and let f be a differentiable real-valued function on M. The *derivative* $\mathbf{v}[f]$ *of* f *with respect to* $\mathbf{v}$ is the common value of $(d/dt)(f\alpha)(0)$ for all curves α in M which have initial velocity $\mathbf{v}$.

Directional derivatives on a surface have exactly the same linear and Leibnizian properties as in the Euclidean case (Theorem 3.3 of Chapter I).

EXERCISES

1. Let $\mathbf{x}$ be the geographical patch in the sphere Σ (Ex. 2.2). Find the coordinate expression $f(\mathbf{x})$ for the following functions on Σ:
   (a) $f(\mathbf{p}) = p_1^2 + p_2^2$,   (b) $f(\mathbf{p}) = (p_1 - p_2)^2 + p_3^2$.

2. Let $\mathbf{x}$ be the parametrization of the torus given in Example 2.6.
   (a) Find the Euclidean coordinates $\alpha_1, \alpha_2, \alpha_3$ of the curve $\alpha(t) = \mathbf{x}(t, t)$.
   (b) Show that α is periodic, and find its period (see p. 20).

3. (a) Prove Corollary 3.4.
   (b) Derive the "chain rule"

$$\mathbf{y}_u = \frac{\partial \bar{u}}{\partial u} \mathbf{x}_u + \frac{\partial \bar{v}}{\partial u} \mathbf{x}_v \qquad \mathbf{y}_v = \frac{\partial \bar{u}}{\partial v} \mathbf{x}_u + \frac{\partial \bar{v}}{\partial v} \mathbf{x}_v$$

   where $\mathbf{x}_u$ and $\mathbf{x}_v$ are evaluated on $(\bar{u}, \bar{v})$.
   (c) Deduce that $\mathbf{y}_u \times \mathbf{y}_v = J\mathbf{x}_u \times \mathbf{x}_v$, where J is the Jacobian of the mapping $\mathbf{x}^{-1}\mathbf{y} = (\bar{u}, \bar{v}): D \to \mathbf{E}^2$.

4. Let $\mathbf{x}$ be a patch in M.
   (a) If $\mathbf{x}_*$ is the derivative map of $\mathbf{x}$ (I.7), show that

$$\mathbf{x}_*(U_1) = \mathbf{x}_u \qquad \mathbf{x}_*(U_2) = \mathbf{x}_v$$

   where U_1, U_2 is the natural frame field on $\mathbf{E}^2$.

(b) If f is a differentiable function on M, prove

$$\mathbf{x}_u[f] = \frac{\partial}{\partial u}\,(f(\mathbf{x})) \qquad \mathbf{x}_v[f] = \frac{\partial}{\partial v}\,(f(\mathbf{x})).$$

5. Prove that:
   (a) $\mathbf{v}$ is tangent to $M\colon z = f(x,\,y)$ at a point $\mathbf{p}$ of M if and only if

   $$v_3 = \frac{\partial f}{\partial x}\,(p_1,\,p_2)v_1 + \frac{\partial f}{\partial y}\,(p_1,\,p_2)v_2.$$

   (b) if $\mathbf{x}$ is a patch in an arbitrary surface M, then $\mathbf{v}$ is tangent to M at $\mathbf{x}(u,\,v)$ if and only if

   $$\mathbf{v} \cdot \mathbf{x}_u(u,\,v) \times \mathbf{x}_v(u,\,v) = 0.$$

6. Let $\mathbf{x}$ and $\mathbf{y}$ be the patches in the unit sphere Σ that are defined on the unit disc $D\colon u^2 + v^2 < 1$ by

   $$\mathbf{x}(u,\,v) = (u,\,v,\,f(u,v)) \qquad \mathbf{y}(u,\,v) = (v,\,f(u,v),\,u)$$

   where $f = \sqrt{1 - u^2 - v^2}$.
   (a) On a sketch of Σ indicate the images $\mathbf{x}(D)$ and $\mathbf{y}(D)$, and the region on which they overlap.
   (b) At which points of D is $\mathbf{y}^{-1}\mathbf{x}$ defined? Find a formula for this function.
   (c) At which points of D is $\mathbf{x}^{-1}\mathbf{y}$ defined? Find a formula for this function.

7. Find a nonvanishing normal vector field on $M\colon z = xy$ and two tangent vector fields that are linearly independent at each point.

8. Let C be the right circular cone parametrized by

   $$\mathbf{x}(u,\,v) = v(\cos u,\,\sin u,\,1).$$

   If α is the curve $\alpha(t) = \mathbf{x}(\sqrt{2}\,t,\,e^t)$
   (a) Express α' in terms of $\mathbf{x}_u$ and $\mathbf{x}_v$.
   (b) Show that at each point of α, the velocity α' bisects the angle between $\mathbf{x}_u$ and $\mathbf{x}_v$. (*Hint:* Verify that $\alpha' \cdot \mathbf{x}_u/\|\,\mathbf{x}_u\,\| = \alpha' \cdot \mathbf{x}_v/\|\,\mathbf{x}_v\,\|$, where $\mathbf{x}_u$ and $\mathbf{x}_v$ are evaluated on $(\sqrt{2}\,t,\,e^t)$.)
   (c) Make a sketch of the cone C showing the curve α.

9. If $\mathbf{z}$ is a nonzero vector normal to M at $\mathbf{p}$, let $\bar{T}_p(M)$ be the plane through $\mathbf{p}$ orthogonal to $\mathbf{z}$ (see page 60). Prove:
   (a) If each tangent vector $\mathbf{v}_p$ to M at $\mathbf{p}$ is replaced by its *tip* $\mathbf{p} + \mathbf{v}$, then $T_p(M)$ becomes $\bar{T}_p(M)$. (Thus $\bar{T}_p(M)$ gives a concrete representation of $T_p(M)$ in $\mathbf{E}^3$.)
   (b) If $\mathbf{x}$ is a patch in M, then $\bar{T}_{\mathbf{x}(u,v)}(M)$ consists of all points $\mathbf{r}$ in $\mathbf{E}^3$ such that $(\mathbf{r} - \mathbf{x}(u,v)) \cdot \mathbf{x}_u(u,\,v) \times \mathbf{x}_v(u,\,v) = 0$.

(c) If M is given implicitly by $g = c$, then $\bar{T}_p(M)$ consists of all points $\mathbf{r}$ in $\mathbf{E}^3$ such that $(\mathbf{r} - \mathbf{p}) \cdot (\nabla g)(\mathbf{p}) = 0$.

10. In each case find an equation of the form $ax + by + cz = d$ for the plane $\bar{T}_p(M)$.

(a) $\mathbf{p} = (0, 0, 0)$ and M is the sphere

$$x^2 + y^2 + (z - 1)^2 = 1.$$

(b) $\mathbf{p} = (1, -2, 3)$ and M is the ellipsoid

$$\frac{x^2}{4} + \frac{y^2}{16} + \frac{z^2}{18} = 1.$$

(c) $\mathbf{p} = \mathbf{x}(2, \pi/4)$, where M is the helicoid parametrized by

$$\mathbf{x}(u, v) = (u \cos v, u \sin v, 2v).$$

11. (*Continuation of Exercise 2*).

(a) If m and n are integers with greatest common divisor 1, show that $\alpha(t) = \mathbf{x}(mt, nt)$ is a simple closed curve on the torus, and find its period.†

(b) If q is an irrational number, show that the curve $\alpha : \mathbf{R} \to T$ such that $\alpha(t) = \mathbf{x}(t, qt)$ is one-to-one.

This curve, called a *winding line* on the torus T, is *dense* in T; that is, given any number $\varepsilon > 0$, α comes within distance ε of every point of T.

12. A Euclidean vector field $Z = \sum z_i U_i$ on M is *differentiable* provided its coordinate functions z_1, z_2, z_3 (on M) are differentiable. If V is a tangent vector field, show that

(a) For every patch $\mathbf{x} : D \to M$, V can be written as

$$V(\mathbf{x}(u, v)) = f(u, v)\mathbf{x}_u(u, v) + g(u, v)\mathbf{x}_v(u, v)$$

(b) V is differentiable if and only if the functions f and g (on D) are differentiable.

The following exercises require some knowledge of point-set topology. They deal with *open sets* in a surface M in $\mathbf{E}^3$, that is, sets $\mathfrak{U}$ in M which contain a neighborhood in M of each of their points.

13. Prove that if $\mathbf{y} : E \to M$ is a proper patch, then $\mathbf{y}$ carries open sets in E to open sets in M. Deduce that if $\mathbf{x} : D \to M$ is an arbitrary patch, then the image $\mathbf{x}(D)$ is an open set in M. (*Hint:* To prove the latter assertion, use Corollary 3.3.)

---

† That is, show $\alpha(t') = \alpha(t)$ if and only if $t' - t$ is a multiple of the period p. Roughly speaking, this means the route of α is O-shaped rather than, say, 8-shaped.

14. Prove that *every* patch $\mathbf{x}: D \to M$ in a surface M in $\mathbf{E}^3$ is proper. (*Hint:* Use Exercise 13. Note that $(\mathbf{x}^{-1}\mathbf{y})\mathbf{y}^{-1}$ is continuous and agrees with $\mathbf{x}^{-1}$ on an open set in $\mathbf{x}(D)$.)

15. If $\mathcal{U}$ is a subset of a surface M in $\mathbf{E}^3$, prove that $\mathcal{U}$ is itself a surface in $\mathbf{E}^3$ if and only if $\mathcal{U}$ is an open set of M.

4   Differential Forms on a Surface

In Chapter I we discussed differential forms on $\mathbf{E}^3$ only in sufficient detail to take care of the Cartan structural equations (Theorem 8.3 of Chapter II). In the next three sections we shall give a rather complete treatment of forms *on a surface*.

Forms are just what we need to describe the geometry of a surface (Chapters VI and VII), but this is only one example of their usefulness. Surfaces and Euclidean spaces are merely special cases of the general notion of *manifold* (Section 8). Every manifold has a differential and integral calculus—expressed in terms of forms—which generalizes the usual elementary calculus on the real line. Thus forms are fundamental to all the many branches of mathematics and its applications that are based on calculus. In the special case of a surface, the calculus of forms is rather easy, but still gives a remarkably accurate picture of the most general case.

Just as for $\mathbf{E}^3$, a 0-*form* f on a surface M is simply a (differentiable) real-valued function on M, and a 1-*form* ϕ on M is a real-valued function on tangent vectors to M that is linear at each point (Definition 5.1 of Chapter I). We did not give a precise definition of 2-forms in Chapter I, but we shall do so now. A 2-form will be a two-dimensional analogue of a 1-form: a real-valued function, not on single tangent vectors, but on *pairs* of tangent vectors. (In this context the term "pair" will always imply that the tangent vectors have the same point of application.)

4.1 Definition   A 2-*form* η on a surface M is a real-valued function on all ordered pairs of tangent vectors $\mathbf{v}, \mathbf{w}$ to M such that

   (1) $\eta(\mathbf{v}, \mathbf{w})$ is linear in $\mathbf{v}$ and in $\mathbf{w}$.

   (2) $\eta(\mathbf{v}, \mathbf{w}) = -\eta(\mathbf{w}, \mathbf{v})$.

Since a surface is two-dimensional, *all p forms with $p > 2$ are zero*, by definition. This fact considerably simplifies the theory of differential forms on a surface.

At the end of this section we will show that our new definitions are consistent with the informal exposition given in Chapter I, Section 6.

Forms are added in the usual pointwise fashion; we add only forms of the same *degree* $p = 0, 1, 2$. Just as we evaluated a 1-form ϕ on a vector field V, we now evaluate a 2-form η on a pair of vector fields V, W to get a real-valued function $\eta(V,W)$ on the surface M. Of course we shall always assume that the forms we deal with are differentiable—that is, convert (differentiable) vector fields into differentiable functions.

Note that the alternation rule (2) in Definition 4.1 implies that

$$\eta(\mathbf{v}, \mathbf{v}) = 0$$

for any tangent vector $\mathbf{v}$. This rule also shows that 2-forms are related to determinants.

4.2 Lemma Let η be a 2-form on a surface M, and let $\mathbf{v}$ and $\mathbf{w}$ be (linearly independent) tangent vectors at some point of M. Then

$$\eta(a\mathbf{v} + b\mathbf{w}, c\mathbf{v} + d\mathbf{w}) = \begin{vmatrix} a & b \\ c & d \end{vmatrix} \eta(\mathbf{v}, \mathbf{w})$$

Proof. Since η is linear in its first variable, its value on the pair of tangent vectors $a\mathbf{v} + b\mathbf{w}$, $c\mathbf{v} + d\mathbf{w}$ is $a\eta(\mathbf{v}, c\mathbf{v} + d\mathbf{w}) + b\eta(\mathbf{w}, c\mathbf{v} + d\mathbf{w})$. Using the linearity of η in its second variable, we get

$$ac\,\eta(\mathbf{v}, \mathbf{v}) + ad\,\eta(\mathbf{v}, \mathbf{w}) + bc\,\eta(\mathbf{w}, \mathbf{v}) + bd\,\eta(\mathbf{w}, \mathbf{w}).$$

Then the alternation rule (2) gives

$$\eta(a\mathbf{v} + b\mathbf{w}, c\mathbf{v} + d\mathbf{w}) = (ad - bc)\,\eta(\mathbf{v}, \mathbf{w}). \qquad \blacksquare$$

Thus the values of a 2-form on *all* pairs of tangent vectors at a point are completely determined by its value on any *one* linearly independent pair. This remark is used frequently in later work.

Wherever they appear, differential forms satisfy certain general properties, established (at least partially) in Chapter I for forms on $\mathbf{E}^3$. To begin with, *the wedge product of a p-form and a q-form is always a $(p + q)$-form.* If p or q is zero, the wedge product is just the usual multiplication by a function. On a surface, the wedge product is always zero if $p + q > 2$. So we need a definition only for the case $p = q = 1$.

4.3 Definition If ϕ and ψ are 1-forms on a surface M, the *wedge product* $\phi \wedge \psi$ is the 2-form on M such that

$$(\phi \wedge \psi)(\mathbf{v}, \mathbf{w}) = \phi(\mathbf{v})\,\psi(\mathbf{w}) - \phi(\mathbf{w})\,\psi(\mathbf{v})$$

for all pairs $\mathbf{v}$, $\mathbf{w}$ of tangent vectors to M.

Note that $\phi \wedge \psi$ really is a 2-form on M, since it is a real-valued function on all pairs of tangent vectors and satisfies the conditions in Definition 4.1. The wedge product has all the usual algebraic properties except commutativity; in general, *if ξ is a p-form and η is a q-form, then*

$$\xi \wedge \eta = (-1)^{pq}\eta \wedge \xi.$$

On a surface the only minus sign occurs in the multiplication of 1-forms, where just as in Chapter I, we have $\phi \wedge \psi = -\psi \wedge \phi$.

The differential calculus of forms is based on the exterior derivative d. For a 0-form (function) f on a surface, the exterior derivative is, as before, the 1-form df such that $df(\mathbf{v}) = \mathbf{v}[f]$. Wherever forms appear, the exterior derivative of a p-form is a $(p + 1)$-form. Thus for surface the only new definition we need is that of the exterior derivative $d\phi$ of a 1-form ϕ.

4.4 Definition   Let ϕ be a 1-form on a surface M. Then the *exterior derivative $d\phi$* of ϕ is the 2-form such that for any patch $\mathbf{x}$ in M,

$$d\phi(\mathbf{x}_u, \mathbf{x}_v) = \frac{\partial}{\partial u}\left(\phi(\mathbf{x}_v)\right) - \frac{\partial}{\partial v}\left(\phi(\mathbf{x}_u)\right).$$

As it stands, this is not yet a valid definition; there is a problem of consistency. What we have actually defined is a form $d_{\mathbf{x}}\phi$ on the image of each patch $\mathbf{x}$ in M. So what we must prove is that on the region where two patches overlap, the forms $d_{\mathbf{x}}\phi$ and $d_{\mathbf{y}}\phi$ are equal. Only then will we have obtained from ϕ a single form $d\phi$ on M.

4.5 Lemma   Let ϕ be a 1-form on M. If $\mathbf{x}$ and $\mathbf{y}$ are patches in M, then $d_{\mathbf{x}}\phi = d_{\mathbf{y}}\phi$ on the overlap of $\mathbf{x}(D)$ and $\mathbf{y}(E)$.

Proof. Because $\mathbf{y}_u$ and $\mathbf{y}_v$ are linearly independent at each point, it suffices by Lemma 4.2 to show that

$$(d_{\mathbf{y}}\phi)(\mathbf{y}_u, \mathbf{y}_v) = (d_{\mathbf{x}}\phi)(\mathbf{y}_u, \mathbf{y}_v).$$

Now as in Corollary 3.4, we write $\mathbf{y} = \mathbf{x}(\bar{u}, \bar{v})$ and, as in an earlier exercise, deduce by the chain rule that

$$
\begin{aligned}
\mathbf{y}_u &= \frac{\partial \bar{u}}{\partial u}\,\mathbf{x}_u + \frac{\partial \bar{v}}{\partial u}\,\mathbf{x}_v \\[2mm]
\mathbf{y}_v &= \frac{\partial \bar{u}}{\partial v}\,\mathbf{x}_u + \frac{\partial \bar{v}}{\partial v}\,\mathbf{x}_v
\end{aligned}
\tag{1}
$$

where $\mathbf{x}_u$ and $\mathbf{x}_v$ are henceforth *evaluated on* $(\bar{u}, \bar{v})$. Then by Lemma 4.2,

$$(d_{\mathbf{x}}\phi)(\mathbf{y}_u, \mathbf{y}_v) = J(d_{\mathbf{x}}\phi)(\mathbf{x}_u, \mathbf{x}_v) \tag{2}$$

where J is the Jacobian $(\partial \bar{u}/\partial u)(\partial \bar{v}/\partial v) - (\partial \bar{u}/\partial v)(\partial \bar{v}/\partial u)$. Thus it is clear from Definition 4.4 that to prove $(d_{\mathbf{y}}\phi)(\mathbf{y}_u, \mathbf{y}_v) = (d_{\mathbf{x}}\phi)(\mathbf{y}_u, \mathbf{y}_v)$, all we need is the equation

$$\frac{\partial}{\partial u}(\phi\mathbf{y}_v) - \frac{\partial}{\partial v}(\phi\mathbf{y}_u) = J\left\{\frac{\partial}{\partial \bar{u}}(\phi\mathbf{x}_v) - \frac{\partial}{\partial \bar{v}}(\phi\mathbf{x}_u)\right\}. \tag{3}$$

It suffices to operate on $(\partial/\partial u)(\phi\mathbf{y}_v)$, for merely reversing u and v will then yield $(\partial/\partial v)(\phi\mathbf{y}_u)$. Since (3) requires us to *subtract* these two derivatives, we can *discard any terms that will cancel* when u and v are everywhere reversed.

Applying ϕ to the second equation in (1) yields

$$\phi(\mathbf{y}_v) = \phi(\mathbf{x}_u)\frac{\partial \bar{u}}{\partial v} + \phi(\mathbf{x}_v)\frac{\partial \bar{v}}{\partial v}.$$

Hence by the chain rule,

$$\frac{\partial}{\partial u}(\phi\mathbf{y}_v) = \frac{\partial}{\partial u}(\phi\mathbf{x}_u)\frac{\partial \bar{u}}{\partial v} + \frac{\partial}{\partial u}(\phi\mathbf{x}_v)\frac{\partial \bar{v}}{\partial v} + \cdots \tag{4}$$

where in accordance with the remark above we have discarded two symmetric terms. Next we use the chain rule—and the same remark—to get

$$\frac{\partial}{\partial u}(\phi\mathbf{y}_v) = \left(\frac{\partial}{\partial \bar{v}}(\phi\mathbf{x}_u)\frac{\partial \bar{v}}{\partial u} + \cdots\right)\frac{\partial \bar{u}}{\partial v} + \left(\frac{\partial}{\partial \bar{u}}(\phi\mathbf{x}_v)\frac{\partial \bar{u}}{\partial u} + \cdots\right)\frac{\partial \bar{v}}{\partial v}. \tag{5}$$

Now reverse u and v in (5) (and also $\bar{u}$ and $\bar{v}$) and subtract from (5) itself. The result is precisely equation (3).   ∎

It is difficult to exaggerate the importance of the exterior derivative. We have already seen in Chapter I that it generalizes the familiar notion of differential of a function, and that it contains the three fundamental derivative operations of classical vector analysis (Exercise 8 of Chapter I, Section 6). In Chapter II it is essential to the Cartan structural equations (Theorem 8.3). Perhaps the clearest statement of its meaning will come in Stokes theorem (6.5), which could actually be used to define the exterior derivative of a 1-form.

On a surface, the exterior derivative of a wedge product displays the same linear and Leibnizian properties (Theorem 6.4 of Chapter I) as in $\mathbf{E}^3$; see Exercise 3. For practical computations these properties are apt to be more efficient than a direct appeal to the definition—compare the discussion of the Euclidean case on page 25. Examples of this technique appear in subsequent exercises.

The most striking property of this notion of derivative is that there are no *second* exterior derivatives: Wherever forms appear, *the exterior derivative applied twice always gives zero*. For a surface, we need only prove this for 0-forms, since even for a 1-form ϕ, the second derivative $d(d\phi)$ is a 3-form, hence is automatically zero.

4.6 Theorem   If f is a (differentiable) real-valued function on M, then $d(df) = 0$.

Proof. Let $\psi = df$, so we must show $d\psi = 0$. It suffices by Lemma 4.2 to prove that for any patch $\mathbf{x}$ in M we have $(d\psi)(\mathbf{x}_u,\mathbf{x}_v) = 0$. Now using Exercise 4 of Section 3 we get

$$\psi(\mathbf{x}_u) = df(\mathbf{x}_u) = \mathbf{x}_u[f] = \frac{\partial}{\partial u}(f\mathbf{x})$$

and similarly

$$\psi(\mathbf{x}_v) = \frac{\partial}{\partial v}(f\mathbf{x}).$$

Hence

$$d\psi(\mathbf{x}_u, \mathbf{x}_v) = \frac{\partial}{\partial u}(\psi\mathbf{x}_v) - \frac{\partial}{\partial v}(\psi\mathbf{x}_u) = \frac{\partial^2(f\mathbf{x})}{\partial u\,\partial v} - \frac{\partial^2(f\mathbf{x})}{\partial v\,\partial u} = 0 \qquad \blacksquare$$

Many computations and proofs reduce to the problem of showing that two forms are equal. As we have seen, to do so it is not necessary to check that the forms have the same value on *all* tangent vectors. In particular, if $\mathbf{x}$ is a coordinate patch, then

(1) for 1-forms on $\mathbf{x}(D)$: $\phi = \psi$ if and only if $\phi(\mathbf{x}_u) = \psi(\mathbf{x}_u)$ and $\phi(\mathbf{x}_v) = \psi(\mathbf{x}_v)$;

(2) for 2-forms on $\mathbf{x}(D)$: $\mu = \nu$ if and only if $\mu(\mathbf{x}_u, \mathbf{x}_v) = \nu(\mathbf{x}_u, \mathbf{x}_v)$.

(To prove these criteria we express arbitrary tangent vectors as linear combinations of $\mathbf{x}_u$ and $\mathbf{x}_v$.) More generally, $\mathbf{x}_u$ and $\mathbf{x}_v$ may be replaced by any two vector fields that are linearly independent at each point.

Let us now check that the rigorous results proved in this section are consistent with the rules of operation stated in Chapter I, Section 6.

4.7 Example   Differential forms on the plane $\mathbf{E}^2$.   Let $u_1 = u$ and $u_2 = v$ be the natural coordinate functions, and U_1, U_2 the natural frame field on $\mathbf{E}^2$. The differential calculus of forms on $\mathbf{E}^2$ is expressed in terms of u_1 and u_2 as follows:

If f is a function, ϕ a 1-form, and η a 2-form, then

(1) $\phi = f_1\,du_1 + f_2\,du_2,$ where $f_i = \phi(U_i)$.

(2) $\eta = g\,du_1\,du_2,$ where $g = \eta(U_1, U_2)$.

(3) for $\psi = g_1\,du_1 + g_2\,du_2$ and ϕ as above,

$$\phi \wedge \psi = (f_1 g_2 - f_2 g_1)du_1\,du_2.$$

(4) $df = \dfrac{\partial f}{\partial u_1}\,du_1 + \dfrac{\partial f}{\partial u_2}\,du_2.$

(5) $d\phi = \left(\dfrac{\partial f_2}{\partial u_1} - \dfrac{\partial f_1}{\partial u_2}\right)du_1\,du_2$ (ϕ as above).

For the proof of these formulas, see Exercise 4.

Similar definitions and coordinate expressions may be established on any Euclidean space. In the case of the real line $\mathbf{E}^1$, the natural frame field

(Definition 2.4 of Chapter I) reduces to the single unit vector field U_1 for which $U_1[f] = df/dt$. All p-forms with $p > 1$ are zero, and if ϕ is a 1-form, then $\phi = \phi(U_1)\, dt$.

Some examples of forms will appear in subsequent exercises; however, many more will occur naturally throughout Chapters VI and VII, where their properties have direct geometric meaning.

EXERCISES

1. If ϕ and ψ are 1-forms on a surface, prove that $\phi \wedge \psi = -\psi \wedge \phi$. Deduce that $\phi \wedge \phi = 0$.

2. A form ϕ such that $d\phi = 0$ is said to be *closed*. A form ϕ such that $\phi = d\xi$ for some form ξ is *exact*. (So if ϕ is a p-form, ξ is necessarily a $(p-1)$-form.) Prove
   (a) Every exact form is closed.
   (b) No 0-form is exact, and on a surface every 2-form is closed.
   (c) Constant functions are closed 0-forms.

3. Prove the Leibnizian formulas
$$d(fg) = df\, g + f\, dg \qquad d(f\phi) = df \wedge \phi + f\, d\phi$$
   where f and g are functions on M, and ϕ is a 1-form (*Hint:* By definition $(f\phi)(v_p) = f(p)\phi(v_p)$; hence $f\phi$ evaluated on x_u is $f(x)\phi(x_u)$.)

4. (a) Prove formulas (1) and (2) in Example 4.7 using the remark preceding Example 4.7. (*Hint:* Show $(du_1\, du_2)(U_1, U_2) = 1$.)
   (b) Derive the remaining formulas using the properties of d and the wedge product.

5. If f is a real-valued function on a surface, and g is a function on the real line, show that
$$v_p[g(f)] = g'(f)v_p[f].$$
   Deduce that
$$d(g(f)) = g'(f)\, df.$$

6. If f, g, and h are functions on a surface M, and ϕ is a 1-form, prove:
   (a) $d(fgh) = gh\, df + fh\, dg + fg\, dh$,
   (b) $d(\phi f) = f\, d\phi - \phi \wedge df \quad (\phi f = f\phi)$,
   (c) $(df \wedge dg)(v, w) = v[f]w[g] - v[g]w[f]$.

7. Suppose that M is covered by open sets $\mathfrak{U}_1, \cdots, \mathfrak{U}_k$, and on each $\mathfrak{U}_i$ there is defined a function f_i such that $f_i - f_j$ is constant on the overlap

of $\mathcal{U}_i$ and $\mathcal{U}_j$. Show that there is a 1-form ϕ on M such that $\phi = df_i$ on each $\mathcal{U}_i$. Generalize to the case of 1-forms ϕ_i such that $\phi_i - \phi_j$ is closed.

8. Let $\mathbf{y} \colon E \to M$ be an arbitrary mapping of an open set of $\mathbf{E}^2$ into a surface M. If ϕ is a 1-form on M, show that the formula

$$d\phi(\mathbf{y}_u, \mathbf{y}_v) = \frac{\partial}{\partial u}(\phi\,\mathbf{y}_v) - \frac{\partial}{\partial v}(\phi\,\mathbf{y}_u)$$

is still valid even when $\mathbf{y}$ is not regular or one-to-one.

(*Hint:* In the proof of Lemma 4.5, check that equation (3) is still valid in this case.)

A patch $\mathbf{x}$ in M establishes a one-to-one correspondence between an open set D of $\mathbf{E}^2$ and an open set $\mathbf{x}(D)$ of M. While we have emphasized the function $\mathbf{x} \colon D \to \mathbf{x}(D)$, there are some advantages to emphasizing instead the inverse function $\mathbf{x}^{-1} \colon \mathbf{x}(D) \to D$.

9. If $\mathbf{x} \colon D \to M$ is a patch in M, let $\tilde{u}$ and $\tilde{v}$ be the coordinate functions of $\mathbf{x}^{-1}$, so $\mathbf{x}^{-1}(\mathbf{p}) = (\tilde{u}(\mathbf{p}), \tilde{v}(\mathbf{p}))$ for all $\mathbf{p}$ in $\mathbf{x}(D)$. Show that
  (a) $\tilde{u}$ and $\tilde{v}$ are differentiable functions on $\mathbf{x}(D)$ such that

$$\tilde{u}(\mathbf{x}(u,v)) = u, \qquad \tilde{v}(\mathbf{x}(u,v)) = v.$$

These functions constitute the *coordinate system* associated with $\mathbf{x}$.
  (b) $d\tilde{u}(\mathbf{x}_u) = 1 \qquad d\tilde{u}(\mathbf{x}_v) = 0$
    $d\tilde{v}(\mathbf{x}_u) = 0 \qquad d\tilde{v}(\mathbf{x}_v) = 1$.
  (c) If ϕ is a 1-form and η is a 2-form, then

$$\phi = f\,d\tilde{u} + g\,d\tilde{v} \qquad \text{where } f(\mathbf{x}) = \phi(\mathbf{x}_u), g(\mathbf{x}) = \phi(\mathbf{x}_v)$$

$$\eta = h\,d\tilde{u}\,d\tilde{v} \qquad \text{where } h(\mathbf{x}) = \eta(\mathbf{x}_u, \mathbf{x}_v).$$

(*Hint:* For (b) use Ex. 4(b) of IV.3.)

10. Identify (or describe) the associated coordinate system $\tilde{u}$, $\tilde{v}$ of
  (a) The polar coordinate patch $\mathbf{x}(u,v) = (u \cos v, u \sin v)$ defined on $D \colon u > 0, 0 < v < 2\pi$.
  (b) The identity patch $\mathbf{x}(u,v) = (u,v)$ in $\mathbf{E}^2$.
  (c) The geographical patch $\mathbf{x}$ in the sphere.

5 Mappings of Surfaces

To define differentiability of a function *from* a surface *to* a surface, we follow the same general scheme used in Section 3, and require that all its coordinate expressions be differentiable.

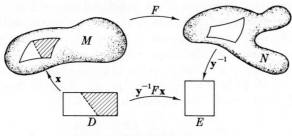

FIG. 4.29

5.1 Definition   A function $F\colon M \to N$ from one surface to another is *differentiable* provided that for each patch $\mathbf{x}$ in M and $\mathbf{y}$ in N the composite function $\mathbf{y}^{-1}F\mathbf{x}$ is Euclidean differentiable (and defined on an open set of $\mathbf{E}^2$). F is then called a *mapping of surfaces*.

Evidently the function $\mathbf{y}^{-1}F\mathbf{x}$ is defined at all points (u, v) of D such that $F(\mathbf{x}(u, v))$ lies in the image of $\mathbf{y}$ (Fig. 4.29). As in Section 3 we deduce from Corollary 3.3 that, in applying this definition, it suffices to check enough patches to cover both M and N.

5.2 Example   (1) Let Σ be the unit sphere in $\mathbf{E}^3$ (center at $\mathbf{0}$) but with *north and south poles removed*, and let C be the cylinder based on the unit circle in the xy plane. So C is in contact with the sphere along the equator. We define a mapping $F\colon \Sigma \to C$ as follows: If $\mathbf{p}$ is a point of Σ, draw the line orthogonally out from the z azis through $\mathbf{p}$, and let $F(\mathbf{p})$ be the point at which this line first meets C, as in Fig. 4.30. To prove that F is a mapping, we use the geographical patch $\mathbf{x}$ in Σ (Example 2.2), and for C we use the patch $\mathbf{y}(u, v) = (\cos u, \sin u, v)$. Now

$$\mathbf{x}(u, v) = (\cos v \cos u, \cos v \sin u, \sin v),$$

so from the definition of F, we get

$$F(\mathbf{x}(u, v)) = (\cos u, \sin u, \sin v).$$

But this point of C is $\mathbf{y}(u, \sin v)$; hence

$$F(\mathbf{x}(u, v)) = \mathbf{y}(u, \sin v).$$

Applying $\mathbf{y}^{-1}$ to both sides of this equation, we find

$$(\mathbf{y}^{-1}F\mathbf{x})(u, v) = (u, \sin v)$$

so $\mathbf{y}^{-1}F\mathbf{x}$ is certainly differentiable. (Actually $\mathbf{x}$ does not entirely cover Σ, but the missing semicircle can be covered by a patch like $\mathbf{x}$.) We conclude that F is a mapping.

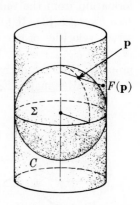

FIG. 4.30

(2) Stereographic projection of the punctured sphere Σ onto the plane. Let Σ be a unit sphere resting on the xy plane at the origin, so the center of Σ is at $(0, 0, 1)$. *Delete the north pole* $\mathbf{n} = (0, 0, 2)$ from Σ. Now imagine that there is a light source at the north pole, and for each point $\mathbf{p}$ of Σ, let $P(\mathbf{p})$ be the shadow of $\mathbf{p}$ in the xy plane (Fig. 4.31). As usual, we identify the xy plane with $\mathbf{E}^2$ by $(p_1, p_2, 0) \leftrightarrow (p_1, p_2)$. Thus we have defined a function P from Σ onto $\mathbf{E}^2$. Evidently P has the form

$$P(p_1, p_2, p_3) = \left(\frac{Rp_1}{r}, \frac{Rp_2}{r} \right),$$

where r and R are the distances from $\mathbf{p}$ and $P(\mathbf{p})$, respectively, to the z axis. But from the similar triangles in Fig. 4.32, we see that $R/2 = r/(2 - p_3)$; hence

$$P(p_1, p_2, p_3) = \left(\frac{2p_1}{2 - p_3}, \frac{2p_2}{2 - p_3} \right).$$

Now if $\mathbf{x}$ is any patch in Σ, the composite function $P\mathbf{x}$ is Euclidean differentiable, so $P \colon \Sigma \to \mathbf{E}^2$ is a mapping.

Just as for mappings of Euclidean space, each mapping of surfaces has a derivative map.

5.3 Definition   Let $F \colon M \to N$ be a mapping of surfaces. The *derivative map* F_* of F assigns to each tangent vector $\mathbf{v}$ to M the tangent vector $F_*(\mathbf{v})$ to N such that: If $\mathbf{v}$ is the initial velocity of a curve α in M, then $F_*(\mathbf{v})$ is the initial velocity of the image curve $F(\alpha)$ in N (Fig. 4.33).

Furthermore, at each point $\mathbf{p}$, the derivative map F_* is a linear transformation from the tangent plane $T_p(M)$ to the tangent plane $T_{F(p)}(N)$ (see Exercise 13). It follows immediately from the definition that F_* preserves velocities of curves: If $\bar{\alpha} = F(\alpha)$ is the image in N of a curve α in M, then $F_*(\alpha') = \bar{\alpha}'$. As in the Euclidean case, we deduce the convenient property that the derivative map of a composition is the composition of the derivative maps (Exercise 14).

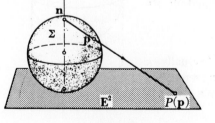

FIG. 4.31

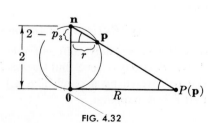

FIG. 4.32

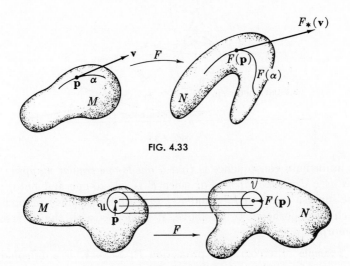

FIG. 4.33

FIG. 4.34

The derivative map of a mapping $F: M \rightarrow N$ may be computed in terms of partial velocities as follows. If $\mathbf{x}: D \rightarrow M$ is a parametrization in M, let $\mathbf{y}$ be the composite mapping $F(\mathbf{x}): D \rightarrow N$ (which need not be a parametrization). Obviously F carries the parameter curves of $\mathbf{x}$ to the corresponding parameter curves of $\mathbf{y}$. Since F_* preserves velocities of curves, it follows at once that

$$F_*(\mathbf{x}_u) = \mathbf{y}_u \qquad F_*(\mathbf{x}_v) = \mathbf{y}_v$$

Since $\mathbf{x}_u$ and $\mathbf{x}_v$ give a basis for the tangent space of M at each point of $\mathbf{x}(D)$, these readily computable formulas completely determine F_*.

The discussion of regular mappings in Section 7 of Chapter I translates easily to the case of a mapping of surfaces $F: M \rightarrow N$. F is *regular* provided all of its derivative maps $F_{*p}: T_p(M) \rightarrow T_{F(p)}(N)$ are one-to-one. Since these tangent planes have the same dimension, we know from linear algebra that the one-to-one requirement is equivalent to F_* being a linear isomorphism. A mapping $F: M \rightarrow N$ that has an inverse mapping $F^{-1}: N \rightarrow M$ is called a *diffeomorphism*. We may think of a diffeomorphism F as smoothly distorting M to produce N. By applying the Euclidean formulation of the inverse function theorem to a coordinate expression $\mathbf{y}^{-1}F\mathbf{x}$ for F, we can deduce this extension of the inverse function theorem (7.10 of Chapter I).

5.4 Theorem   Let $F: M \rightarrow N$ be a mapping of surfaces, and suppose that $F_{*p}: T_p(M) \rightarrow T_{F(p)}(N)$ is a linear isomorphism at some one point $\mathbf{p}$ of M. Then there exists a neighborhood $\mathfrak{u}$ of $\mathbf{p}$ in M such that the restriction of F to $\mathfrak{u}$ is a diffeomorphism onto a neighborhood υ of $F(\mathbf{p})$ in N (Fig. 4.34).

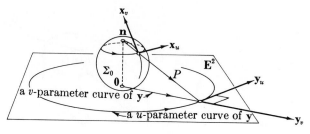

FIG. 4.35

An immediate consequence is that *a one-to-one regular mapping F of M onto N is a diffeomorphism*. For since F is one-to-one and onto, it has a unique inverse function F^{-1}, and F^{-1} is a (differentiable) mapping, since on each neighborhood $\mathcal{V}$, as above, it coincides with the inverse of the diffeomorphism $\mathcal{U} \to \mathcal{V}$.

5.5 Example   Stereographic projection $P: \Sigma_0 \to \mathbf{E}^2$ is a diffeomorphism.

It is clear from Example 5.2 that P is a one-to-one mapping from the punctured sphere Σ_0 to the plane $\mathbf{E}^2$. Thus we need only show that P_* is one-to-one at each point. A minor modification of the geographical patch in Example 2.2 yields a parametrization

$$\mathbf{x}(u,v) = (\cos v \cos u, \cos v \sin u, 1 + \sin v)$$

of all of Σ_0 except the south pole, located at the origin $\mathbf{0}$.

Now its geometric definition shows that P carries the u-parameter curves of $\mathbf{x}$ (circles of latitude) to circles in the plane, centered at the origin, and that P carries the v-parameter curves (meridians of longitude) to straight lines radiating out from the origin, as shown in Fig. 4.35.

Indeed these two families of image curves are just the parameter curves of $\mathbf{y} = P(\mathbf{x})$, and from the formula for P in Example 5.2 we get

$$\mathbf{y}(u,v) = P(\mathbf{x}(u,v)) = \left(\frac{2 \cos v \cos u}{1 - \sin v}, \frac{2 \cos v \sin u}{1 - \sin v} \right)$$

Since $P_*(\mathbf{x}_u) = \mathbf{y}_u$, $P_*(\mathbf{x}_v) = \mathbf{y}_v$, the regularity of P_* may be proved by computing $\mathbf{y}_u$ and $\mathbf{y}_v$, which turn out to be orthogonal and nonzero, hence linearly independent. (At the south pole $\mathbf{0}$ a different proof is required, since $\mathbf{x}$ is not a parametrization there (see Exercise 15).) We conclude that P is a diffeomorphism.

Differential forms have the remarkable property that they can be moved from one surface to another by means of an arbitrary mapping.† Let us experiment with a 0-form, that is, a real-valued function f. If $F: M \to N$ is

† This is not the case with vector fields.

a mapping of surfaces and f is a function on M, there is simply no reasonable general way to move f over to a function on N. But if instead f is a function on N, the problem is easy; we pull f back to the composite function $f(F)$ on M. The corresponding pull-back for 1-forms and 2-forms is accomplished as follows.

5.6 Definition   Let $F: M \to N$ be a mapping of surfaces.

(1) If ϕ is a 1-form on N, let $F^*\phi$ be the 1-form on M such that

$$(F^*\phi)(\mathbf{v}) = \phi(F_*\mathbf{v})$$

for all tangent vectors $\mathbf{v}$ to M.

(2) If η is a 2-form on N, let $F^*\eta$ be the 2-form on M such that

$$(F^*\eta)(\mathbf{v}, \mathbf{w}) = \eta(F_*\mathbf{v}, F_*\mathbf{w})$$

for all pairs of tangent vectors $\mathbf{v}$, $\mathbf{w}$ on M (Fig. 4.36).

When we are dealing with a function f in its role as a 0-form, we shall sometimes write F^*f instead of $f(F)$, in accordance with the notation for the pullback of 1-forms and 2-forms.

The essential operations on forms are sum, wedge product, and exterior derivative; all are preserved by mappings.

5.7 Theorem   Let $F: M \to N$ be a mapping of surfaces, and let ξ and η be forms on N. Then

(1) $F^*(\xi + \eta) = F^*\xi + F^*\eta$,

(2) $F^*(\xi \wedge \eta) = F^*\xi \wedge F^*\eta$,

(3) $F^*(d\xi) = d(F^*\xi)$.

Proof. In (1), ξ and η are both assumed to be p-forms (degree $p = 0, 1, 2$) and the proof is a routine computation. In (2), we must allow ξ and η to have different degrees. When, say, ξ is a function f, the given formula means simply $F^*(f\eta) = f(F)F^*(\eta)$. In any case, the proof of (2) is also a straight-forward computation. But (3) is more interesting. The easier case when ξ is a function is left as an exercise (Exercise 8), and we address ourselves to the difficult case when ξ is a 1-form.

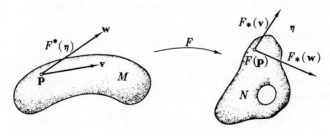

FIG. 4.36

It suffices to show that for every patch $\mathbf{x}: D \to M$

$$(d(F^*\xi))(\mathbf{x}_u, \mathbf{x}_v) = (F^*(d\xi))(\mathbf{x}_u, \mathbf{x}_v).$$

Let $\mathbf{y} = F(\mathbf{x})$, and recall that $F_*(\mathbf{x}_u) = \mathbf{y}_u$ and $F_*(\mathbf{x}_v) = \mathbf{y}_v$. Thus using the definitions of d and F^*, we get

$$d(F^*\xi)(\mathbf{x}_u, \mathbf{x}_v) = \frac{\partial}{\partial u}\{(F^*\xi)(\mathbf{x}_v)\} - \frac{\partial}{\partial v}\{(F^*\xi)(\mathbf{x}_u)\}$$

$$= \frac{\partial}{\partial u}\{\xi(F_*\mathbf{x}_v)\} - \frac{\partial}{\partial v}\{\xi(F_*\mathbf{x}_u)\}$$

$$= \frac{\partial}{\partial u}(\xi(\mathbf{y}_v)) - \frac{\partial}{\partial v}(\xi(\mathbf{y}_u)).$$

Even if $\mathbf{y}$ is not a patch, Exercise 8 of Section 4 shows that this last expression is still equal to $d\xi(\mathbf{y}_u, \mathbf{y}_v)$. But

$$d\xi(\mathbf{y}_u, \mathbf{y}_v) = d\xi(F_*\mathbf{x}_u, F_*\mathbf{x}_v) = (F^*(d\xi))(\mathbf{x}_u, \mathbf{x}_v).$$

Thus we conclude that $d(F^*\xi)$ and $F^*(d\xi)$ have the same value on $\mathbf{x}_u, \mathbf{x}_v$. ∎

The elegant formulas in Theorem 5.7 are the key to the deeper study of mappings. In Chapter VI we shall apply them to the connection forms of frame fields to get fundamental information about the *geometry* of mappings of surfaces.

EXERCISES

1. Let M and N be surfaces in $\mathbf{E}^3$. If $F: \mathbf{E}^3 \to \mathbf{E}^3$ is a mapping such that the image $F(M)$ of M is contained in N, then the restriction of F to M is a function $F \mid M: M \to N$. Prove that $F \mid M$ is a mapping of surfaces. (*Hint:* Use Theorem 3.2.)

2. Let Σ be the sphere of radius r with center at the origin of $\mathbf{E}^3$. Describe the effect of the following mappings $F: \Sigma \to \Sigma$ on the meridians and parallels of Σ.
   (a) $F(\mathbf{p}) = -\mathbf{p}$.   (b) $F(p_1, p_2, p_3) = (p_3, p_1, p_2)$.
   (c) $F(p_1, p_2, p_3) = \left(\dfrac{p_1 + p_2}{\sqrt{2}}, \dfrac{p_1 - p_2}{\sqrt{2}}, -p_3\right)$.

3. Let M be a *simple surface*, that is, one which is the image of a single proper patch $\mathbf{x}: D \to \mathbf{E}^3$. If $\mathbf{y}: D \to N$ is any mapping into a surface N, show that the function $F: M \to N$ such that

$$F(\mathbf{x}(u, v)) = \mathbf{y}(u, v) \qquad \text{for all } (u, v) \text{ in } D$$

is a mapping of surfaces. (*Hint:* Write $F = \mathbf{y}\mathbf{x}^{-1}$, and use Corollary 3.3.)

4. Use Exercise 3 to construct a mapping of the helicoid H (Ex. 7, IV.2) onto the torus T (Example 2.6) such that the rulings of H are carried into meridians of T.

5. If Σ is the sphere $\| \mathbf{p} \| = r$, the function $A : \Sigma \to \Sigma$ such that $A(\mathbf{p}) = -\mathbf{p}$ is called the *antipodal mapping* of Σ. Prove that A is a diffeomorphism and that $A_*(\mathbf{v}_p) = (-\mathbf{v})_{-p}$.

6. Let $\mathbf{x} : D \to M$ be a coordinate patch in a surface M. For any form ψ on M, the form $\mathbf{x}^*(\psi)$ on D is called a *coordinate expression* for ψ. (When ψ is a 0-form, that is, a function, then $\mathbf{x}^*(\psi) = \psi(\mathbf{x})$, so this terminology is consistent with that of IV.3.)

If ϕ is a 1-form and ν a 2-form, prove

(a) $\mathbf{x}^*(\phi) = \phi(\mathbf{x}_u)du + \phi(\mathbf{x}_v)dv.$ (b) $\mathbf{x}^*(\nu) = \nu(\mathbf{x}_u, \mathbf{x}_v)du\,dv.$

(c) $\mathbf{x}^*(d\phi) = \left(\dfrac{\partial}{\partial u}(\phi\,\mathbf{x}_v) - \dfrac{\partial}{\partial v}(\phi\,\mathbf{x}_u) \right) du\,dv.$

(In practice, instead of substituting in the formula (c), it is usually easier to apply the exterior derivative to the formula (a).)

7. (*Continuation*). Let $\mathbf{x}$ be the geographical patch in the sphere Σ.

(a) If ϕ is the 1-form on Σ such that $\phi(\mathbf{v}_p) = p_1v_2 - p_2v_1$, show that $\phi(\mathbf{x}_u) = r^2 \cos^2 v$ and $\phi(\mathbf{x}_v) = 0$. Then find the coordinate expressions for ϕ and $d\phi$.

(b) Prove that the formula $\nu(\mathbf{v}_p, \mathbf{w}_p) = p_3(v_1w_2 - v_2w_1)$ defines a 2-form on Σ and find its coordinate expression.

8. Let $F : M \to N$ be a mapping, and g a function on N.

(a) Prove that F preserves directional derivatives in this sense: If $\mathbf{v}$ is a tangent vector to M, then $\mathbf{v}[g(F)] = (F_*\mathbf{v})[g]$.

(b) Deduce that $F^*(dg) = d(F^*(g))$.

9. If $\mathbf{x} : D \to M$ is a parametrization, prove that the restriction of $\mathbf{x}$ to a sufficiently small neighborhood of a point (u_0, v_0) in D is a patch in M. (Thus a parametrization may be cut into patches.)

10. If $G : P \to M$ is a regular mapping *onto* M, and $H : P \to N$ is an arbitrary mapping, then the formula $F(G(\mathbf{p})) = H(\mathbf{p})$ is *consistent* provided $G(\mathbf{p}) = G(\mathbf{q})$ implies $H(\mathbf{p}) = H(\mathbf{q})$ for all $\mathbf{p}, \mathbf{q}$ in P. Prove that in this case F is a well-defined (differentiable) mapping.

$$M \xrightarrow{\quad F \quad} N$$
$$G \nwarrow \quad \nearrow H$$
$$P$$

We shall frequently apply this result in the case where G is a parametrization of M.

11. Let $\mathbf{x}: \mathbf{E}^2 \to T$ be the parametrization of the torus given in Example 2.6. In each case below, show that the formula

$$F(\mathbf{x}(u, v)) = \mathbf{x}(f(u, v), g(u, v))$$

is consistent (Exercise 10), and describe the resulting mapping $F: T \to T$. (For example, give its effect on the meridians and parallels of T.)

(a) $f = 3u, g = v$.   (c) $f = v, g = u$.
(b) $f = u + \pi, g = v + 2\pi$.   (d) $f = u + v, g = u - v$.

Which of these mappings are diffeomorphisms?

12. Let $F: M \to N$ be a mapping. Let $\mathbf{x}$ be a patch in M and let $\mathbf{y} = F(\mathbf{x})$. (Note that $\mathbf{y}$ lies in N, but need not be a patch.) If

$$\alpha(t) = \mathbf{x}(a_1(t), a_2(t))$$

is a curve in M, prove that the image $\bar{\alpha} = F(\alpha)$ in N has velocity

$$\bar{\alpha}' = \frac{da_1}{dt}\, \mathbf{y}_u(a_1, a_2) + \frac{da_2}{dt}\, \mathbf{y}_v(a_1, a_2).$$

13. Deduce from Exercise 12:
(a) The invariance property needed to justify definition (5.3) of F_*.
(b) The fact that derivative maps $F_*: T_p(M) \to T_{F(p)}(N)$ are linear transformations.

14. Given mappings $M \xrightarrow{F} N \xrightarrow{G} P$, let $GF: M \to P$ be the composite function. Show that
(a) GF is a mapping,   (b) $(GF)_* = G_* F_*$,   (c) $(GF)^* = F^* G^*$,
that is, for any form ξ on N, $F^*(G^*\xi) = (GF)^*(\xi)$. (Note the reversal of factors for $(GF)^*$; forms travel in the opposite direction from points and tangent vectors.)

15. For stereographic projection $P: \Sigma_0 \to \mathbf{E}^2$, show that the derivative map at the origin $\mathbf{0}$ is essentially just an identity mapping. (*Hint:* Express P near $\mathbf{0}$ in terms of a Monge patch.)

16. (a) Prove that the inverse mapping of stereographic projection $P: \Sigma_0 \to \mathbf{E}^2$ is given by the formula

$$P^{-1}(u, v) = \frac{(4u, 4v, 2f)}{f + 4},$$

where $f = u^2 + v^2$. (Show that both PP^{-1} and $P^{-1}P$ are identity mappings.)

(b) Deduce that the entire sphere Σ can be covered by only two patches. (The scheme in IV.1 requires six.)

6 Integration of Forms

Differential forms have yet another role in the calculus, which the reader probably noticed when they first appeared in Chapter I. In, say, a double integral $\iint f(u, v)\, du\, dv$, it is a *2-form* on $\mathbf{E}^2$ that appears after the integral signs. In a sense, it is only on Euclidean space that forms are actually integrated. But we can easily extend this notion of integration to forms on an arbitrary surface—by pulling them back to Euclidean space and then integrating.

Consider first the one-dimensional case. By a *curve segment* (or *1-segment*) in a surface M we shall mean a "curve" $\alpha\colon [a,\, b] \to M$ defined on a closed interval in the real line $\mathbf{E}^1$. (Differentiability for α means that it can be extended to a genuine curve on a larger *open* interval as required by Definition 4.1 of Chapter I.)

Now let ϕ be a 1-form on M. The pull-back $\alpha^*\phi$ of ϕ to the interval $[a,\, b]$ has the expression $f(t)\, dt$, where by the remarks following Example 4.7,

$$f(t) = (\alpha^*\phi)(U_1(t)) = \phi(\alpha_*(U_1(t))) = \phi(\alpha'(t)).$$

Thus the scheme described above yields the following definition.

6.1 Definition   Let ϕ be a 1-form on M, and let $\alpha\colon [a,\, b] \to M$ be a 1-segment (Fig. 4.37). Then the *integral of ϕ over α* is

$$\int_\alpha \phi = \int_{[a,b]} \alpha^*\phi = \int_a^b \phi(\alpha'(t))\, dt.$$

In engineering and physics, the integral $\int_\alpha \phi$ is called a *line integral*, and it has a wide variety of uses. For example, let us suppose that a vector field V on a surface is a *force field*, so for each point $\mathbf{p}$ of M, $V(\mathbf{p})$ is a force exerted at $\mathbf{p}$. Returning to our original idea of curve, we further suppose that $\alpha\colon [a,\, b] \to \mathbf{M}$ describes the motion of a mass point—with $\alpha(t)$ its position at time t. What is the total amount of *work* W needed to move α from $\mathbf{p} = \alpha(a)$ to $\mathbf{q} = \alpha(b)$? The discussion of velocity in Chapter I,

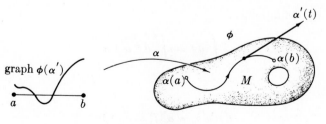

FIG. 4.37

Section 4, shows that for Δt small, the route of the curve α from $\alpha(t)$ to $\alpha(t + \Delta t)$ is approximately the straight line segment described by $\Delta t\, \alpha'(t)$. Now the moving point is opposed only by the component of force tangent to α, that is,

$$V(\alpha) \cdot \frac{\alpha'}{\| \alpha' \|} = \| V(\alpha) \| \cos \vartheta$$

(Fig. 4.38). Thus the work done *against* the force during time Δt is (approximately) *force* $- V(\alpha\,(t)) \cdot [\alpha'(t)/\| \alpha'(t)\|]$ times *distance* $\| \alpha'(t)\|\ \Delta t$. Adding these contributions over the whole time interval $[a, b]$ and taking the usual limit, we get

$$W = -\int_a^b V(\alpha(t)) \cdot \alpha'(t)\ dt.$$

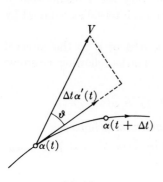

FIG. 4.38

To express this more simply, we introduce the *dual* 1-form ϕ such that for each tangent vector $\mathbf{w}$ at $\mathbf{p}$, $\phi(\mathbf{w}) = \mathbf{w} \cdot V(\mathbf{p})$. Then, by Definition 6.1, the total work is just

$$W = -\int_\alpha \phi.$$

We emphasize that this notion of line integral—like everything we are doing with forms—applies without change if the surface M is replaced by a Euclidean space or, indeed, by any *manifold* (Section 8).

When the 1-form ϕ is an exterior derivative df, the line integral $\int_\alpha \phi$ has an interesting property which generalizes the fundamental theorem of calculus.

6.2 Theorem Let f be a function on M, and let $\alpha: [a, b] \to M$ be a 1-segment in M from $\mathbf{p} = \alpha(a)$ to $\mathbf{q} = \alpha(b)$. Then

$$\int_\alpha df = f(\mathbf{q}) - f(\mathbf{p}).$$

Proof. By definition,

$$\int_\alpha df = \int_a^b df(\alpha')\ dt.$$

But

$$df(\alpha') = \alpha'[f] = \frac{d}{dt}\,(f\alpha).$$

Hence by the fundamental theorem of calculus,

$$\int_{\alpha} df = \int_{a}^{b} \frac{d}{dt}\,(f\alpha)\,dt = f(\alpha(b)) - f(\alpha(a)) = f(\mathbf{q}) - f(\mathbf{p}). \quad\blacksquare$$

The integral $\int_{\alpha} df$ is thus said to be *path-independent*. In the language used above, if the force field V has dual 1-form df, the work done depends not on where the point $\alpha(t)$ moves, but only on where it starts and finishes. In particular, if it follows a closed curve, $\mathbf{p} = \alpha(a) = \alpha(b) = \mathbf{q}$, it does no (total) work at all.

Mathematically we look at the preceding theorem roughly as follows: the "boundary" of the segment α from $\mathbf{p}$ to $\mathbf{q}$ is $\mathbf{q} - \mathbf{p}$, where the purely formal minus sign indicates that α goes *from* $\mathbf{p}$ and *to* $\mathbf{q}$. Then the integral of df over α equals the "integral" of f over the boundary $\mathbf{q} - \mathbf{p}$; that is, $f(\mathbf{q}) - f(\mathbf{p})$. This interpretation will be justified by the analogous theorem (6.5), in dimension 2.

Now a two-dimensional interval is just a closed rectangle $R\colon a \leqq u \leqq b$, $c \leqq v \leqq d$ in $\mathbf{E}^2$. And a 2-*segment* in M is a differentiable mapping $\mathbf{x}\colon R \to M$ of a closed rectangle into M (Fig. 4.39). (As before, differentiability means one can extend $\mathbf{x}$ differentiably to an open set containing R.)

Although we use the patch notation $\mathbf{x}$, we do not assume that $\mathbf{x}$ is either regular or one-to-one. The partial velocities $\mathbf{x}_u$ and $\mathbf{x}_v$ are still available, however, even when $\mathbf{x}$ is not a patch.

If η is a 2-form on M, then the pullback $\mathbf{x}^*\eta$ of η has, using Example 4.7, the coordinate expression $h\,du\,dv$, where

$$h = (\mathbf{x}^*\eta)(U_1,\,U_2) = \eta(\mathbf{x}_* U_1,\,\mathbf{x}_* U_2) = \eta(\mathbf{x}_u,\,\mathbf{x}_v).$$

Thus by strict analogy with Definition 6.1 we establish

6.3 Definition   Let η be a 2-form on M, and let $\mathbf{x}\colon R \to M$ be a 2-segment. Then the *integral of η over* $\mathbf{x}$ is

$$\iint_{\mathbf{x}} \eta = \iint_{R} \mathbf{x}^*\eta = \int_{a}^{b} \int_{c}^{d} \eta(\mathbf{x}_u,\,\mathbf{x}_v)\,du\,dv.$$

The physical applications of this notion of integral are perhaps richer

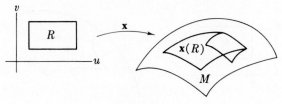

FIG. 4.39

than those of Definition 6.1, but we must proceed without delay to the two-dimensional analogue of Theorem 6.2.

6.4 Definition　Let $\mathbf{x} \colon R \to M$ be a 2-segment in M with R the closed rectangle $a \leqq u \leqq b,\ c \leqq v \leqq d$ (Fig. 4.40). The *edge curves* (or *edges*) of $\mathbf{x}$ are the 1-segments $\alpha,\ \beta,\ \gamma,\ \delta$ such that

$$\alpha(u) = \mathbf{x}(u, c)$$

$$\beta(v) = \mathbf{x}(b, v)$$

$$\gamma(u) = \mathbf{x}(u, d)$$

$$\delta(v) = \mathbf{x}(a, v)$$

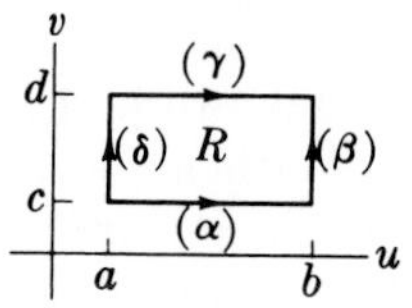

FIG. 4.40

Then the *boundary* $\partial \mathbf{x}$ of the 2-segment $\mathbf{x}$ is the formal expression

$$\partial \mathbf{x} = \alpha + \beta - \gamma - \delta.$$

These four curve segments are what we get by considering the function $\mathbf{x} \colon R \to M$ only on the four line segments that comprise the boundary of the rectangle R. The formal minus signs before γ and δ in $\partial \mathbf{x}$ remind us that γ and δ must be "reversed" to give a consistent trip around the rim of R, and thus of $\mathbf{x}$ (Fig. 4.41). Then if ϕ is a 1-form on M, the *integral of ϕ over the boundary of* $\mathbf{x}$ is defined to be

$$\int_{\partial \mathbf{x}} \phi = \int_{\alpha} \phi + \int_{\beta} \phi - \int_{\gamma} \phi - \int_{\delta} \phi.$$

The two-dimensional analogue of Theorem 6.2 is then

6.5 Theorem　(Stokes' Theorem)　If ϕ is a 1-form on M, and $\mathbf{x} \colon R \to M$ is a 2-segment, then

$$\iint_{\mathbf{x}} d\phi = \int_{\partial \mathbf{x}} \phi.$$

Proof. We shall work on the double integral and show that it turns out

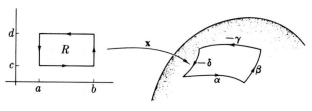

FIG. 4.41

to be the integral of ϕ over the boundary of $\mathbf{x}$. Combining Definitions 6.3 and 4.4 we have

$$\iint_{\mathbf{x}} d\phi = \iint_{R} (d\phi)\,(\mathbf{x}_u\,,\,\mathbf{x}_v)\,du\,dv = \iint_{R} \left(\frac{\partial}{\partial u}\,(\phi\mathbf{x}_v) - \frac{\partial}{\partial v}\,(\phi\mathbf{x}_u) \right) du\,dv$$

Let $f = \phi(\mathbf{x}_u)$ and $g = \phi(\mathbf{x}_v)$; this equation then becomes

$$\iint_{\mathbf{x}} d\phi = \iint_{R} \frac{\partial g}{\partial u}\,du\,dv - \iint_{R} \frac{\partial f}{\partial v}\,du\,dv \tag{1}$$

Now we treat these double integrals as iterated integrals. Suppose the rectangle R is given by the inequalities $a \leqq u \leqq b, c \leqq v \leqq d$. Then integrating first with respect to u, we find

$$\iint_{R} \frac{\partial g}{\partial u}\,du\,dv = \int_{c}^{d} I(v)\,dv, \quad \text{where} \quad I(v) = \int_{a}^{b} \frac{\partial g}{\partial u}\,(u,v)\,du.$$

In the partial integral defining $I(v)$, v is constant, so the integrand is just the ordinary derivative with respect to u. Thus the fundamental theorem of calculus applies to give

$$I(v) = g(b,v) - g(a,v).$$

(Fig. 4.42). Hence

$$\iint_{R} \frac{\partial g}{\partial u}\,du\,dv = \int_{c}^{d} g(b,v)\,dv - \int_{c}^{d} g(a,v)\,dv. \tag{2}$$

Again we work on the first integral. Now by definition

$$g(b,v) = \phi(\mathbf{x}_v(b,v)).$$

But $\mathbf{x}_v(b,v)$ is precisely the velocity $\beta'(v)$ of the "right side" curve β in $\partial \mathbf{x}$. Hence by Definition 6.1,

$$\int_{c}^{d} g(b,v)\,dv = \int_{c}^{d} \phi(\beta'(v))\,dv = \int_{\beta} \phi.$$

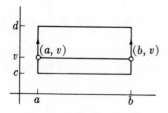

FIG. 4.42

A similar argument shows that the second integral in (2) is $\int_\delta \phi$; hence

$$\iint_R \frac{\partial g}{\partial u} \, du \, dv = \int_\beta \phi - \int_\delta \phi. \tag{3}$$

In the same way—but integrating first with respect to v!—we find

$$\iint_R \frac{\partial f}{\partial v} \, du \, dv = \int_\gamma \phi - \int_\alpha \phi \tag{4}$$

Assembling the information in (1), (3), and (4) we obtain the required result:

$$\iint_x d\phi = \left\{ \int_\beta \phi - \int_\delta \phi \right\} - \left\{ \int_\gamma \phi - \int_\alpha \phi \right\} = \int_{\partial x} \phi \qquad \blacksquare$$

Stokes' theorem may be considered a two-dimensional formulation of the fundamental theorem of calculus; it ranks as one of the most useful results in mathematics. Alternative formulations of the theorem and extensive applications may be found in texts on advanced calculus or applied mathematics; we shall use it to study the geometry of surfaces.

The line integral $\int_\alpha \phi$ is not particularly sensitive to reparametrization of α; all that matters is the *direction* in which the route of α is traversed. The following lemma uses the notation of Exercise 10, II.2.

6.6 Lemma   Let $\alpha(h) \colon [a, b] \to M$ be a reparametrization of a curve segment $\alpha \colon [c, d] \to M$. For any 1-form ϕ on M

$$\int_{\alpha(h)} \phi = \begin{cases} \displaystyle\int_\alpha \phi & \text{if } h \text{ is orientation-preserving} \\[2ex] -\displaystyle\int_\alpha \phi & \text{if } h \text{ is orientation-reversing} \end{cases}$$

Proof. Since $\alpha(h)$ has velocity

$$\alpha(h)' = \frac{dh}{dt} \, \alpha'(h),$$

we have

$$\int_{\alpha(h)} \phi = \int_a^b \phi(\alpha(h)') \, du = \int_a^b \phi(\alpha'(h)) \frac{dh}{du} \, du$$

Now we use the theorem on change of variables in an integral. If h is orientation-preserving, then $h(a) = c$ and $h(b) = d$, so the integral above becomes

$$\int_c^d \phi(\alpha') \, du = \int_\alpha \phi.$$

But in the orientation-reversing case, $h(a) = d$ and $h(b) = c$, which gives

$$\int_d^c \phi(\alpha') \, du = -\int_c^d \phi(\alpha') \, du = -\int_\alpha \phi \qquad \blacksquare$$

This lemma lets us give a concrete interpretation to the formal minus signs in the boundary $\partial \mathbf{x} = \alpha + \beta - \gamma - \delta$ of a 2-segment $\mathbf{x}$. For any curve $\xi\colon [t_0, t_1] \to M$, let $-\xi$ be any orientation-*reversing* reparametrization of ξ, say

$$(-\xi)(t) = \xi(t_0 + t_1 - t).$$

Thus by the lemma,

$$\int_{-\xi} \phi = -\int_\xi \phi$$

and if $\mathbf{x}$ is a 2-segment, then

$$\int_{\partial \mathbf{x}} \phi = \int_\alpha \phi + \int_\beta \phi - \int_\gamma \phi - \int_\delta \phi = \int_\alpha \phi + \int_\beta \phi + \int_{-\gamma} \phi + \int_{-\delta} \phi.$$

EXERCISES

1. If $\alpha = (\alpha_1, \alpha_2)$ is a curve in $\mathbf{E}^2$ and ϕ is a 1-form, prove this computational rule for finding $\phi(\alpha') \, dt$: Substitute $u = \alpha_1(t)$ and $v = \alpha_2(t)$ in a coordinate expression $\phi = f(u, v) \, du + g(u, v) \, dv$.

2. Consider the curve segment $\alpha\colon [-1, 1] \to \mathbf{E}^2$ such that $\alpha(t) = (t, t^2)$.
   (a) If $\phi = v^2 \, du + 2uv \, dv$, compute $\int_\alpha \phi$.
   (b) Find a function f such that $df = \phi$ and check Theorem 6.2 in this case.

3. Let ϕ be a 1-form on a surface M.
   (a) If ϕ is *closed*, show that $\int_{\partial \mathbf{x}} \phi = 0$ for every 2-segment in M.
   (b) If ϕ is *exact*, show (more generally) that

$$\sum_i \int_{\alpha_i} \phi = 0$$

for any "cycle" of curve segments $\alpha_1, \cdots,$
$\alpha_k \ (\alpha_{k+1} = \alpha_1)$ such that α_i ends at the starting point of α_{i+1} (Fig. 4.43). (*Closed* means $d\phi = 0$, *exact* means $\phi = df$; see Ex. 2 of IV.4.)

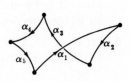

FIG. 4.43

4. The 1-form

$$\psi = \frac{u\,dv - v\,du}{u^2 + v^2}$$

is well-defined on the plane $\mathbf{E}^2$ with the origin $(0, 0)$ removed. Show:

(a) ψ is closed, but not exact. (*Hint:* Integrate around the unit circle and use Exercise 3.)

(b) if ψ is restricted to, say, the right half-plane, $u > 0$, then ψ becomes exact.

5. (*Continuation*). It follows from Ex. 12 of II.1 that every curve α in $\mathbf{E}^2$ that does not pass through the origin can be written in the polar form

$$\alpha(t) = (r(t) \cos \vartheta(t), r(t) \sin \vartheta(t)).$$

Prove that for every closed curve α, $(1/2\pi) \int_\alpha \psi$ is an integer. This integer is called the *winding number of* α; it represents the total algebraic number of times α has gone around the origin in the counterclockwise direction.

6. Let $\mathbf{x}$ be a patch in a surface M. For a curve segment

$$\alpha(t) = \mathbf{x}(a_1(t), a_2(t)), \qquad a \le t \le b,$$

show that

$$\int_\alpha \phi = \int_a^b \left(\phi(\mathbf{x}_u) \frac{da_1}{dt} + \phi(\mathbf{x}_v) \frac{da_2}{dt} \right) dt$$

where $\mathbf{x}_u$ and $\mathbf{x}_v$ are evaluated on (a_1, a_2). (This generalizes Ex. 1, since we can use the identity patch $\mathbf{x}(u, v) = (u, v)$ on $\mathbf{E}^2$.)

7. Let α be the closed curve

$$\alpha(t) = \mathbf{x}(mt, nt), \qquad 0 \le t \le 2\pi$$

in the torus T (see Ex. 11 of IV.3.) Compute

(a) $\int_\alpha \xi$, where ξ is the 1-form on T such that $\xi(\mathbf{x}_u) = 1$, and $\xi(\mathbf{x}_v) = 0$.

(b) $\int_\alpha \eta$, where η is the 1-form such that $\eta(\mathbf{x}_u) = 0$ and $\eta(\mathbf{x}_v) = 1$.

If γ is an arbitrary closed curve, $\int_\gamma \xi / 2\pi$ gives the total number of times γ travels around the torus in the general direction of the parallels, while $\int_\gamma \eta / 2\pi$ gives a similar measurement for the direction of the meridians. This suggests the commonly used notation $\xi = d\vartheta$, $\eta = d\varphi$, where ϑ and φ are the (multivalued!) longitude and latitude functions on T; however, see Exercise 13.

8. Let $F: M \to N$ be a mapping. Prove:

(a) If α is a curve segment in M and ϕ is a 1-form on N, then

$$\int_\alpha F^*\phi = \int_{F(\alpha)} \phi.$$

(b) If $\mathbf{x}$ is a 2-segment in M and ν is a 2-form on N, then

$$\int_{\mathbf{x}} F^*\nu = \int_{F(\mathbf{x})} \nu.$$

9. Let $\mathbf{x}: R \to \Sigma$ be the 2-segment in the sphere Σ obtained by restricting the geographical patch

$$\mathbf{x}(u, v) = (r \cos v \cos u, r \cos v \sin u, r \sin v)$$

to the rectangle $R: 0 \leq u, v \leq \pi/2$. Find explicit formulas for the edge curves α, β, γ, δ of $\mathbf{x}$, and show these curves and the image $\mathbf{x}(R)$ on a sketch of Σ.

10. Let $\mathbf{x}: R \to M$ be a 2-segment defined on the rectangle

$$R: 0 \leq u, v \leq 1.$$

If ϕ is the 1-form on M such that

$$\phi(\mathbf{x}_u) = u + v \quad \text{and} \quad \phi(\mathbf{x}_v) = uv,$$

compute $\iint_{\mathbf{x}} d\phi$ and $\int_{\partial \mathbf{x}} \phi$ and check the results by Stokes' theorem. (*Hint:* $\mathbf{x}^*(d\phi) = d(\mathbf{x}^*\phi) = (v - 1)du\ dv$.)

11. Same as Exercise 10, except that $R: 0 \leq u \leq \pi/2, 0 \leq v \leq \pi$, and $\phi(\mathbf{x}_u) = u \cos v, \phi(\mathbf{x}_v) = v \sin u$.

12. A closed curve α in M is *homotopic to a constant* provided there is a 2-segment $\mathbf{x}: R \to M$ for which (a) α is, in fact, the α edge curve of $\mathbf{x}$, (b) $\beta = \delta$, and (c) γ is a constant curve (Fig. 4.44). (Suppose $R: a \leq u \leq b, c \leq v \leq d$. Then as v_0 varies from c to d, the closed u-parameter curve, $v = v_0$, of $\mathbf{x}$ varies smoothly from α to the constant curve γ.) Prove that every closed curve in $\mathbf{E}^2$ is homotopic to constant.

FIG. 4.44

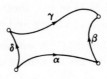

FIG. 4.45

13. Let ϕ be a closed 1-form and α a closed curve. Prove that $\int_\alpha \phi = 0$ if either

(a) ϕ is exact, or (b) α is homotopic to a constant.

Deduce that on the torus T meridians and parallels are not homotopic to constants, and the closed forms ξ and η (Exercise 7) are not exact.

A surface M for which *every* closed curve is homotopic to a constant is said to be *simply connected*. Thus the plane is simply connected (Ex. 12), but a torus, or a plane with even a single point removed, is not (Ex. 7 and 5). Roughly speaking, a simply connected surface has no holes in it, and any four curves α, β, γ, δ linked together as in Fig. 4.45 are in fact the boundary curves of some 2-segment.† Use this assertion to prove:

14. On a simply connected surface, the integral of a *closed* 1-form ϕ is path-independent. (That is, $\int_\alpha \phi$ is the same for all α with the same end points.)

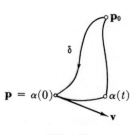

15. On a simply connected surface, every *closed* 1-form is *exact*. (*Hint:* Fix a point $\mathbf{p}_0$ in M and define $f(\mathbf{p}) = \int_\delta \phi$ for any curve segment from $\mathbf{p}_0$ to $\mathbf{p}$ (Fig. 4.46). To show $df(\mathbf{v}) = \phi(\mathbf{v})$ for a tangent vector $\mathbf{v}$ at $\mathbf{p}$, prove that if α is a curve with initial velocity $\mathbf{v}$, then

$$f(\alpha(t)) = f(\mathbf{p}) + \int_0^t \phi(\alpha'(u))\, du.)$$

FIG. 4.46

7 Topological Properties of Surfaces

We now discuss some of the very basic properties that a surface may possess.

7.1 Definition A surface M is *connected* provided that for any two points $\mathbf{p}$ and $\mathbf{q}$ of M there is a curve segment in M from $\mathbf{p}$ to $\mathbf{q}$.

Thus a connected surface M is all in one piece, since one can travel from any point in M to any other without leaving M. Most of the surfaces we have discussed so far have been connected; the surface $M: z^2 - x^2 - y^2 = 1$ (hyperboloid of two sheets) is not connected. Connectedness is a mild and reasonable condition and might well be included in the definition of surface.

7.2 Definition A surface M is *compact* provided that M can be covered by the images of a finite number of 2-segments in M.

Roughly speaking, compactness means that the surface is finite in size. For example, spheres are compact, since if we use the formula for $\mathbf{x}(u, v)$ in Example 2.2 on the *closed* rectangle

$$R: -\pi \leqq u \leqq \pi, \qquad -\pi/2 \leqq v \leqq \pi/2,$$

† For a systematic account of simple connectedness, see pp. 157–165 of Lefschetz [8], where it is shown that spheres are simply connected.

then Σ is the image of this single 2-segment. Similarly the torus of revolution (Example 2.6), or *any* closed surface of revolution, is compact.

The proof of the following lemma uses this fundamental fact: If f is a continuous real-valued function defined on a closed rectangle

$$R: a \leqq u \leqq b, \qquad c \leqq v \leqq d,$$

then f takes on its maximum at some point of R.

7.3 Lemma If f is a continuous function on a compact surface M, then f takes on maximum at some point of M. (Obviously we can also replace *maximum* by *minimum*.)

Proof. By definition there exist a finite number of 2-segments

$$\mathbf{x}_i: R_i \to M \ (1 \leqq i \leqq k)$$

whose images cover all of M. Since each $\mathbf{x}_i$ is differentiable, it is also continuous, so each composite function $f\mathbf{x}_i: R_i \to \mathbf{R}$ is continuous. Thus by the remark above, for each index i, there is a point (u_i, v_i) in R_i where the function $f\mathbf{x}_i$ takes its maximum. Let, say, $f(\mathbf{x}_1(u_1, v_1))$ be the largest of this finite number k of maximum values. We assert that f takes on its maximum value at the point $\mathbf{m} = \mathbf{x}_1(u_1, v_1)$. In fact, we shall prove that if $\mathbf{p}$ is any point of M, then $f(\mathbf{m}) \geqq f(\mathbf{p})$. Since the 2-segments $\mathbf{x}_1, \cdots, \mathbf{x}_k$ cover M, there is an index i such that $\mathbf{p} = \mathbf{x}_i(u, v)$. But then by the preceding construction,

$$f(\mathbf{m}) = f(\mathbf{x}_1(u_1, v_1)) \geqq f(\mathbf{x}_i(u_i, v_i)) \geqq f(\mathbf{x}_i(u, v)) = f(\mathbf{p}). \quad \blacksquare$$

This very useful result can be applied to prove noncompactness. For example, no cylinder C (as in Example 1.5) is compact, since the coordinate function z on C gives the height $z(\mathbf{p})$ of each point $\mathbf{p}$ above the xy plane, and thus has no maximum value on C.

However, Definition 7.2 is a little trickier than it looks. Consider, for example, the open unit disc $D: x^2 + y^2 < 1$ in the xy plane. Now, D is a surface and has finite area π. But D is not compact: It suffices to note that the continuous function $f = (1 - x^2 - y^2)^{-1}$ does not have a maximum on D. In general, a compact surface cannot have any open edges, as D does. It must be smoothly closed up everywhere—as well as finite in size—like a sphere or torus.

Roughly speaking, an orientable surface is one that is not twisted. Of the many equivalent formulations of orientability the one that follows is perhaps the simplest.

7.4 Definition A surface M is *orientable* provided there exists a 2-form μ on M which is nonzero at each point of M.

(A 2-form is *zero* at a point $\mathbf{p}$ if it is zero on every pair of tangent vectors at **p**.) Thus the plane $\mathbf{E}^2$ is orientable, since $du\ dv$ is a nonvanishing 2-form. Although simple, this definition is somewhat mysterious, so we shall prove a more intuitive criterion.

7.5 Theorem   A surface M in $\mathbf{E}^3$ is orientable if and only if there exists a normal vector field Z on M that is nonzero at each point of M.

Proof. We use the cross product of $\mathbf{E}^3$ to convert normal vector fields into 2-forms, and vice versa. For Z as above, define a 2-form μ on M as follows: For any pair $\mathbf{v}$, $\mathbf{w}$ of tangent vectors to M at **p**, let

$$\mu(\mathbf{v}, \mathbf{w}) = Z(\mathbf{p}) \cdot \mathbf{v} \times \mathbf{w}.$$

Standard properties of the cross product show that μ is, in fact, a nonvanishing 2-form on M. Thus M is orientable.

Conversely, suppose that M is orientable, with μ a nonvanishing 2-form. If $\mathbf{v}$, $\mathbf{w}$ is a linearly independent pair of tangent vectors at **p**, then

$$\mu(\mathbf{v}, \mathbf{w}) \neq 0,$$

for otherwise, μ would be zero at **p**. Now define

$$Z(\mathbf{p}) = \frac{\mathbf{v} \times \mathbf{w}}{\mu(\mathbf{v}, \mathbf{w})}.$$

This formula has the remarkable property that it is independent of the choice of $\mathbf{v}$, $\mathbf{w}$ at **p**. Explicitly, for any other such pair $\overline{\mathbf{v}}$, $\overline{\mathbf{w}}$, it follows from Lemma 4.2 and the analogous formula for cross products that

$$\frac{\overline{\mathbf{v}} \times \overline{\mathbf{w}}}{\mu(\overline{\mathbf{v}}, \overline{\mathbf{w}})} = \frac{\mathbf{v} \times \mathbf{w}}{\mu(\mathbf{v}, \mathbf{w})}.$$

We have thus obtained a well-defined Euclidean vector field on all of M. Again the properties of the cross product show that Z is everywhere normal to M, but never zero.   ∎

Thus it follows from Lemma 3.8 that every surface in $\mathbf{E}^3$ that can be defined implicitly is orientable. For example, all cylinders, surfaces of revolution, and spheres (in fact, all quadric surfaces) are orientable. However, nonorientable surfaces do exist in $\mathbf{E}^3$. The simplest example is the famous Möbius band M, which can be made from a strip of paper by giving it a half twist, then gluing its ends together. (The formal construction of a particular Möbius band is given in Exercise 7.) M is nonorientable, since every normal vector field Z on M must somewhere be zero. To see this, let γ be a closed curve as indicated in Fig. 4.47, with $\gamma(0) = \gamma(1) = \mathbf{p}$. If we assume that Z is never zero, then the twist in M forces the contradiction $Z(\gamma(1)) = -Z(\gamma(0))$, since the function $t \to Z(\gamma(t))$ is differentiable (that is, Z varies smoothly as it moves around γ).

The three properties discussed in this section—connectedness, compactness, and orientability—are *topological properties:* It is possible to define them using only open sets and continuous functions, with no differentiability considerations at all. Using these more general definitions, a more conceptual proof can be given of the following result.

7.6 Theorem   Let M and N be surfaces in $\mathbf{E}^3$ such that M is contained in N. If M is compact, and N is connected, then $M = N$.

(If N were not connected, it might consist of two surfaces, of which M is one. Similarly the result fails if M is not compact; consider the case of an open disc M in the xy plane N.)

Proof. Exercise 15 of Section 3 shows that M is an open set of N. We assume that M does not fill all of N, and deduce a contradiction. By assumption there is a point $\mathbf{n}$ of N that is not in M. Let $\mathbf{m}$ be a point of M. Since N is connected, there is a curve segment α in N from

$$\alpha(0) = \mathbf{m} \qquad \text{to} \qquad \alpha(1) = \mathbf{n}.$$

Let t^* be the least upper bound of those numbers t such that $\alpha(t)$ is in M. We assert that *the point* $\mathbf{p}^* = \alpha(t^*)$ *is in* M (Fig. 4.48).

To prove this, consider the real-valued function f on M such that at each

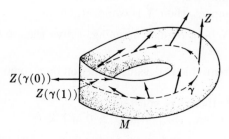

FIG. 4.47

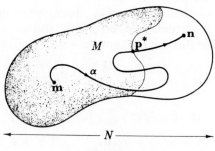

FIG. 4.48

point $\mathbf{p}$ of M, $f(\mathbf{p})$ is the Euclidean distance $d(\mathbf{p}^*, \mathbf{p})$ from $\mathbf{p}^*$ to $\mathbf{p}$. Now $f \geqq 0$ is continuous, in the sense that for each patch $\mathbf{x}$ in M the composite function $f(\mathbf{x})$ is continuous. Since M is compact, Lemma 7.3 applies to show that f takes a minimum at some point of M. By definition of least upper bound, there are numbers $t < t^*$, arbitrarily close to t^*, such that $\alpha(t)$ is in M. Since α is continuous (being differentiable) the corresponding distances $d(\mathbf{p}^*, \alpha(t))$ become arbitrarily small; hence the minimum value of f can only be zero. Thus the only possible point at which f can be a minimum is $\mathbf{p}^*$ itself, which means that $\mathbf{p}^*$ is in the domain M of f.

Since M is an open set of N, it follows that every point of N near enough to $\mathbf{p}^*$ is also in M. Thus if $t > t^*$ is near enough to t^*, $\alpha(t)$ must still be in M—a contradiction to the definition of t^*.   ∎

EXERCISES

1. Decide which of the following surfaces are compact and which are connected:
    (a) A sphere with one point removed.   (c) The surface in Fig. 4.10.
    (b) The region $z > 0$ in $M: z = xy$.   (d) $M: x^2 + y^4 + z^6 = 1$.
    (e) A torus with the curve $\alpha(t) = \mathbf{x}(t, t)$ removed. (See Ex. 2 of IV.3.)

2. Let F be a mapping of a surface M *onto* a surface N. Prove:
    (a) If M is connected, then N is connected.
    (b) If M is compact, then N is compact.

3. Let $F: M \to N$ be a regular mapping. Prove that if N is orientable, then M is orientable.

4. Let f be a differentiable real-valued function on a connected surface M. Prove that:
    (a) If $df = 0$, then f is constant.
    (b) If f is never zero, then either $f > 0$ or $f < 0$.

5. (a) Prove that a connected orientable surface has exactly two unit normal vector fields, which are negatives of each other. We denote these by $\pm U$. (*Hint:* Use Ex. 4.)
    (b) If M is a nonorientable surface, prove that any point of M is contained in a connected orientable region. (Thus, even on a nonorientable surface, unit normal vector fields exist *locally*.)

6. Let $F: M \to N$ be a regular mapping. Prove this generalization of Theorem 7.6: If M is compact and N is connected, then F carries M *onto* N.

7. A Möbius band M (Fig. 4.47) can be constructed as a ruled surface

$$x(u, v) = \beta(u) + v\delta(u), \qquad -\tfrac{1}{2} \leqq v \leqq \tfrac{1}{2},$$

where

$$\beta(u) = (\cos u, \sin u, 0)$$

$$\delta(u) = \left(\cos \frac{u}{2}\right)\beta(u) + \left(\sin \frac{u}{2}\right) U_3.$$

(The ruling L makes only a half-turn as it traverses the circle β once.)
(a) Compute

$$E = \frac{v^2}{4} + [1 + v \cos (u/2)]^2, \qquad F = 0, \qquad G = 1,$$

and deduce as in Exercise 2 of Section 2 that $\mathbf{x}$ is regular.

(b) Show the u-parameter curve, $v = \frac{1}{4}$, on a sketch of M. Prove that the u-parameter curves are closed and (β excepted) have period 4π.

8. Let M^* be the surface obtained by removing the central circle β from the Möbius band in the last exercise. Is M^* connected? Orientable?

9. (*Counterexamples*). Give examples to show that the following are all false:
(a) Converses of (a) and (b) of Exercise 2.
(b) Exercise 3 with F not regular.
(c) Converse of Exercise 3.

10. A surface M in $\mathbf{E}^3$ is *closed in* $\mathbf{E}^3$ provided the points of $\mathbf{E}^3$ not in M constitute an open set of $\mathbf{E}^3$. (If $\mathbf{p}$ is not in M, there is an ε-neighborhood of $\mathbf{p}$ that does not meet M.) Show that:
(a) Every surface in $\mathbf{E}^3$ that can be described in the implicit form $M: g = c$ is closed in $\mathbf{E}^3$.
(b) Every compact surface in $\mathbf{E}^3$ is closed in $\mathbf{E}^3$.

11. (*Boundedness*). A surface M in $\mathbf{E}^3$ is *bounded* provided there is a number R such that $\|\mathbf{p}\| \leq R$ for all points $\mathbf{p}$ of M. (Thus M lies inside a sphere.) Prove that a compact surface in $\mathbf{E}^3$ is bounded.

The last two exercises have shown that a compact surface in $\mathbf{E}^3$ is closed and bounded; the converse follows from a fundamental topological theorem.

12. Prove Theorem 7.6 assuming M is merely closed in $\mathbf{E}^3$ (instead of compact).

13. In each case, decide whether the surface $M: g = 1$ is compact, or connected:
(a) $g = x^2 - y^3 + z^4$.   (c) $g = z^2 + x^2 y^2$.
(b) $g = x^4 - y + z$.   (d) $g = (x^2 + y^2 - 4)^2 + (z - 4)^2$.

14. Prove that every surface of revolution M is connected and orientable, but that M is compact if and only if its profile curve C is closed.

8 Manifolds

Surfaces in $\mathbf{E}^3$ are a matter of everyday experience, so it is reasonable to try to investigate them mathematically. But examining this concept with a critical eye, we may well ask if there could not be surfaces in $\mathbf{E}^4 \cdots$ or $\mathbf{E}^n \cdots$ or even surfaces that are not in any Euclidean space at all. To devise a definition for such a surface, we must rely not on our direct experience of the real world, but on our mathematical experience of surfaces in $\mathbf{E}^3$. Thus we shall strip away from the basic definition (1.2) every feature that involves $\mathbf{E}^3$ in any way. What is left will be just a *surface*.

To begin with, a surface will be a set M: a collection of any objects whatsoever, *not* necessarily points of $\mathbf{E}^3$. An *abstract patch in M* will now be just a one-to-one function $\mathbf{x}: D \rightarrow M$ from an open set D of $\mathbf{E}^2$ into the set M. There is, as yet, no way to say what it means for such a function to be differentiable. But all we need to get a workable definition of surface is the smooth overlap condition (Corollary 3.3). To *prove* this now is a logical impossibility, so in the usual fashion of mathematics, we make it an axiom.

8.1 Definition A surface is a set M furnished with a certain collection $\mathcal{P}$ of abstract patches in M such that:

(1) The images of the patches in the collection $\mathcal{P}$ cover M.

(2) For any two patches $\mathbf{x}$, $\mathbf{y}$ in the collection $\mathcal{P}$, the composite functions $\mathbf{y}^{-1}\mathbf{x}$ and $\mathbf{x}^{-1}\mathbf{y}$ are Euclidean differentiable (and defined on open sets of $\mathbf{E}^2$).

This definition generalizes Definition 1.2: A surface-in-$\mathbf{E}^3$ *is* a surface. But there are vast numbers of surfaces that can never be found in $\mathbf{E}^3$.

8.2 Example *The projective plane $\bar{\Sigma}$.* Starting from the unit sphere Σ in $\mathbf{E}^3$, we construct the projective plane $\bar{\Sigma}$ by identifying *antipodal points* of Σ; that is, by considering $\mathbf{p}$ and $-\mathbf{p}$ to be the same point (Fig. 4.49). For-

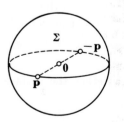

FIG. 4.49

mally, this means the set $\bar{\Sigma}$ consists of all antipodal pairs $\{\mathbf{p}, -\mathbf{p}\}$ of points of the sphere. (Order is not important here; that is, $\{\mathbf{p}, -\mathbf{p}\} = \{-\mathbf{p}, \mathbf{p}\}$.)

To make $\bar{\Sigma}$ a surface—and later to study it—we use two functions: the antipodal mapping $A(\mathbf{p}) = -\mathbf{p}$ of the sphere Σ, and the *projection* $P(\mathbf{p}) = \{\mathbf{p}, -\mathbf{p}\}$ of Σ onto $\bar{\Sigma}$. Note that $PA = P$, and $(\ddagger)$: $P(\mathbf{p}) = P(\mathbf{q})$ if and only if either $\mathbf{q} = \mathbf{p}$ or $\mathbf{q} = -\mathbf{p}$.

Let us call a patch $\mathbf{x}$ in Σ "small" if the Euclidean distance between any two of its points is less than 1. If $\mathbf{x}: D \rightarrow \Sigma$ is a small patch, then the composite function $P(\mathbf{x}): D \rightarrow \bar{\Sigma}$ is one-to-one, and is thus an abstract patch in $\bar{\Sigma}$. The collection of all such abstract patches makes $\bar{\Sigma}$ a surface—the first condition in Definition 8.1 is clear, and we merely outline the proof of the second.

Suppose that $P(\mathbf{x})$ and $P(\mathbf{y})$ overlap in $\bar{\Sigma}$; that is, their images have a point in common. If $\mathbf{x}$ and $\mathbf{y}$ overlap in Σ, show that $(P\mathbf{y})^{-1}(P\mathbf{x}) = \mathbf{y}^{-1}\mathbf{x}$, which by Corollary 3.3 is differentiable and defined on an open set. (Hint: Use smallness and $\ddagger$.) On the other hand, if $\mathbf{x}$ and $\mathbf{y}$ do not overlap, replace $\mathbf{y}$ by $A(\mathbf{y})$. Then $\mathbf{x}$ and $A(\mathbf{y})$ *do* overlap, so the previous argument applies.

Conclusion: The projective plane $\bar{\Sigma}$ is a surface.†

To emphasize the distinction between a surface in $\mathbf{E}^3$ and the general concept of surface defined above, we shall sometimes call the latter an *abstract surface*. Note that $\mathbf{E}^2$ is an abstract surface if it is furnished with the single patch $\mathbf{x}(u, v) = (u, v)$.

To get as many patches as possible in an abstract surface M, it is customary to enlarge the given patch collection $\mathcal{P}$ to include all abstract patches in M that overlap smoothly with those in $\mathcal{P}$. In working with M, these are the only patches we can use. We emphasize that abstract surfaces M_1 and M_2 with the *same* set of points are nevertheless different surfaces if their (enlarged) patch collections $\mathcal{P}_1$ and $\mathcal{P}_2$ are different.

There is essentially only one problem to solve in establishing the calculus of an abstract surface M, and that is to define the *velocity* of a curve in M. For everything else—differentiable functions, curves themselves, tangent vectors, tangent vector fields, differential forms, and so on—the *definitions* and *theorems* given for surfaces in $\mathbf{E}^3$ apply without change. (It is necessary to tinker with a few *proofs* to take care of the new Definition 8.3, but no serious problems arise.) The velocity of a curve fails us in the abstract case, since before it consisted of tangent vectors to $\mathbf{E}^3$, and now $\mathbf{E}^3$ is gone.

It makes not the slightest difference what we define the velocity $\alpha'(t)$ to be—provided the new definition leads to the same essential properties as before. The directional derivative property (Lemma 4.6 of Chapter I) is what is needed.

† The terminology in this example derives from projective geometry, however, Exercise 2 shows that $\bar{\Sigma}$ can more aptly be described as a twisted sphere.

8.3 Definition   Let $\alpha: I \to M$ be a curve in an abstract surface M. For each t in I, the *velocity vector* $\alpha'(t)$ is the function such that

$$\alpha'(t)[f] = \frac{d(f\alpha)}{dt}(t)$$

for every differentiable real-valued function f on M.

Thus $\alpha'(t)$ is a real-valued function whose domain is the set of all differentiable functions on M. This is all we need to generalize the calculus for surfaces in $\mathbf{E}^3$ to the case of an abstract surface.

The reader may feel he has gone far enough in the direction of abstraction, but in one more step we shall have gone all the way.

We now have a calculus for $\mathbf{E}^n$ (Chapter I) and a calculus for surfaces. These are strictly analogous, but analogies in mathematics (although useful at first) are, in the long run, annoying. What we need is a *single* calculus, of which these two will be special cases. The most general object on which calculus can be conducted is called a *manifold*. It is simply an abstract surface of arbitrary dimension n.

8.4 Definition   *An n-dimensional manifold M* is a set furnished with a collection $\mathcal{P}$ of *abstract patches* (one-to-one functions $\mathbf{x}: D \to M$, D an open set in $\mathbf{E}^n$) such that

(1) M is covered by the images of the patches in the collection $\mathcal{P}$.

(2) For any two patches $\mathbf{x}$, $\mathbf{y}$ in the collection $\mathcal{P}$, the composite functions $\mathbf{y}^{-1}\mathbf{x}$ and $\mathbf{x}^{-1}\mathbf{y}$ are Euclidean-differentiable (and defined on open sets in $\mathbf{E}^n$).

Thus *a surface* (Definition 8.1) *is the same thing as a two-dimensional manifold*. The Euclidean space $\mathbf{E}^n$ is a very special n-dimensional manifold; its patch collection consists only of the identity function.

To keep this definition as close as possible to that of a surface in $\mathbf{E}^3$, we have deviated slightly from the standard definition of manifold in which it is usually the inverse functions $\mathbf{x}^{-1}: \mathbf{x}(D) \to D$ that are axiomatized.

The calculus of an arbitrary n-dimensional manifold M is defined in the same way as in the special case, $n = 2$, of an abstract surface. Differentiable functions, tangent vectors, vector fields, and mappings are gotten exactly as before: We need only replace $i = 1, 2$ by $i = 1, 2, \cdots, n$. Differential forms on a manifold M have the same general properties as in the case $n = 2$, which we have explored in Sections 4, 5, and 6. But there are p-forms for $0 \leq p \leq n$, so when the dimension n of M is large, the situation becomes rather more complicated than for $n = 2$, and more sophisticated techniques are called for.

Whenever calculus appears in mathematics and its applications, manifolds will also be found, and higher-dimensional manifolds turn out to be

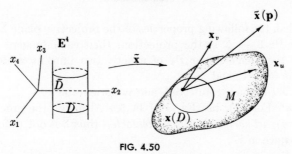

FIG. 4.50

important in problems (both pure and applied) that initially seem to involve only dimensions 2 or 3. For example, we now describe a four-dimensional manifold that has already appeared, implicitly at least, in this chapter.

8.5 Example   *The tangent bundle of a surface.*   If M is a surface, let $T(M)$ be the set of all tangent vectors to M at all points of M. (For the sake of concreteness we shall think of M as a surface in $\mathbf{E}^3$, but M could just as well be an abstract surface, or indeed a manifold of any dimension.) Now M itself has dimension 2 and each tangent plane $T_p(M)$ has dimension 2, so $T(M)$ will turn out to have dimension 4. To get the patch collection $\tilde{\mathcal{P}}$ that will make the set $T(M)$ a manifold, we shall derive from each patch $\mathbf{x}$ in M a patch $\tilde{\mathbf{x}}$ in $T(M)$. Given $\mathbf{x}: D \to M$, let $\tilde{\mathbf{D}}$ be the open set in $\mathbf{E}^4$ consisting of all point (p_1, p_2, p_3, p_4) for which (p_1, p_2) is in D. Then let $\tilde{\mathbf{x}}: \tilde{\mathbf{D}} \to T(M)$ be the function such that

$$\tilde{\mathbf{x}}(p_1, p_2, p_3, p_4) = p_3 \mathbf{x}_u(p_1, p_2) + p_4 \mathbf{x}_v(p_1, p_2).$$

(In Fig. 4.50 we identify $\mathbf{E}^2$ with the $x_1 x_2$ plane of $\mathbf{E}^4$ and deal as best we can with dimension 4.)

Using Exercise 3 of Section 3 and the proof of Lemma 3.6, it is not difficult to check that (1) each such function $\tilde{\mathbf{x}}$ is one-to-one, hence is a patch in $T(M)$, in the sense of Definition 8.4, and (2) the collection $\tilde{\mathcal{P}}$ of all such patches satisfies the two conditions in Definition 8.4. Thus $T(M)$ is a four-dimensional manifold called the *tangent bundle* of M.

EXERCISES

1. Show that a surface M is nonorientable if there is a closed curve $\alpha: [0, 1] \to M$ and a vector field Y on α such that
   (a) Y and α' are linearly independent at each point.
   (b) $Y(1) = -Y(0)$.

2. Establish the following properties of the projective plane $\bar{\Sigma}$:
   (a) If $P\colon \Sigma \to \bar{\Sigma}$ is the projection, then each tangent vector to $\bar{\Sigma}$ is the image under P_* of exactly two tangent vectors to Σ—of the form $\mathbf{v}_p$ and $(-\mathbf{v})_{-p}$.
   (b) $\bar{\Sigma}$ is compact, connected, and nonorientable.
   (*Hint:* For (a), use Ex. 5 of IV.5.) Although the proof is difficult, *every compact surface in* $\mathbf{E}^3$ *is orientable*—thus $\bar{\Sigma}$ is not diffeomorphic to any surface in $\mathbf{E}^3$.

3. Prove that the tangent bundle (8.5) is a manifold. (If $\mathbf{x}$ and $\mathbf{y}$ are overlapping patches in M, find an explicit formula for $\tilde{\mathbf{y}}^{-1}\tilde{\mathbf{x}}$.)

4. If M is the image of a single patch $\mathbf{x}\colon \mathbf{E}^2 \to M$, show that the tangent bundle $T(M)$ is diffeomorphic to $\mathbf{E}^4$.

5. (*Plane with two origins*). Let M consist of all ordered pairs of real numbers (u, v) and one additional point $\mathbf{0}^*$. Let $\mathbf{x}$ and $\mathbf{y}$ be the functions from $\mathbf{E}^2$ to M such that

$$\mathbf{x}(u, v) = \mathbf{y}(u, v) = (u, v) \quad \text{if } (u, v) \neq (0, 0),$$

but

$$\mathbf{x}(0, 0) = \mathbf{0} = (0, 0) \quad \text{and} \quad \mathbf{y}(0, 0) = \mathbf{0}^*.$$

Prove that:
   (a) The abstract patches $\mathbf{x}$ and $\mathbf{y}$ make M a surface.
   (b) M is connected.
   (c) The function $F\colon M \to M$ is a mapping, where $F(\mathbf{0}) = \mathbf{0}^*$ and $F(\mathbf{0}^*) = \mathbf{0}$, but $F(\mathbf{p}) = \mathbf{p}$ for all other points of M.
   Surfaces such as this one are troublesome to deal with; we eliminate them by adding an additional hypothesis to Definition 8.1: For any points $\mathbf{p} \neq \mathbf{q}$ of M there exist abstract patches $\mathbf{x}$ and $\mathbf{y}$ in $\mathcal{P}$ such that $\mathbf{p}$ is in $\mathbf{x}(D)$, $\mathbf{q}$ is in $\mathbf{y}(E)$, and $\mathbf{x}(D)$ and $\mathbf{y}(E)$ do not meet (*Hausdorff axiom*).

6. Let V be a vector field on a surface M. A curve α in M is an *integral curve* of V provided $\alpha'(t) = V(\alpha(t))$ for all t. Thus an integral curve has at each point the velocity prescribed by V. If $\alpha(0) = \mathbf{p}$, we say that α starts at $\mathbf{p}$.
   (a) In the special case $M = \mathbf{E}^2$, show that the curve $t \to (a_1(t), a_2(t))$ is an integral curve of $\mathbf{v} = f_1 U_1 + f_2 U_2$ starting at (a, b) if and only if

$$\begin{cases} \dfrac{da_1}{dt} = f_1(a_1, a_2) \\[2mm] \dfrac{da_2}{dt} = f_2(a_1, a_2) \end{cases} \quad \text{and} \quad \begin{cases} a_1(0) = a \\[2mm] a_2(0) = b \end{cases}$$

The theory of differential equations predicts a unique solution for such systems.

(b) Find the integral curve of

$$\mathbf{v} = -u^2 U_1 + uv U_2 \qquad \text{on } \mathbf{E}^2$$

which starts at the point $(1, -1)$. (*Hint:* The differential equations in this case can be solved by elementary methods. Use the arbitrary constants in the solution to make the starting point $(1, -1)$.)

7. Prove that every vector field V on a surface M has an integral curve starting at any given point. (*Hint:* Pick a patch $\mathbf{x}$ in M, with $\mathbf{x}(a, b) = \mathbf{p}$, and let $\mathbf{v}$ be the vector field on $\mathbf{E}^2$ such that $\mathbf{x}_*(\mathbf{v}) = V$.)

8. Prove that every surface of revolution is diffeomorphic to either a torus or a circular cylinder. (Similarly an augmented surface of revolution—Ex. 12 of IV.1—is diffeomorphic to either a plane or a sphere.)

9. (*Cartesian products*). If M and N are surfaces, let $M \times N$ be the set of all ordered pairs $(\mathbf{p}, \mathbf{q})$, with $\mathbf{p}$ in M and $\mathbf{q}$ in N. If $\mathbf{x}: D \to M$ and $\mathbf{y}: E \to N$ are patches, let $D \times E$ be the region in $\mathbf{E}^4$ consisting of all points (u, v, u_1, v_1) with (u, v) in D, (u_1, v_1) in E. Then define $\mathbf{x} \times \mathbf{y}: D \times E \to M \times N$ by

$$(\mathbf{x} \times \mathbf{y})(u, v, u_1, v_1) = (\mathbf{x}(u,v), \mathbf{y}(u_1, v_1)).$$

Prove that the collection $\mathcal{P}$ of all such abstract patches makes $M \times N$ a manifold (of dimension 4). $M \times N$ is called the *Cartesian product* of M and N.

The same scheme works for any two manifolds—for example, $\mathbf{E}^1 \times \mathbf{E}^1$ is precisely $\mathbf{E}^2$.

10. If M is an abstract surface, a *proper imbedding* of M in $\mathbf{E}^3$ is a one-to-one regular mapping $F: M \to \mathbf{E}^3$ such that the inverse function $F^{-1}: F(M) \to M$ is continuous. Prove that the image $F(M)$ of a proper imbedding is a surface in $\mathbf{E}^3$ (Definition 1.2) and is diffeomorphic to M.

If $F: M \to \mathbf{E}^3$ is merely regular, then F is an *immersion* of M in $\mathbf{E}^3$ and the image $F(M)$ is sometimes called an "immersed surface," even though it need not satisfy Definition 1.2.

9 Summary

In this chapter we have progressed from the familiar notion of surface in $\mathbf{E}^3$ to the general notion of *manifold*. Let us now reverse this process: An n-dimensional manifold M is a space that—near each point—is like the Euclidean space $\mathbf{E}^n$. Every manifold has a calculus consisting of differentiable functions, tangent vectors, vector fields, mappings, and, above all, differential forms. The simplest manifold of dimension n is $\mathbf{E}^n$ itself. A two-dimensional manifold is called a surface. Some surfaces appear in $\mathbf{E}^3$, some do not. This theory all comes from the usual elementary calculus on the best-known manifold of all, the real line. But the calculus of *every* manifold behaves in the same general way.

CHAPTER V

Shape Operators

In Chapter II we measured the shape of a curve in $\mathbf{E}^3$ by its curvature and torsion functions. Now we consider the analogous measurement problem for surfaces. It turns out that the shape of a surface M in $\mathbf{E}^3$ is described infinitesimally by a certain linear operator S defined on each of the tangent planes of M. As with curves, to say that two surfaces in $\mathbf{E}^3$ have the same shape means simply that they are congruent. And just as with curves, we shall justify our infinitesimal measurements by proving that two surfaces with "the same" shape operators are, in fact, congruent. The *algebraic* invariants (determinant, trace, $\cdots$) of its shape operators thus have *geometric* meaning for the surface M. We shall investigate this matter in detail and find efficient ways to compute these invariants, which we test on a number of geometrically interesting surfaces.

From now on the notation $M \subset \mathbf{E}^3$ means a connected surface M in $\mathbf{E}^3$ as defined in Chapter IV.

1 The Shape Operator of $M \subset \mathbf{E}^3$

Suppose that Z is a Euclidean vector field (Definition 3.7 of Chapter IV) on a surface M in $\mathbf{E}^3$. Although Z is defined only at points of M, the covariant derivative $\nabla_v Z$ (Chapter II, Section 5) still makes sense *as long as* $\mathbf{v}$ *is tangent to* M. As usual, $\nabla_v Z$ is the rate of change of Z in the $\mathbf{v}$ direction, and there are two main ways to compute it.

Method 1. Let α be a curve in M that has initial velocity $\alpha'(0) = \mathbf{v}$. Let Z_α be the restriction of Z to α, that is, the vector field $t \to Z(\alpha(t))$ on α (Fig. 5.1). Then

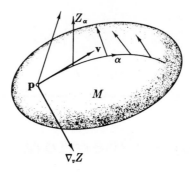

FIG. 5.1

$$\nabla_v Z = (Z_\alpha)'(0)$$

where the derivative is that of Chapter II, Section 2.

Method 2. Express Z in terms of the natural frame field of $\mathbf{E}^3$ by

$$Z = \sum z_i U_i.$$

Then

$$\nabla_v Z = \sum \mathbf{v}[z_i] U_i$$

where the directional derivative is that of Chapter IV, Section 3.

It is easy to check that these two methods give consistent results. Note that even if Z is a tangent vector field, the covariant derivative $\nabla_v Z$ need not be tangent to M.

It follows immediately from Theorem 7.5 of Chapter IV that if M is an *orientable* surface in $\mathbf{E}^3$, then there is a unit normal vector field U on M. In fact, if Z is a nonvanishing normal vector field, then $U = Z/\|Z\|$ is still normal, and has unit length. Since M is now assumed to be connected, *there are exactly two unit normal vector fields U and $-U$ defined on the whole surface M.* But even when M is not orientable, unit normals U and $-U$ are still available *on some neighborhood* of each point of $\mathbf{p}$ of M (see Exercises for Chapter IV, Section 7).

We are now in a position to find a mathematical measurement of the shape of a surface in $\mathbf{E}^3$.

1.1 Definition If $\mathbf{p}$ is a point of M, then for each tangent vector $\mathbf{v}$ to M at $\mathbf{p}$, let

$$S_p(\mathbf{v}) = -\nabla_v U$$

where U is a unit normal vector field on a neighborhood of $\mathbf{p}$ in M. S_p is called the *shape operator* of M at $\mathbf{p}$ (derived from U).† (Fig. 5.2.)

† The minus sign artificially introduced in this definition will sharply reduce the total number of minus signs needed later on.

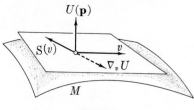

FIG. 5.2

The tangent plane of M at any point $\mathbf{q}$ consists of all Euclidean vectors orthogonal to $U(\mathbf{q})$. Thus the rate of change $\nabla_v U$ of U in the $\mathbf{v}$ direction tells how the tangent planes of M are varying in the $\mathbf{v}$ direction—and this gives an infinitesimal description of the way M itself is curving in $\mathbf{E}^3$.

Note that if U is replaced by $-U$, then S_p changes to $-S_p$.

1.2 Lemma   For each point $\mathbf{p}$ of $M \subset \mathbf{E}^3$, the shape operator is a linear operator

$$S_p \colon T_p(M) \to T_p(M)$$

on the tangent plane of M at $\mathbf{p}$.

Proof. In Definition 1.1, U is a unit vector field, so $U \cdot U = 1$. Thus by a Leibnizian property of covariant derivatives,

$$0 = \mathbf{v}[U \cdot U] = 2\nabla_v U \cdot U(\mathbf{p}) = -2S_p(\mathbf{v}) \cdot U(\mathbf{p})$$

where $\mathbf{v}$ is tangent to M at $\mathbf{p}$. Since U is also a *normal* vector field, it follows that $S_p(\mathbf{v})$ is tangent to M at $\mathbf{p}$. Thus S_p is a function from $T_p(M)$ to $T_p(M)$. (It is to emphasize this that we use the term "operator" instead of "transformation.")

The linearity of S_p is a consequence of a linearity property of covariant derivatives.

$$S_p(a\mathbf{v} + b\mathbf{w}) = -\nabla_{av+bw}U = -(a\nabla_v U + b\nabla_w U)$$

$$= aS_p(\mathbf{v}) + bS_p(\mathbf{w}). \qquad \blacksquare$$

At each point $\mathbf{p}$ of $M \subset \mathbf{E}^3$ there are actually two shape operators $\pm S_p$ derived from the two unit normals $\pm U$ near $\mathbf{p}$. We shall refer to all of these, collectively, as *the shape operator S of M*. Thus if a choice of unit normal is not specified, there is a (relatively harmless) ambiguity of sign.

1.3 Example   Shape operators of some surfaces in $\mathbf{E}^3$.   (1) Let Σ be the sphere of radius r consisting of all points $\mathbf{p}$ of $\mathbf{E}^3$ with $\| \mathbf{p} \| = r$. Let U be the "outward normal" on Σ. Now as U moves away from any point $\mathbf{p}$ in the direction $\mathbf{v}$, evidently U topples forward in the exact direction of $\mathbf{v}$ itself (Fig. 5.3). Thus $S(\mathbf{v})$ must have the form $-c\mathbf{v}$.

In fact, using *gradients* as in Example 3.9 of Chapter IV, we find

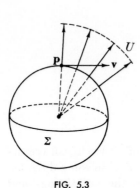

FIG. 5.3

$$U = \frac{1}{r} \sum x_i U_i.$$

But then

$$\nabla_v U = \frac{1}{r} \sum v[x_i]\, U_i(\mathbf{p}) = \frac{\mathbf{v}}{r}.$$

Thus $S(\mathbf{v}) = -\mathbf{v}/r$ for all $\mathbf{v}$. So the shape operator S is merely scalar multiplication by $-1/r$. This uniformity in S reflects the roundness of spheres: They bend the same way in all directions at all points.

(2) Let P be a plane in $\mathbf{E}^3$. A unit normal vector field U on P is evidently *parallel* in $\mathbf{E}^3$ (constant Euclidean coordinates) (Fig. 5.4). Hence

$$S(\mathbf{v}) = -\nabla_v U = 0$$

for all tangent vectors $\mathbf{v}$ to P. Thus the shape operator is identically zero —to be expected, since planes do not bend at all.

(3) Let C be the circular cylinder $x^2 + y^2 = r^2$ in $\mathbf{E}^3$. At any point $\mathbf{p}$ of C, let $\mathbf{e}_1$ and $\mathbf{e}_2$ be unit tangent vectors, with $\mathbf{e}_1$ tangent to the ruling of the cylinder through $\mathbf{p}$, and $\mathbf{e}_2$ tangent to the cross-sectional circle.

Use the outward normal U as indicated in Fig. 5.5.

Now, when U moves from $\mathbf{p}$ in the $\mathbf{e}_1$ direction, it stays parallel to itself just as on a plane; hence $S(\mathbf{e}_1) = 0$. When U moves in the $\mathbf{e}_2$ direction, it topples forward exactly as on a sphere of radius r; hence $S(\mathbf{e}_2) = -\mathbf{e}_2/r$. In this way S describes the "half-flat, half-round" shape of a cylinder.

(4) The *saddle surface* $M: z = xy$. For the moment we investigate S only at $\mathbf{p} = (0,0,0)$ in M. Since the x and y axes of $\mathbf{E}^3$ lie in M, the vectors $\mathbf{u}_1 = (1,0,0)$ and $\mathbf{u}_2 = (0,1,0)$ are tangent to M at $\mathbf{p}$. We use the "upward" unit normal U, which at $\mathbf{p}$ is $(0,0,1)$. Along the x axis, U stays

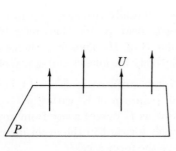

FIG. 5.4

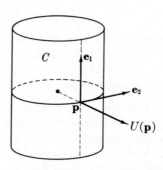

FIG. 5.5

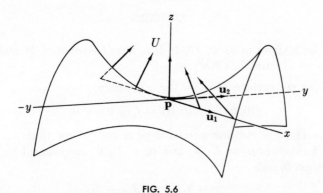

FIG. 5.6

orthogonal to the x axis, and as we proceed in the $\mathbf{u}_1$ direction, U swings from left to right (Fig. 5.6). In fact, a routine computation (Exercise 3) shows that $\nabla_{u_1} U = -\mathbf{u}_2$. Similarly we find $\nabla_{u_2} U = -\mathbf{u}_1$.

Thus the shape operator of M at $\mathbf{p}$ is given by the formula

$$S(a\mathbf{u}_1 + b\mathbf{u}_2) = b\mathbf{u}_1 + a\mathbf{u}_2.$$

These examples clarify the analogy between the shape operator of a surface and the curvature and torsion of a curve. In the case of a curve, there is only one direction to move, and κ and τ measure the rate of change of the unit vector fields T and B (hence N). For a surface only one unit vector field is intrinsically determined—the unit normal U. Furthermore, at each point, there are now a whole plane of directions in which U can move, so that rates of change of U are measured, not numerically, but by the linear operators S.

1.4 Lemma For each point $\mathbf{p}$ of $M \subset \mathbf{E}^3$ the shape operator

$$S: T_pM \to T_pM$$

is a *symmetric* linear operator; that is,

$$S(\mathbf{v}) \cdot \mathbf{w} = S(\mathbf{w}) \cdot \mathbf{v}$$

for any pair of tangent vectors to M at $\mathbf{p}$.

We postpone the proof of this crucial fact to Section 4, where it occurs naturally in the course of general computations.

From the viewpoint of linear algebra, a symmetric linear operator on a two-dimensional vector space is a very simple object indeed. For a shape operator, its characteristic values and vectors, its trace and determinant, all turn out to have geometric meaning of first importance for the surface $M \subset \mathbf{E}^3$.

EXERCISES

1. Let α be a curve in $M \subset \mathbf{E}^3$. If U is a unit normal of M restricted to the curve α, show that $S(\alpha') = -U'$.

2. Consider the surface $M: z = f(x, y)$, where

$$f(0, 0) = f_x(0, 0) = f_y(0, 0) = 0.$$

(The subscripts indicate partial derivatives.) Show that
(a) The vectors $\mathbf{u}_1 = U_1(0)$ and $\mathbf{u}_2 = U_2(0)$ are tangent to M at the origin $\mathbf{0}$, and

$$U = \frac{-f_x U_1 - f_y U_2 + U_3}{\sqrt{1 + f_x{}^2 + f_y{}^2}}$$

is a unit normal vector field on M.
(b) $S(\mathbf{u}_1) = f_{xx}(0, 0)\mathbf{u}_1 + f_{xy}(0, 0)\mathbf{u}_2$
   $S(\mathbf{u}_2) = f_{yx}(0, 0)\mathbf{u}_1 + f_{yy}(0, 0)\mathbf{u}_2.$
(*Note:* The square root in the denominator is no real problem here because of the special character of f at $(0, 0)$. In general, direct computation of S is difficult, and in Section 4 we shall establish indirect ways of getting at it.)

3. (*Continuation*). In each case, express $S(a\mathbf{u}_1 + b\mathbf{u}_2)$ in terms of $\mathbf{u}_1$ and $\mathbf{u}_2$, and determine the rank of S at $\mathbf{0}$ (rank S is dimension of image S: 0, 1, or 2).
(a) $z = xy.$ (c) $z = (x + y)^2.$
(b) $z = 2x^2 + y^2.$ (d) $z = xy^2.$

4. Let M be a surface in $\mathbf{E}^3$ oriented by a unit normal vector field

$$U = g_1 U_1 + g_2 U_2 + g_3 U_3.$$

Then the *Gauss mapping* $G: M \rightarrow \Sigma$ of M sends each point $\mathbf{p}$ to the point $(g_1(\mathbf{p}), g_2(\mathbf{p}), g_3(\mathbf{p}))$ of the unit sphere Σ. Pictorially: Move $U(\mathbf{p})$ to the origin by parallel motion; there it points to $G(\mathbf{p})$ (Fig. 5.7).

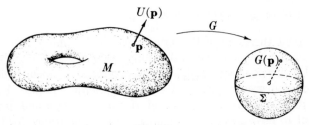

FIG. 5.7

Thus G completely describes the turning of U as it traverses M.

For each of the following surfaces, describe the image $G(M)$ of the Gauss mapping in the sphere Σ (use either normal):

(a) Cylinder, $x^2 + y^2 = r^2$.

(b) Cone, $z = \sqrt{x^2 + y^2}$.

(c) Plane, $x + y + z = 0$.

(d) Sphere, $(x - 1)^2 + y^2 + (z + 2)^2 = 1$.

5. Let $G: T \to \Sigma$ be the Gauss mapping of the torus T (as in IV. 2.6) derived from its outward unit normal U. What are the image curves under G of the meridians and parallels of T? Which points of Σ are the image of exactly two points of T?

6. Let $G: M \to \Sigma$ be the Gauss mapping of the saddle surface $M: z = xy$ derived from the unit normal U obtained as in Exercise 2. What is the image under G of one of the straight lines, y constant, in M? How much of the sphere is covered by the entire image $G(M)$?

7. Show that the shape operator of M is (minus) the derivative of its Gauss mapping: If S and $G: M \to \Sigma$ both derive from U, then $S(\mathbf{v})$ and $-G_*(\mathbf{v})$ are parallel for every tangent vector $\mathbf{v}$ to M.

8. An orientable surface has two Gauss mappings derived from its two unit normals. Show that they differ only by the antipodal mapping of Σ (Ex. 5 of IV. 5). Define a Gauss-type mapping for a nonorientable surface in $\mathbf{E}^3$.

9. If V is a tangent vector field on M (with unit normal U), then by the pointwise principle, $S(V)$ is the tangent vector field on M whose value at each point $\mathbf{p}$ is $S_p(V(\mathbf{p}))$. Show that

$$S(V) \cdot W = \nabla_V W \cdot U.$$

Deduce that the symmetry of S is equivalent to the assertion that the *bracket*

$$[V, W] = \nabla_V W - \nabla_W V$$

of two tangent vector fields is again a tangent vector field.

2 Normal Curvature

Throughout this section we shall work in a region of $M \subset \mathbf{E}^3$ which has been *oriented* by the choice of a unit normal vector field U, and we use the shape operator S derived from U.

The shape of a surface in $\mathbf{E}^3$ influences the shape of the curves in M.

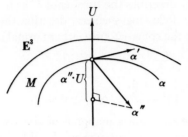

FIG. 5.8

2.1 Lemma   If α is a curve in $M \subset \mathbf{E}^3$, then

$$\alpha'' \cdot U = S(\alpha') \cdot \alpha'.$$

Proof. Since α is in M, its velocity α' is always tangent to M. Thus $\alpha' \cdot U = 0$, where as in Section 1 we restrict U to the curve α. Differentiation yields

$$\alpha'' \cdot U + \alpha' \cdot U' = 0.$$

But from Section 1, we know that $S(\alpha') = -U'$. Hence

$$\alpha'' \cdot U = -U' \cdot \alpha' = S(\alpha') \cdot \alpha'. \qquad \blacksquare$$

Geometric interpretation: at each point $\alpha'' \cdot U$ is the component of acceleration α'' normal to the surface M (Fig. 5.8). The lemma shows that this component depends only on the velocity α' and the shape operator of M. Thus *all curves in M with a given velocity $\mathbf{v}$ at point $\mathbf{p}$ will have the same normal component of acceleration at $\mathbf{p}$, namely, $S(\mathbf{v}) \cdot \mathbf{v}$.* This is the component of acceleration *which the bending of M in $\mathbf{E}^3$ forces them to have.*

Thus if we standardize $\mathbf{v}$ by reducing it to a unit vector $\mathbf{u}$, we get a measurement of the way M is bent in the $\mathbf{u}$ direction.

2.2 Definition   Let $\mathbf{u}$ be a unit vector tangent to $M \subset \mathbf{E}^3$ at a point $\mathbf{p}$. Then the number $k(\mathbf{u}) = S(\mathbf{u}) \cdot \mathbf{u}$ is called the *normal curvature of M in the $\mathbf{u}$ direction.*

To make the term *direction* precise, we define a *tangent direction to M at* $\mathbf{p}$ to be a one-dimensional subspace L of $T_p(M)$, that is, a line through the zero vector (located for intuitive purposes at $\mathbf{p}$) (Fig. 5.9). Any nonzero tangent vector at $\mathbf{p}$ determines a direction L, but we prefer to use one of

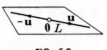

FIG. 5.9

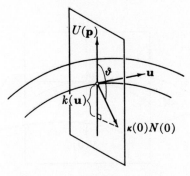

FIG. 5.10

the two unit vectors $\pm \mathbf{u}$ in L. Note that

$$k(\mathbf{u}) = S(\mathbf{u}) \cdot \mathbf{u} = S(-\mathbf{u}) \cdot (-\mathbf{u}) = k(-\mathbf{u}).$$

Thus, although we evaluate k on unit vectors, it is, in effect, a real-valued function on the set of all tangent directions to M.

Given a unit tangent vector to M at $\mathbf{p}$, let α be a (unit speed) curve in M with initial velocity $\alpha'(0) = \mathbf{u}$. Using the Frenet apparatus of α, the preceding lemma gives

$$k(\mathbf{u}) = S(\mathbf{u}) \cdot \mathbf{u} = \alpha''(0) \cdot U(\mathbf{p}) = \kappa(0)N(0) \cdot U(\mathbf{p})$$

$$= \kappa(0) \cos \vartheta.$$

Thus the normal curvature of M in the $\mathbf{u}$ direction is $\kappa(0) \cos \vartheta$, where $\kappa(0)$ is the curvature of α at $\alpha(0) = \mathbf{p}$, and ϑ is the angle between the principal normal $N(0)$ and the surface normal $U(\mathbf{p})$, as in Fig. 5.10.

Given $\mathbf{u}$, there is a natural way to choose the curve so that ϑ is 0 or π. In fact, if P is the plane determined by $\mathbf{u}$ and $U(\mathbf{p})$, then P cuts from M (near $\mathbf{p}$) a curve σ called the *normal section* of M in the $\mathbf{u}$ direction. If we give σ unit-speed parametrization with $\sigma'(0) = \mathbf{u}$, it is easy to see that $N(0) = \pm U(\mathbf{p})$. ($\sigma''(0) = \kappa(0)N(0)$ is orthogonal to $\sigma'(0) = \mathbf{u}$ and tangent to P.) Thus for a normal section in the $\mathbf{u}$ direction (Fig. 5.11),

$$k(\mathbf{u}) = \kappa_\sigma(0)N(0) \cdot U(\mathbf{p}) = \pm\kappa_\sigma(0).$$

Thus it is possible to make a reasonable estimate of the normal curvatures in various directions on a surface $M \subset \mathbf{E}^3$ by picturing what the corresponding normal sections would look like. We know that the principal normal N of a curve tells in which direction it is turning. Thus the preceding discussion gives geometric meaning to the sign of the normal curvature $k(\mathbf{u})$ (relative to our fixed choice of U).

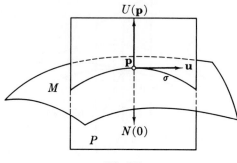

FIG. 5.11

(1) If $k(\mathbf{u}) > 0$, then $N(0) = U(\mathbf{p})$, so the normal section σ is bending toward $U(\mathbf{p})$ at $\mathbf{p}$ (Fig. 5.12). Thus in the $\mathbf{u}$ direction the surface M is bending *toward* $U(\mathbf{p})$.

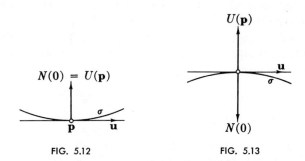

FIG. 5.12   FIG. 5.13

(2) If $k(\mathbf{u}) < 0$, then $N(0) = -U(\mathbf{p})$, so the normal section σ is bending *away from* $U(\mathbf{p})$ at $\mathbf{p}$. Thus in the $\mathbf{u}$ direction M is bending away from $U(\mathbf{p})$ (Fig. 5.13).

(3) If $k(\mathbf{u}) = 0$, then $\kappa_\sigma(0) = 0$ ($N(0)$ is undefined). Here the normal section σ is not turning at $\sigma(0) = \mathbf{p}$. We cannot conclude that in the $\mathbf{u}$ direction M is not bending at all, since κ might be zero only at $\sigma(0) = \mathbf{p}$. But we can conclude that its rate of bending is unusually small.

In different directions at a fixed point $\mathbf{p}$, the surface may bend in quite different ways. For example, consider the saddle surface $z = xy$ in Example 1.3. If we identify the tangent plane of M at $\mathbf{p} = (0,0,0)$ with the xy plane of $\mathbf{E}^3$, then clearly the normal curvature in the direction of the x and y axes is zero, since the normal sections are straight lines. However, Fig. 5.6 shows that in the tangent direction given by the line $y = x$, the normal curvature is positive, for the normal section is a parabola bending upward. ($U(\mathbf{p}) = (0,0,1)$ is "upward.") But in the direction of the line $y = -x$, normal curvature is negative, since this parabola bends downward.

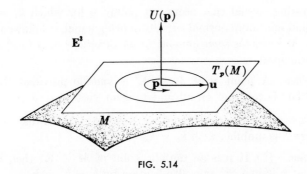

$$U(\mathbf{p})$$

$$\mathbf{E}^3$$

$$T_p(M)$$

$$\mathbf{p} \qquad \mathbf{u}$$

$$M$$

FIG. 5.14

Let us now fix a point $\mathbf{p}$ of $M \subset \mathbf{E}^3$ and imagine that a unit tangent vector $\mathbf{u}$ at $\mathbf{p}$ revolves, sweeping out the unit circle in the tangent plane $T_p(M)$. From the corresponding normal sections, we get a moving picture of the way M is bending in *every* direction at $\mathbf{p}$ (Fig. 5.14).

2.3 Definition   Let $\mathbf{p}$ be a point of $M \subset \mathbf{E}^3$. The maximum and minimum values of the normal curvature $k(\mathbf{u})$ of M at $\mathbf{p}$ are called the *principal curvatures* of M at $\mathbf{p}$, and are denoted by k_1 and k_2. The directions in which these extreme values occur are called *principal directions* of M at $\mathbf{p}$. Unit vectors in these directions are called *principal vectors* of M at $\mathbf{p}$.

Using the normal-section scheme discussed above, it is often fairly easy to pick out the directions of maximum and minimum bending. For example, if we use the outward normal (U) on a circular cylinder C, then the normal sections of all bend away from U, so $k(\mathbf{u}) \leqq 0$. Furthermore, it is reasonably clear that the maximum value $k_1 = 0$ occurs only in the direction of a ruling, minimum value $k_2 < 0$ occurs only in the direction tangent to a cross section, as in Fig. 5.15.

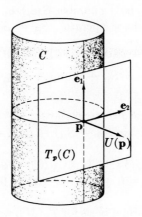

$$C$$

$$\mathbf{e}_1$$

$$\mathbf{e}_2$$

$$\mathbf{p}$$

$$U(\mathbf{p})$$

$$T_p(C)$$

FIG. 5.15

An interesting special case occurs at points **p** for which $k_1 = k_2$. The maximum and minimum normal curvature being equal, it follows that $k(\mathbf{u})$ is constant: *M bends the same amount in all directions at* **p** (and all directions are principal).

2.4 Definition   A point **p** of $M \subset E^3$ is *umbilic* provided the normal curvature $k(\mathbf{u})$ is constant on all unit tangent vectors **u** at **p**.

For example, what we found in (1) of Example 1.3 was that every point of the sphere Σ is umbilic, with $k_1 = k_2 = -1/r$.

2.5 Theorem   (1) If **p** is an umbilic point of $M \subset E^3$, then the shape operator S at **p** is just sc ar multiplication by $k = k_1 = k_2$.

(2) If **p** is a nonumbilic point, $k_1 \neq k_2$, then there are exactly two principal directions, and these are orthogonal. Furthermore, if $\mathbf{e}_1$ and $\mathbf{e}_2$ are principal vectors in these directions, then

$$S(\mathbf{e}_1) = k_1\mathbf{e}_1 \qquad S(\mathbf{e}_2) = k_2\mathbf{e}_2.$$

In short, the principal curvatures of M at **p** are the *characteristic values* of S, and the principal vectors of M at **p** are the *characteristic vectors* of S.

Proof. Suppose that k takes on its maximum value k_1 at $\mathbf{e}_1$, so

$$k_1 = k(\mathbf{e}_1) = S(\mathbf{e}_1) \cdot \mathbf{e}_1.$$

Let $\mathbf{e}_2$ be merely a unit tangent vector orthogonal to $\mathbf{e}_1$ (presently we shall show that it is also a principal vector.)

If **u** is any unit tangent vector at **p**, we write

$$\mathbf{u} = \mathbf{u}(\vartheta) = c\mathbf{e}_1 + s\mathbf{e}_2$$

where $c = \cos \vartheta$, $s = \sin \vartheta$ (Fig. 5.16). Thus normal curvature k at **p** becomes a function on the real line: $k(\vartheta) = k(\mathbf{u}(\vartheta))$.

For $1 \leq i, j \leq 2$, let S_{ij} be the number $S(\mathbf{e}_i) \cdot \mathbf{e}_j$. Note that $S_{11} = k_1$, and by the symmetry of the shape operator, $S_{12} = S_{21}$. We compute

$$\begin{aligned} k(\vartheta) &= S(c\mathbf{e}_1 + s\mathbf{e}_2) \cdot (c\mathbf{e}_1 + s\mathbf{e}_2) \\ &= c^2 S_{11} + 2sc S_{12} + s^2 S_{22}. \end{aligned} \tag{1}$$

Hence

$$\frac{dk}{d\vartheta}(\vartheta) = 2sc(S_{22} - S_{11}) + 2(c^2 - s^2)S_{12}. \tag{2}$$

If $\vartheta = 0$, then $c = 1$ and $s = 0$, so $\mathbf{u}(0) = \mathbf{e}_1$. Thus, by assumption, $k(\vartheta)$ is a maximum at $\vartheta = 0$, so $(dk/d\vartheta)(0) = 0$. It follows immediately from (2) that $S_{12} = 0$.

Since $\mathbf{e}_1, \mathbf{e}_2$ is an orthonormal basis for $T_p(M)$, we deduce by orthonormal expansion that

$$S(\mathbf{e}_1) = S_{11}\mathbf{e}_1 \qquad S(\mathbf{e}_2) = S_{22}\mathbf{e}_2. \tag{3}$$

Now if $\mathbf{p}$ is umbilic, then $S_{22} = k(\mathbf{e}_2)$ is the same as $S_{11} = k(\mathbf{e}_1) = k_1$, so (3) shows that S is scalar multiplication by $k_1 = k_2$.

If $\mathbf{p}$ is *not* umbilic, we look back at (1), which has become

$$k(\vartheta) = c^2 k_1 + s^2 S_{22}. \tag{4}$$

Since k_1 is the maximum value of $k(\vartheta)$, and $k(\vartheta)$ is now nonconstant, it follows that $k_1 > S_{22}$. But then (4) shows: (a) the maximum value k_1 is taken on *only* when $c = \pm 1$, $s = 0$, that is, in the $\mathbf{e}_1$ direction, and (b) the minimum value k_2 is S_{22}, and is taken on *only* when $c = 0$, $s = \pm 1$, that is, in the $\mathbf{e}_2$ direction. This proves the second assertion in the theorem, since (3) now reads:

$$S(\mathbf{e}_1) = k_1\mathbf{e}_1, \qquad S(\mathbf{e}_2) = k_2\mathbf{e}_2. \qquad \blacksquare$$

Contained in the preceding proof is Euler's formula for the normal curvature of M in *all* directions at $\mathbf{p}$.

2.6 Corollary Let k_1, k_2 and $\mathbf{e}_1$, $\mathbf{e}_2$ be the principal curvatures and vectors of $M \subset \mathbf{E}^3$ at $\mathbf{p}$. Then if $\mathbf{u} = \cos\vartheta\,\mathbf{e}_1 + \sin\vartheta\,\mathbf{e}_2$, the normal curvature of M in the $\mathbf{u}$ direction is (Fig. 5.16)

$$k(\mathbf{u}) = k_1 \cos^2\vartheta + k_2 \sin^2\vartheta.$$

Here is another way to show how the principal curvatures k_1 and k_2 control the shape of M near an arbitrary point $\mathbf{p}$. Since the position of M in $\mathbf{E}^3$ is of no importance, we can assume that (1) $\mathbf{p}$ is at the origin of $\mathbf{E}^3$, (2) the tangent plane $\bar{T}_p(M)$ is the xy plane of $\mathbf{E}^3$, and (3) the x and y axes are the principal directions. *Near* $\mathbf{p}$, M can be expressed as $M: z = f(x,y)$, as shown in Fig. 5.17, and the idea is to construct an *approximation* of M near $\mathbf{p}$ by using only terms up to quadratic in the Taylor expansion of the

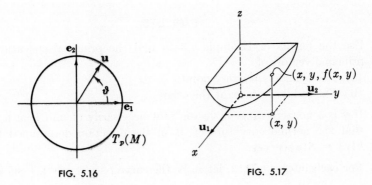

FIG. 5.16   FIG. 5.17

function f. Now (1) and (2) imply $f^0 = f_x^0 = f_y^0 = 0$, where the subscripts indicate partial derivatives, and the superscript zero denotes evaluation at $x = 0, y = 0$. Thus the quadratic approximation of f near $(0,0)$ reduces to

$$f(x,y) \sim \tfrac{1}{2}(f_{xx}^0 x^2 + 2f_{xy}^0 xy + f_{yy}^0 y^2).$$

In Exercise 2 of Section 1 we found that for the tangent vectors

$$\mathbf{u}_1 = (1,0,0) \quad \text{and} \quad \mathbf{u}_2 = (0,1,0)$$

at $\mathbf{p} = 0$

$$S(\mathbf{u}_1) = -\nabla_{u_1} U = f_{xx}^0 \mathbf{u}_1 + f_{xy}^0 \mathbf{u}_2$$

$$S(\mathbf{u}_2) = -\nabla_{u_2} U = f_{xy}^0 \mathbf{u}_1 + f_{yy}^0 \mathbf{u}_2.$$

By condition (3) above, $\mathbf{u}_1$ and $\mathbf{u}_2$ are *principal* vectors, so it follows from Theorem 2.5 that $k_1 = f_{xx}^0$, $k_2 = f_{yy}^0$, and $f_{xy}^0 = 0$.

Substituting these values in the quadratic approximation of f, we conclude that *the shape of M near $\mathbf{p}$ is approximately the same as that of the surface*

$$\hat{M}: z = \tfrac{1}{2}(k_1 x^2 + k_2 y^2)$$

near $\mathbf{0}$. $\hat{M}$ is called the *quadratic approximation of M near $\mathbf{p}$.* It is an analogue for surfaces of a Frenet approximation of a curve.

From Definition 2.2 through Corollary 2.6 we have been concerned with the geometry of $M \subset \mathbf{E}^3$ near one of its points $\mathbf{p}$. These results thus apply simultaneously to all the points of the oriented region $\mathcal{O}$ on which, by our initial assumption, the unit normal U is defined. In particular then, we have actually defined principal curvature *functions* k_1 and k_2 on $\mathcal{O}$. At each point $\mathbf{p}$ of $\mathcal{O}$, $k_1(\mathbf{p})$ and $k_2(\mathbf{p})$ are the principal curvatures of M at $\mathbf{p}$. We emphasize that these functions are only defined "modulo sign": If U is replaced by $-U$, they become $-k_1$ and $-k_2$.

EXERCISES

1. Use the results of Example 1.3 to find the principal curvatures and principal vectors of
   (a) The cylinder, at every point.
   (b) The saddle surface, at the origin.

2. If $\mathbf{v}$ is a nonzero tangent vector (not necessarily of unit length), show that the normal curvature of M in the direction determined by $\mathbf{v}$ is $k(\mathbf{v}) = S(\mathbf{v}) \cdot \mathbf{v}/\mathbf{v} \cdot \mathbf{v}$.

3. For each integer $n \geq 2$, let α_n be the curve $t \to (r \cos t, r \sin t, \pm t^n)$

in the cylinder $M: x^2 + y^2 = r^2$. These curves all have the same velocity at $t = 0$; test Lemma 2.1 by showing that they all have the same normal component of acceleration at $t = 0$.

4. For each of the following surfaces, find the quadratic approximation near the origin:
   (a) $z = \exp(x^2 + y^2) - 1$.
   (b) $z = \log \cos x - \log \cos y$.
   (c) $z = (x + 3y)^3$.

5. Justify the first sentence in the proof of Theorem 2.5: Show that k has a maximum value.

3 Gaussian Curvature

In the preceding section we found the geometrical meaning of the characteristic values and vectors of the shape operator. Now we examine the determinant and trace of S.

3.1 Definition   The *Gaussian curvature* of $M \subset \mathbf{E}^3$ is the real-valued function $K = \det S$ on M. Explicitly, for each point $\mathbf{p}$ of M, the Gaussian curvature $K(\mathbf{p})$ of M at $\mathbf{p}$ is the determinant of the shape operator S of M at $\mathbf{p}$.

The *mean curvature* of $M \subset \mathbf{E}^3$ is the function $H = \frac{1}{2}$ trace S. Gaussian and mean curvature are expressed in terms of principal curvature by

3.2 Lemma $K = k_1 k_2, H = (k_1 + k_2)/2$.

Proof. The determinant (and trace) of a linear operator may be defined as the common value of the determinant (and trace) of all its matrices. If $\mathbf{e}_1$ and $\mathbf{e}_2$ are principal vectors at a point $\mathbf{p}$, then by Theorem 2.5, we have $S(\mathbf{e}_1) = k_1(\mathbf{p})\mathbf{e}_1$ and $S(\mathbf{e}_2) = k_2(\mathbf{p})\mathbf{e}_2$. Thus the matrix of S at $\mathbf{p}$ with respect to $\mathbf{e}_1$, $\mathbf{e}_2$ is

$$\begin{pmatrix} k_1(\mathbf{p}) & 0 \\ 0 & k_2(\mathbf{p}) \end{pmatrix}$$

This immediately gives the required result.   ∎

A significant fact about the Gaussian curvature: It is independent of the choice of the unit normal U. If U is changed to $-U$, then the signs of *both* k_1 and k_2 change, so $K = k_1 k_2$ is unaffected. This is obviously not the case with mean curvature $H = (k_1 + k_2)/2$, which has the same ambiguity of sign as the principal curvatures themselves.

The normal section method in Section 2 lets us tell, by inspection, approximately what the principal curvatures of M are at each point. Thus we

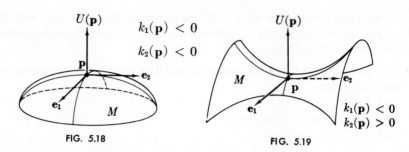

FIG. 5.18   FIG. 5.19

get a reasonable idea of what the Gaussian curvature $K = k_1 k_2$ is at each
point $\mathbf{p}$ by merely *looking* at the surface M. In particular, we can usually
tell what the sign of $K(\mathbf{p})$ is—and this sign has an important geometric
meaning, which we now illustrate.

3.3 Remark *The sign of Gaussian curvature at a point* $\mathbf{p}$.

(1) *Positive.* If $K(\mathbf{p}) > 0$, then by Lemma 3.2, the principal curvatures
$k_1(\mathbf{p})$ and $k_2(\mathbf{p})$ have the same sign. By Corollary 2.6, either $k(\mathbf{u}) > 0$ for
all unit vectors $\mathbf{u}$ at $\mathbf{p}$ or $k(\mathbf{u}) < 0$. Thus M *is bending away from its
tangent plane* $T_p(M)$ *in all tangent directions at* $\mathbf{p}$. (Fig. 5.18).

The quadratic approximation of M near $\mathbf{p}$ is the paraboloid

$$2z = k_1(\mathbf{p})x^2 + k_2(\mathbf{p})y^2.$$

(2) *Negative.* If $K(\mathbf{p}) < 0$, then by Lemma 3.2 the principal curvatures
$k_1(\mathbf{p})$ and $k_2(\mathbf{p})$ have opposite signs. Thus the quadratic approximation of
M near $\mathbf{p}$ is a hyperboloid, so M also is saddle-shaped *near* $\mathbf{p}$ (Fig. 5.19).

(3) *Zero.* If $K(\mathbf{p}) = 0$, then by Lemma 3.2 there are two cases:

(a) Only one principal curvature is zero, say

$$k_1(\mathbf{p}) \neq 0, \qquad k_2(\mathbf{p}) = 0.$$

(b) Both principal curvatures are zero:

$$k_1(\mathbf{p}) = k_2(\mathbf{p}) = 0.$$

In case (a) the quadratic approximation is the cylinder $2z = k_1(\mathbf{p})x^2$, so
M is trough-shaped near $\mathbf{p}$ (Fig. 5.20).

In case (b), the quadratic approximation reduces simply to the plane
$z = 0$, so we get no information about the shape of M near $\mathbf{p}$.

A torus of revolution T provides a good example of these different cases.
At points on the outer half $\mathcal{O}$ of T, the torus bends away from its tangent
plane as one can see from Fig. 5.21; hence $K > 0$ on $\mathcal{O}$. But near each point
$\mathbf{p}$ of the inner half $\mathcal{I}$, T is saddle-shaped and cuts through $T_p(M)$. Hence
$K < 0$ on $\mathcal{I}$.

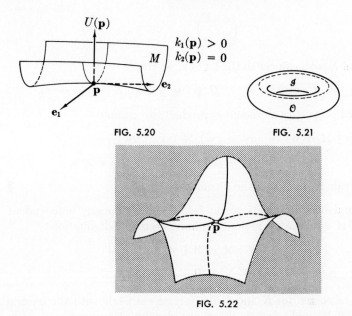

FIG. 5.20   FIG. 5.21

FIG. 5.22

Near each point on the two circles (top and bottom) which separate $\mathcal{O}$ and $\mathcal{G}$, the torus is trough-shaped; hence $K = 0$ there. (A quantitative check of these qualitative results is given in Section 6.)

In case 3 (b) above, where both principal curvatures vanish, **p** is called a *planar* point of M. (There are no planar points on the torus.) For example, the central point **p** of a *monkey saddle*, say

$$M: z = x(x + \sqrt{3}y)(x - \sqrt{3}y),$$

is planar. Here three hills and valleys meet, as shown in Fig. 5.22. Thus **p** *must* be a planar point—the shape of M near **p** is too complicated for the other three possibilities in Remark 3.3.

We consider now some ways to compute Gaussian and mean curvature.

3.4 Lemma If **v** and **w** are linearly independent tangent vectors at a point **p** of $M \subset \mathbf{E}^3$, then

$$S(\mathbf{v}) \times S(\mathbf{w}) = K(\mathbf{p})\mathbf{v} \times \mathbf{w}$$

$$S(\mathbf{v}) \times \mathbf{w} + \mathbf{v} \times S(\mathbf{w}) = 2H(\mathbf{p})\mathbf{v} \times \mathbf{w}.$$

Proof. Since **v**, **w** is a basis for the tangent plane $T_p(M)$, we can write

$$S(\mathbf{v}) = a\mathbf{v} + b\mathbf{w}$$

$$S(\mathbf{w}) = c\mathbf{v} + d\mathbf{w}.$$

Thus

$$\begin{pmatrix} a & b \\ c & d \end{pmatrix}$$

is the matrix of S with respect to the basis $\mathbf{v}$, $\mathbf{w}$. Hence

$$K(\mathbf{p}) = \det S = ad - bc \qquad H(\mathbf{p}) = \tfrac{1}{2} \text{ trace } S = \tfrac{1}{2}(a + d).$$

Using standard properties of the cross product, we compute

$$S(\mathbf{v}) \times S(\mathbf{w}) = (a\mathbf{v} + b\mathbf{w}) \times (c\mathbf{v} + d\mathbf{w})$$

$$= (ad - bc)\,\mathbf{v} \times \mathbf{w} = K(\mathbf{p})\,\mathbf{v} \times \mathbf{w}$$

and a similar calculation gives the formula for $H(\mathbf{p})$.　　　■

Thus if V and W are tangent vector fields that are linearly independent at each point of an oriented region, we have vector field equations

$$S(V) \times S(W) = K\,V \times W$$

$$S(V) \times W + V \times S(W) = 2H\,V \times W.$$

These may be solved for K and H by dotting each side with the normal vector field $V \times W$, and using the Lagrange identity (Exercise 6). We then find

$$K = \frac{\begin{vmatrix} SV \cdot V & SV \cdot W \\ SW \cdot V & SW \cdot W \end{vmatrix}}{\begin{vmatrix} V \cdot V & V \cdot W \\ W \cdot V & W \cdot W \end{vmatrix}} \qquad H = \frac{\begin{vmatrix} SV \cdot V & SV \cdot W \\ W \cdot V & W \cdot W \end{vmatrix} + \begin{vmatrix} V \cdot V & V \cdot W \\ SW \cdot V & SW \cdot W \end{vmatrix}}{2\begin{vmatrix} V \cdot V & V \cdot W \\ W \cdot V & W \cdot W \end{vmatrix}}$$

(The denominators are never zero, since the independence of V and W is equivalent to $(V \times W) \cdot (V \times W) > 0$.) In particular the functions K and H are *differentiable*.

Once K and H are known, it is a simple matter to find k_1 and k_2.

3.5 Corollary On an oriented region $\mathcal{O}$ in M, the principal curvature functions are

$$k_1, k_2 = H \pm \sqrt{H^2 - K}.$$

Proof. To verify the formula it suffices to substitute

$$K = k_1 k_2 \quad \text{and} \quad H = (k_1 + k_2)/2,$$

and note that

$$H^2 - K = \frac{(k_1 + k_2)^2}{4} - k_1 k_2 = \frac{(k_1 - k_2)^2}{4}$$
　　　　　　　　　　　　　　　　　　　　　　　　　　　　　■

A more enlightening *derivation* (Exercise 4) uses the characteristic polynomial of S.

This formula shows only that k_1 and k_2 are *continuous* functions on $\mathcal{O}$; they need not be differentiable since the square-root function is badly behaved at zero. The identity in the proof shows that $H^2 - K$ is zero only at umbilic points, however, so k_1 *and* k_2 *are differentiable on any oriented region free of umbilics.*

A natural way to single out special types of surfaces in $\mathbf{E}^3$ is by restrictions on Gaussian and mean curvature.

3.6 Definition A surface M in $\mathbf{E}^3$ is *flat* provided its Gaussian curvature is zero, and *minimal* provided its mean curvature is zero.

As expected, a plane is flat, for by Example 1.3 its shape operators are all zero, so $K = \det S = 0.$ On a circular cylinder, (3) of Example 1.3 shows that S is *singular* at each point $\mathbf{p}$, that is, has rank less than the dimension of the tangent plane $T_p(M)$. Thus, although S itself is never zero, its determinant is always zero, so cylinders are also flat. This terminology seems odd at first for a surface so obviously curved, but it will be amply justified in later work.

Note that minimal surfaces have Gaussian curvature $K \leqq 0$, because if $H = (k_1 + k_2)/2 = 0$, then $k_1 = -k_2$, so $K = k_1 k_2 \leqq 0$.

Another notable class of surfaces consists of those with *constant* Gaussian curvature. As mentioned earlier, Example 1.3 shows that a sphere of radius r has $k_1 = k_2 = -1/r$ (for U outward). Thus the sphere Σ has constant positive curvature $K = 1/r^2$: The smaller the sphere, the larger its curvature.

We shall find many examples of these various special types of surface as we proceed through this chapter.

EXERCISES

1. Show that there are no umbilics on a surface with $K < 0$, and that if $K \leqq 0$, umbilic points are planar.

2. Let $\mathbf{u}_1$ and $\mathbf{u}_2$ be orthonormal tangent vectors at a point $\mathbf{p}$ of M. What geometric information can be deduced from each of the following conditions on S at $\mathbf{p}$?
   (a) $S(\mathbf{u}_1) \cdot \mathbf{u}_2 = 0.$ (c) $S(\mathbf{u}_1) \times S(\mathbf{u}_2) = 0.$
   (b) $S(\mathbf{u}_1) + S(\mathbf{u}_2) = 0.$ (d) $S(\mathbf{u}_1) \cdot S(\mathbf{u}_2) = 0.$

3. (*Mean curvature*). Prove that
   (a) the average value of the normal curvature in *any* two orthogonal directions at $\mathbf{p}$ is $H(\mathbf{p})$. (The analogue for K is false.)

(b) $$H(\mathbf{p}) = (1/2\pi) \int_0^{2\pi} k(\vartheta)\, d\vartheta,$$

where $k(\vartheta)$ is the normal curvature, as in Corollary 2.6.

4. The *characteristic polynomial* of an arbitrary linear operator S is

$$p(k) = \det(A - kI),$$

where A is any matrix of S.

(a) Show that the characteristic polynomial of the shape operator is $k^2 - 2Hk + K$.

(b) Every linear operator satisfies its characteristic equation; that is, $p(S)$ is the zero operator when S is formally substituted in $p(k)$. Prove this in the case of the shape operator by showing that

$$S\mathbf{v}\cdot S\mathbf{w} - 2H S\mathbf{v}\cdot\mathbf{w} + K\mathbf{v}\cdot\mathbf{w} = 0$$

for any pair of tangent vectors to M.

The real-valued functions

$$I(\mathbf{v},\mathbf{w}) = \mathbf{v}\cdot\mathbf{w}, \quad II(\mathbf{v},\mathbf{w}) = S\mathbf{v}\cdot\mathbf{w},$$

and

$$III(\mathbf{v},\mathbf{w}) = S^2\mathbf{v}\cdot\mathbf{w} = S\mathbf{v}\cdot S\mathbf{w},$$

defined for all pairs of tangent vectors to an oriented surface, are traditionally called the *first*, *second*, and *third fundamental forms* of M. They are not differential forms; in fact, they are symmetric in $\mathbf{v}$ and $\mathbf{w}$ rather than alternate. The shape operator does not appear explicitly in the classical treatment of this subject; it is replaced by the second fundamental form.

5. (*Dupin curves*). For a point $\mathbf{p}$ of an oriented region of M, let C_0 be the intersection of M near $\mathbf{p}$ with its tangent plane $\bar{T}_p(M)$; specifically, C_0 consists of those points of M *near* $\mathbf{p}$ which lie in the plane through $\mathbf{p}$ orthogonal to $U(\mathbf{p})$. C_0 may be approximated by substituting for M its quadratic approximation $\hat{M}$; thus C_0 is approximated by the curve

$$\hat{C}_0\colon k_1 x^2 + k_2 y^2 = 0, \qquad \text{near } (0,0).$$

(a) Describe $\hat{C}_0$ in each of the three cases $K(\mathbf{p}) > 0$, $K(\mathbf{p}) < 0$, and $K(\mathbf{p}) = 0$ (not planar).

(b) Repeat (a) with C_0 replaced by C_ε and $C_{-\varepsilon}$, where the tangent plane has been replaced by the two parallel planes at distance $\pm\varepsilon$ from it.

(c) This scheme fails for planar points since the quadratic approximation becomes $\hat{M}\colon z = 0$. For the monkey saddle, sketch C_0, C_ε, and $C_{-\varepsilon}$.

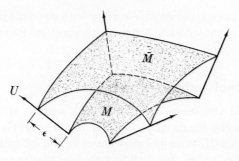

FIG. 5.23

6. If $\mathbf{v}$, $\mathbf{w}$, $\mathbf{a}$, and $\mathbf{b}$ are vectors in $\mathbf{E}^3$, prove the *Lagrange identity*

$$(\mathbf{v} \times \mathbf{w}) \cdot (\mathbf{a} \times \mathbf{b}) = \begin{vmatrix} \mathbf{v} \cdot \mathbf{a} & \mathbf{v} \cdot \mathbf{b} \\ \mathbf{w} \cdot \mathbf{a} & \mathbf{w} \cdot \mathbf{b} \end{vmatrix}.$$

7. (*Parallel surfaces*). Let M be a surface oriented by U; for a fixed number ε (positive or negative) let $F: M \to \mathbf{E}^3$ be the mapping such that

$$F(\mathbf{p}) = \mathbf{p} + \varepsilon U(\mathbf{p}).$$

(a) If $\mathbf{v}$ is tangent to M at $\mathbf{p}$, show that $\bar{\mathbf{v}} = F_*(\mathbf{v})$ is $\mathbf{v} - \varepsilon S(\mathbf{v})$. Deduce that

$$\bar{\mathbf{v}} \times \bar{\mathbf{w}} = J(\mathbf{p}) \, \mathbf{v} \times \mathbf{w},$$

where

$$J = 1 - 2\varepsilon H + \varepsilon^2 K = (1 - \varepsilon k_1)(1 - \varepsilon k_2).$$

If the function J does not vanish on M (M is compact and $|\varepsilon|$ small), this shows that F is a regular mapping, so the image

$$\bar{M} = F(M)$$

is at least an immersed surface in $\mathbf{E}^3$ (Ex. 10 of IV.8). $\bar{M}$ is said to be *parallel* to M at distance ε (Fig. 5.23).

(b) Show that the canonical isomorphisms of $\mathbf{E}^3$ make U a unit normal on $\bar{M}$ for which $\bar{S}(\bar{\mathbf{v}}) = S(\mathbf{v})$.

(c) Derive the following formulas for the Gaussian and mean curvatures of M:

$$\bar{K}(F) = K/J; \qquad \bar{H}(F) = (H - \varepsilon K)/J.$$

8. (*Continuation*)

(a) Check the results in (c) in the case of a sphere of radius r oriented by the outward normal U. Describe the mapping $F = F_\varepsilon$ when ε is 0, $-r$, and $-2r$.

(b) Starting from an orientable surface with constant positive Gaussian curvature, construct a surface with constant mean curvature.

4 Computational Techniques

We have defined the shape operator S of a surface M in $\mathbf{E}^3$ and found geometrical meaning for its main algebraic invariants: Gaussian curvature K, mean curvature H, principal curvatures k_1 and k_2, and (at each point) principal vectors $\mathbf{e}_1$ and $\mathbf{e}_2$. We shall now see how to express these invariants in terms of patches in M.

If $\mathbf{x} \colon D \to M$ is a patch in $M \subset \mathbf{E}^3$, we have already used the three real-valued functions

$$E = \mathbf{x}_u \cdot \mathbf{x}_u, \qquad F = \mathbf{x}_u \cdot \mathbf{x}_v = \mathbf{x}_v \cdot \mathbf{x}_u, \qquad G = \mathbf{x}_v \cdot \mathbf{x}_v.$$

on D. Here $E > 0$ and $G > 0$ are the squares of the speeds of the u- and v-parameter curves of $\mathbf{x}$, and F measures the *coordinate angle* ϑ between $\mathbf{x}_u$ and $\mathbf{x}_v$, since

$$F = \mathbf{x}_u \cdot \mathbf{x}_v = \| \mathbf{x}_u \| \, \| \mathbf{x}_v \| \cos \vartheta = \sqrt{EG} \cos \vartheta.$$

(Fig. 5.24). E, F, and G are the "warping functions" of the patch $\mathbf{x}$: They measure the way $\mathbf{x}$ distorts the flat region D in $\mathbf{E}^2$ in order to apply it to the curved region $\mathbf{x}(D)$ in M. These functions completely determine the dot product of tangent vectors at points of $\mathbf{x}(D)$, for if

$$\mathbf{v} = v_1 \mathbf{x}_u + v_2 \mathbf{x}_v \qquad \text{and} \qquad \mathbf{w} = w_1 \mathbf{x}_u + w_2 \mathbf{x}_u,$$

then

$$\mathbf{v} \cdot \mathbf{w} = E v_1 w_1 + F(v_1 w_2 + v_2 w_1) + G v_2 w_2.$$

(In such equations we understand that $\mathbf{x}_u$, $\mathbf{x}_v$, E, F, and G are evaluated at (u, v) where $\mathbf{x}(u, v)$ is the point of application of $\mathbf{v}$ and $\mathbf{w}$.)

Now $\mathbf{x}_u \times \mathbf{x}_v$ is a function on D whose value at each point (u, v) of D is a vector orthogonal to both $\mathbf{x}_u(u, v)$ and $\mathbf{x}_v(u, v)$—and hence normal to M at $\mathbf{x}(u, v)$. Furthermore, by Lemma 1.8 of Chapter II,

$$\| \mathbf{x}_u \times \mathbf{x}_v \|^2 = EG - F^2.$$

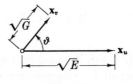

FIG. 5.24

Since $\mathbf{x}$ is, by definition, regular, this real-valued function on D is never zero. Thus we can construct the *unit normal function*

$$U = \frac{\mathbf{x}_u \times \mathbf{x}_v}{\| \mathbf{x}_u \times \mathbf{x}_v \|}$$

on D, which assigns to each (u, v) in D a unit normal vector at $\mathbf{x}(u, v)$. We emphasize that in this context, U, like $\mathbf{x}_u$ and $\mathbf{x}_v$, is not a vector field on $\mathbf{x}(D)$, but merely a vector-valued function on D. Nevertheless we may regard the system $\mathbf{x}_u$, $\mathbf{x}_v$, U as a kind of defective frame field. At least U has unit length and is orthogonal to both $\mathbf{x}_u$ and $\mathbf{x}_v$, even though $\mathbf{x}_u$ and $\mathbf{x}_v$ are generally not orthonormal.

In this context, covariant derivatives are usually computed along the parameter curves of $\mathbf{x}$, where by the discussion in Section 1, they reduce to partial differentiation with respect to u and v. As in the case of $\mathbf{x}_u$ and $\mathbf{x}_v$, these partial derivative are again denoted by subscripts u and v. If

$$\mathbf{x}(u, v) = (x_1(u, v), x_2(u, v), x_3(u, v)),$$

then just as for $\mathbf{x}_u$ and $\mathbf{x}_v$ on page 134, we have

$$\mathbf{x}_{uu} = \left(\frac{\partial^2 x_1}{\partial u^2}, \frac{\partial^2 x_2}{\partial u^2}, \frac{\partial^2 x_3}{du^2} \right)_{\mathbf{x}}$$

$$\mathbf{x}_{vu} = \left(\frac{\partial^2 x_1}{\partial u \partial v}, \frac{\partial^2 x_2}{\partial u \partial v}, \frac{\partial^2 x_3}{\partial u \partial v} \right)_{\mathbf{x}}$$

$$\mathbf{x}_{vv} = \left(\frac{\partial^2 x_1}{\partial v^2}, \frac{\partial^2 x_2}{\partial v^2}, \frac{\partial^2 x_3}{\partial v^2} \right)_{\mathbf{x}}.$$

Evidently $\mathbf{x}_{uu}$ and $\mathbf{x}_{vv}$ give the accelerations of the u- and v-parameter curves. Since order of partial differentiation is immaterial, $\mathbf{x}_{uv} = \mathbf{x}_{vu}$, which gives both the covariant derivative of $\mathbf{x}_u$ in the $\mathbf{x}_v$ direction, and $\mathbf{x}_v$ in the $\mathbf{x}_u$ direction.

Now if S is the shape operator derived from U, we define three more real-valued functions on D:

$$\ell = S(\mathbf{x}_u) \cdot \mathbf{x}_u$$

$$m = S(\mathbf{x}_u) \cdot \mathbf{x}_v = S(\mathbf{x}_v) \cdot \mathbf{x}_u$$

$$n = S(\mathbf{x}_v) \cdot \mathbf{x}_v.$$

Because $\mathbf{x}_u$, $\mathbf{x}_v$ gives a basis for the tangent space of M at each point of $\mathbf{x}(D)$, it is clear that these functions uniquely determine the shape operator. Since this basis is generally not orthonormal, ℓ, m, and n do not lead to simple expression for $S(\mathbf{x}_u)$ and $S(\mathbf{x}_v)$ in terms of $\mathbf{x}_u$ and $\mathbf{x}_v$. In the formulas preceding Corollary 3.5, however, they *do* provide simple expressions for Gaussian and mean curvature.

4.1 Corollary If $\mathbf{x}$ is a patch in $M \subset \mathbf{E}^3$, then

$$K(\mathbf{x}) = \frac{\ell n - m^2}{EG - F^2}, \qquad H(\mathbf{x}) = \frac{G\ell + En - 2Fm}{2(EG - F^2)}.$$

Proof. At a point $\mathbf{p}$ of $\mathbf{x}(D)$, the formulas on page 206 express $K(\mathbf{p})$ and $H(\mathbf{p})$ in terms of tangent vectors $V(\mathbf{p})$ and $W(\mathbf{p})$ at $\mathbf{p}$. If $V(\mathbf{p})$ and $W(\mathbf{p})$ are replaced by the tangent vectors $\mathbf{x}_u(u, v)$ and $\mathbf{x}_v(u, v)$ at $\mathbf{x}(u, v)$, we find the required formulas for $K(\mathbf{x}(u, v))$ and $H(\mathbf{x}(u, v))$. ∎

When the patch $\mathbf{x}$ is clear from context, we shall usually abbreviate the composite functions $K(\mathbf{x})$ and $H(\mathbf{x})$ to merely K and H.

By a device like that used in Lemma 2.1 we can find a simple way to *compute ℓ, m, and n*—and thereby K and H. For example, since $U \cdot \mathbf{x}_u = 0$, partial differentiation with respect to v—that is, ordinary differentiation along v-parameter curves—yields

$$0 = \frac{\partial}{\partial v}(U \cdot \mathbf{x}_u) = U_v \cdot \mathbf{x}_u + U \cdot \mathbf{x}_{uv}.$$

(Recall that U_v is the covariant derivative of the vector field $v \to U(u_0, v)$ on each v-parameter curve, $u = u_0$.) Since $\mathbf{x}_v$ gives the velocity vectors of such curves, Exercise 1 of Section 1 shows that $U_v = -S(\mathbf{x}_v)$. Thus the preceding equation becomes

$$S(\mathbf{x}_v) \cdot \mathbf{x}_u = U \cdot \mathbf{x}_{uv}.$$

(Fig. 5.25). Three similar equations may be found by replacing u by v, and v by u. In particular,

$$S(\mathbf{x}_u) \cdot \mathbf{x}_v = U \cdot \mathbf{x}_{vu} = U \cdot \mathbf{x}_{uv} = S(\mathbf{x}_v) \cdot \mathbf{x}_u.$$

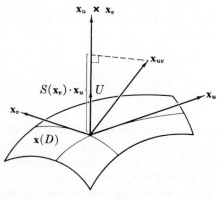

FIG. 5.25

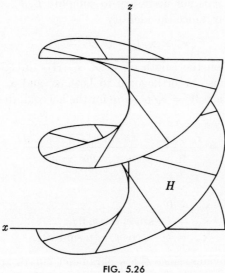

FIG. 5.26

Again, since $\mathbf{x}_u$ and $\mathbf{x}_v$ give a basis for the tangent space at each point, this is sufficient to prove that S *is symmetric* (Lemma 1.4).

4.2 Lemma   If $\mathbf{x}$ is a patch in $M \subset \mathbf{E}^3$, then

$$\ell = S(\mathbf{x}_u) \cdot \mathbf{x}_u = U \cdot \mathbf{x}_{uu}$$

$$m = S(\mathbf{x}_u) \cdot \mathbf{x}_v = U \cdot \mathbf{x}_{uv}$$

$$n = S(\mathbf{x}_v) \cdot \mathbf{x}_v = U \cdot \mathbf{x}_{vv}.$$

The first equation in each case is just definition, and *u and v may be reversed* in the formulas for m.

4.3 Example   *Computation of Gaussian and mean curvature*
(1) *Helicoid* (Exercise 7 of Section 2, Chapter IV). This surface H, shown in Fig. 5.26, is covered by a single patch

$$\mathbf{x}(u, v) = (u \cos v, u \sin v, bv), \qquad b \neq 0,$$

for which

$$\mathbf{x}_u = (\cos v, \sin v, 0) \qquad E = 1$$

$$\mathbf{x}_v = (-u \sin v, u \cos v, b) \qquad F = 0$$

$$G = b^2 + u^2.$$

Hence

$$\mathbf{x}_u \times \mathbf{x}_v = (b \sin v, -b \cos v, u).$$

To find K alone it is not necessary to compute E, F, and G, but it is wise to do so anyway, since the identity

$$\| \mathbf{x}_u \times \mathbf{x}_v \| = \sqrt{EG - F^2}$$

then gives a check on the length of $\mathbf{x}_u \times \mathbf{x}_v$. (Its direction may also be checked, since it must be orthogonal to both $\mathbf{x}_u$ and $\mathbf{x}_v$.) If we denote $\| \mathbf{x}_u \times \mathbf{x}_v \|$ by W, then $W = \sqrt{b^2 + u^2}$ for the helicoid, so the unit normal function is

$$U = \frac{\mathbf{x}_u \times \mathbf{x}_v}{W} = \frac{(b \sin v, -b \cos v, u)}{\sqrt{b^2 + u^2}}.$$

Next we find

$$\mathbf{x}_{uu} = 0$$
$$\mathbf{x}_{uv} = (-\sin v, \cos v, 0)$$
$$\mathbf{x}_{vv} = (-u \cos v, -u \sin v, 0).$$

Here $\mathbf{x}_{uu} = 0$ is obvious, since the u-parameter curves are straight lines. The v-parameter curves are helices, and this formula for the acceleration $\mathbf{x}_{vv}$ was found already in Chapter II. Now by Lemma 4.2,

$$\ell = \mathbf{x}_{uu} \cdot \frac{(\mathbf{x}_u \times \mathbf{x}_v)}{W} = 0$$

$$m = \mathbf{x}_{uv} \cdot \frac{(\mathbf{x}_u \times \mathbf{x}_v)}{W} = -\frac{b}{W}$$

$$n = \mathbf{x}_{vv} \cdot \frac{(\mathbf{x}_u \times \mathbf{x}_v)}{W} = 0.$$

Hence by Corollary 4.1 and the results above,

$$K = \frac{\ell n - m^2}{EG - F^2} = \frac{-(b/W)^2}{W^2} = \frac{-b^2}{W^4} = \frac{-b^2}{(b^2 + u^2)^2}$$

$$H = \frac{G\ell + En - 2Fm}{2(EG - F^2)} = 0.$$

Thus the helicoid is a minimal surface with Gaussian curvature

$$-1 \leqq K < 0.$$

The minimum value $K = -1$ occurs on the central axis ($u = 0$) of the helicoid, and $K \to 0$ as distance $|u|$ from the axis increases to infinity.

(2) *The saddle surface* $M: z = xy$ (Example 1.3). This time we use the Monge patch $\mathbf{x}(u, v) = (u, v, uv)$ and with the same format as above, compute

$$\mathbf{x}_u = (1, 0, v) \qquad E = 1 + v^2$$

$$\mathbf{x}_v = (0, 1, u) \qquad F = uv$$

$$G = 1 + u^2$$

$$U = (-v, -u, 1)/W \qquad W = \sqrt{1 + u^2 + v^2}$$

$$\mathbf{x}_{uu} = 0 \qquad \ell = 0$$

$$\mathbf{x}_{uv} = (0, 0, 1) \qquad m = 1/W$$

$$\mathbf{x}_{vv} = 0 \qquad n = 0.$$

Hence

$$K = \frac{-1}{(1 + u^2 + v^2)^2}, \qquad H = \frac{-uv}{(1 + u^2 + v^2)^{3/2}}.$$

Strictly speaking, these functions are $K(\mathbf{x})$ and $H(\mathbf{x})$ defined on the domain $\mathbf{E}^2$ of $\mathbf{x}$. But it is easy to express K and H directly as functions on M by using the cylindrical coordinate functions $r = \sqrt{x^2 + y^2}$ and z. Note from Fig. 5.27 that

$$r(\mathbf{x}(u, v)) = \sqrt{u^2 + v^2}$$

and

$$z(\mathbf{x}(u, v)) = uv;$$

hence on M:

$$K = \frac{-1}{(1 + r^2)^2}, \qquad H = \frac{-z}{(1 + r^2)^{3/2}}.$$

Thus the Gaussian curvature of M depends only on distance to the z

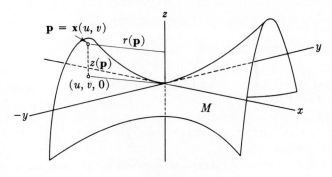

FIG. 5.27

axis, rising from $K = -1$ (at the origin) toward zero as r goes to infinity, while H varies more radically.

Like all simple (that is, one-patch) surfaces, the helicoid and saddle surface are orientable, since computations as above provide a unit normal on the whole surface. Thus the principal curvature functions k_1 and k_2 are defined unambiguously on each surface. These can always be found from K and H by Corollary 3.5. Since the helicoid is a minimal surface, we get the simple result

$$k_1, k_2 = \frac{\pm b}{(b^2 + u^2)}.$$

For the saddle surface,

$$k_1, k_2 = \frac{-z \pm \sqrt{1 + r^2 + z^2}}{(1 + r^2)^{3/2}}.$$

Techniques for computing principal vectors are left to the exercises.

There is a different computational approach which depends on having an *explicit* formula $Z = \sum z_i U_i$ for a nonvanishing normal vector field Z on M. The main case is a surface given in the form $M : g = c$, for there we know from Chapter IV, Section 3, that the gradient

$$\nabla g = \sum \frac{\partial g}{\partial x_i} U_i$$

is such a vector field—thus we may use any convenient scalar multiple of ∇g as Z. Let S be the shape operator derived from the unit normal

$$U = Z/\| Z \|.$$

If V is a tangent vector field on M, then by Method 2 in Section 1, we find

$$\nabla_V Z = \sum V[z_i] U_i.$$

Using a Leibnizian property of such derivatives,

$$\nabla_V U = \nabla_V \frac{Z}{\| Z \|} = \frac{(\nabla_V Z)}{\| Z \|} + V\left[\frac{1}{\| Z \|}\right] Z$$

(Fig. 5.28). What is important here is that $V[1/\| Z \|]Z$ is a *normal* vector field; we do not care which one it is, so we denote it merely by $-N_V$. Thus

$$S(V) = -\nabla_V U = \frac{-(\nabla_V Z)}{\| Z \|} + N_V.$$

Note that if W is another tangent vector field on M, then $N_V \times N_W = 0$, while products such as $N_V \times Y$ are tangent to M for *any* Euclidean vector

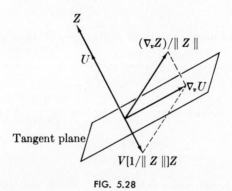

FIG. 5.28

field Y on M. Thus it is a routine matter to deduce the following lemma from Lemma 3.4.

4.4 Lemma   Let Z be a nonvanishing normal vector field on M. If V and W are tangent-vector fields such that $V \times W = Z$, then

$$K = \frac{(Z \cdot \nabla_V Z \times \nabla_W Z)}{\|Z\|^4}$$

$$H = -Z \cdot \frac{(\nabla_V Z \times W + V \times \nabla_W Z)}{2\|Z\|^3}.$$

To compute, say, the Gaussian curvature of a surface $M\colon g = c$ using patches, one must begin by explicitly finding enough of them to cover all of M; a complete computation of K may thus be tedious, even when g is a rather simple function. The following example shows to advantage the approach just described.

4.5 Example   Curvature of the ellipsoid

$$M\colon g = \frac{x^2}{a^2} + \frac{y^2}{b^2} + \frac{z^2}{c^2} = 1.$$

We write $g = \sum x_i^2/a_i^2$, and use the (nonvanishing) normal vector field

$$Z = \tfrac{1}{2} \nabla g = \sum \frac{x_i}{a_i^2} U_i.$$

Now if $V = \sum v_i U_i$ is a tangent vector field on M,

$$\nabla_V Z = \sum \frac{V[x_i]}{a_i^2} U_i = \sum \frac{v_i}{a_i^2} U_i$$

since

$$V[x_i] = dx_i(V) = v_i.$$

Similar results for another tangent vector field W then yield

$$Z \cdot \nabla_V Z \times \nabla_W Z = \begin{vmatrix} \dfrac{x_1}{a_1{}^2} & \dfrac{x_2}{a_2{}^2} & \dfrac{x_3}{a_3{}^2} \\[6pt] \dfrac{v_1}{a_1{}^2} & \dfrac{v_2}{a_2{}^2} & \dfrac{v_3}{a_3{}^2} \\[6pt] \dfrac{w_1}{a_1{}^2} & \dfrac{w_2}{a_2{}^2} & \dfrac{w_3}{a_3{}^2} \end{vmatrix} = \dfrac{1}{a_1{}^2 a_2{}^2 a_3{}^2} X \cdot V \times W.$$

where X is the special vector field $\sum x_i U_i$ which was used in Example 3.9 in Chapter IV.

It is always possible to choose V and W so that $V \times W = Z$. But then

$$X \cdot V \times W = X \cdot Z = \sum \frac{x_i{}^2}{a_i{}^2} = 1.$$

Thus by Lemma 4.4 we have found

$$K = \frac{1}{a_1{}^2 a_2{}^2 a_3{}^2 \parallel Z \parallel^4} \qquad \text{where} \qquad \parallel Z \parallel^4 = \left(\sum \frac{x_i{}^2}{a_i{}^4} \right)^2.$$

For *any* oriented surface in $\mathbf{E}^3$, its *support function* h assigns to each point $\mathbf{p}$ the orthogonal distance $h(\mathbf{p}) = \mathbf{p} \cdot U(\mathbf{p})$ from the origin to the tangent plane $T_p(M)$, as shown in Fig. 5.29 for the ellipsoid. Using the vector field X (whose value at $\mathbf{p}$ is the tangent vector $\mathbf{p}_p$), we find for the ellipsoid that

$$h = X \cdot U = X \cdot \frac{Z}{\parallel Z \parallel} = \frac{1}{\parallel Z \parallel}.$$

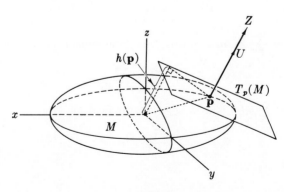

FIG. 5.29

Thus a more intuitive expression for the Gaussian curvature of the ellipsoid is

$$K = \frac{h^4}{a^2b^2c^2}.$$

Note that when $a = b = c = r$, the ellipsoid is a sphere and this formula becomes $K = 1/r^2$.

The computational results in this section, though stated for surfaces, still apply to immersed surfaces (Exercise 10 of Chapter IV, Section 8). In particular, the formulas in Corollary 4.1 make sense for an *arbitrary* regular mapping $\mathbf{x} \colon D \to \mathbf{E}^3$. The theoretical justification of this added generality is outlined in Chapter VII, Section 7.

EXERCISES

1. Show that the sphere of radius r has $K = 1/r^2$ by applying the methods of this section to the geographical patch

$$\mathbf{x}(u, v) = (r \cos v \cos u, \, r \cos v \sin u, \, r \sin v).$$

2. For a Monge patch, $\mathbf{x}(u, v) = (u, v, f(u, v))$, show that

$$E = 1 + f_u^{\,2} \qquad \ell = f_{uu}/W$$

$$F = f_u f_v \qquad\quad m = f_{uv}/W$$

$$G = 1 + f_v^{\,2} \qquad n = f_{vv}/W$$

where

$$W = (1 + f_u^{\,2} + f_v^{\,2})^{1/2}.$$

Find formulas for K and H.

3. (*Continuation*). Deduce that the image of $\mathbf{x}$ is flat if and only if

$$f_{uu}f_{vv} - f_{uv}^2 = 0;$$

minimal if and only if

$$(1 + f_u^{\,2})f_{vv} + (1 + f_v^{\,2})\,f_{uu} - 2f_u f_v f_{uv} = 0.$$

4. Show that the image of the patch

$$\mathbf{x}(u, v) = (u, v, \log \cos v - \log \cos u)$$

is a minimal surface with Gaussian curvature

$$K = \frac{-\sec^2 u \, \sec^2 v}{W^4},$$

where

$$W^2 = 1 + \tan^2 u + \tan^2 v.$$

5. Express the curvature K of the monkey saddle $M: z = x^3 - 3xy^2$ (Fig. 4.47) in terms of $r = \sqrt{x^2 + y^2}$. Is this surface minimal?

6. Find the Gaussian curvature of the elliptic and hyperbolic paraboloid,

$$M: z = \frac{x^2}{a^2} + \varepsilon \frac{y^2}{b^2}, \qquad (\varepsilon = \pm 1).$$

7. Show that the curve segment

$$\alpha(t) = \mathbf{x}(a_1(t), a_2(t)), \qquad a \leq t \leq b$$

has length

$$L(\alpha) = \int_a^b (E a_1'^2 + 2F a_1' a_2' + G a_2'^2)^{1/2} \, dt,$$

where E, F, and G are evaluated on a_1, a_2.

8. Prove that the coordinate angle ϑ of a patch $\mathbf{x}: D \to M$, $0 < \vartheta < \pi$, is a *differentiable* function on D. (*Hint:* Use the Schwarz inequality in II.1.)

9. (a) A patch $\mathbf{x}$ in M is *orthogonal* provided $F = 0$ (so $\mathbf{x}_u$ and $\mathbf{x}_v$ are orthogonal at each point). Show that in this case

$$S(\mathbf{x}_u) = \frac{\ell}{E} \mathbf{x}_u + \frac{m}{G} \mathbf{x}_v$$

$$S(\mathbf{x}_v) = \frac{m}{E} \mathbf{x}_u + \frac{n}{G} \mathbf{x}_v.$$

   (b) A patch $\mathbf{x}$ is *principal* provided $F = m = 0$. Prove that $\mathbf{x}_u$ and $\mathbf{x}_v$ are principal vectors at each point, with corresponding principal curvatures ℓ/E and n/G.

10. Prove that a tangent vector

$$\mathbf{v} = v_1 \mathbf{x}_u + v_2 \mathbf{x}_v$$

is a principal vector if and only if

$$\begin{vmatrix} v_2^2 & -v_1 v_2 & v_1^2 \\ E & F & G \\ \ell & m & n \end{vmatrix} = 0.$$

(*Hint:* $\mathbf{v}$ is principal if and only if the normal vector $S(\mathbf{v}) \times \mathbf{v}$ is zero.)

11. Show that on the saddle surface M (4.3) the two vector fields

$$(\sqrt{1 + u^2}, \ \pm \sqrt{1 + v^2}, \ v\sqrt{1 + u^2} \pm u\sqrt{1 + v^2})$$

are principal at each point. Check that they are orthogonal and tangent to M.

12. (*Enneper's minimal surface*). This is the immersed surface given by

$$\mathbf{x}(u, v) = \left(u - \frac{u^3}{3} + uv^2, \ v - \frac{v^3}{3} + u^2 v, \ u^2 - v^2 \right).$$

Prove that this immersed surface is minimal and that $\mathbf{x}$ is not one-to-one. (*Hint:* For $H = 0$ it suffices to prove

$$E = G, F = 0, \text{ and } \mathbf{x}_{uu} + \mathbf{x}_{vv} = 0.)$$

13. (*Patch criterion for umbilics*). Show that the point $\mathbf{x}(u, v)$ is umbilic if and only if there is a number k such that $\ell = kE$, $m = kF$, and $n = kG$ at (u, v) (k is then the principal curvature $k_1 = k_2$).

14. If $\mathbf{v} = v_1 \mathbf{x}_u + v_2 \mathbf{x}_v$ is tangent to M at $\mathbf{x}(u, v)$, the normal curvature in the direction determined by $\mathbf{v}$ is

$$k(\mathbf{v}) = \frac{\ell v_1^2 + 2m v_1 v_2 + n v_2^2}{E v_1^2 + 2F v_1 v_2 + G v_2^2}$$

where the various functions are evaluated at (u, v).

15. Find the umbilic points (if any) on the following surfaces:
    (a) Saddle (Example 4.3).
    (b) Monkey saddle (Exercise 5).
    (c) Elliptic paraboloid (Exercise 6).

16. (*Tubes*). If β is a unit-speed curve in $\mathbf{E}^3$ with $\kappa > 0$, let

$$\mathbf{x}(u, v) = \beta(u) + \varepsilon (\cos v \, N(u) + \sin v \, B(u)).$$

Thus the v-parameter curves are circles of (constant) radius ϵ in planes orthogonal to β. Show that
    (a) $\mathbf{x}$ is regular if ε is small enough; so $\mathbf{x}$ is an immersed surface called the *tube* of radius ϵ around β.
    (b) $U = \cos v \, N(u) + \sin v \, B(u)$ is a unit normal function on the tube.

(c) $K = \dfrac{-\kappa(u) \cos v}{\varepsilon(1 - \kappa(u) \, \varepsilon \cos v)}$

(*Hint:* Use $S(\mathbf{x}_u) \times S(\mathbf{x}_v) = K \mathbf{x}_u \times \mathbf{x}_v$.)

17. Show that the elliptic hyperboloids of one and two sheets (Ex. 10 of IV.2) have Gaussian curvature $K = -h^4/a^2 b^2 c^2$ and $K = h^4/a^2 b^2 c^2$, re-

spectively, and that both support functions h are given by the same formula as for the ellipsoid (4.5).

18. If h is the support function of an oriented surface $M \subset \mathbf{E}^3$, show that
    (a) A point $\mathbf{p}$ of M is a critical point of h if and only if $\mathbf{p} \cdot S(\mathbf{v}) = 0$ for all tangent vectors $\mathbf{v}$ to M at $\mathbf{p}$. (*Hint:* Write h as $X \cdot U$, where $X = \sum x_i U_i$.)
    (b) When $K(\mathbf{p}) \neq 0$, $\mathbf{p}$ is a critical point of h if and only if $\mathbf{p}$ (considered as a vector) is orthogonal to M at $\mathbf{p}$.

19. Use the preceding exercises to find the Gaussian curvature intervals of the ellipsoid and the elliptic hyperboloids of one and two sheets. (Ex. 10 of IV. 2) Assume $a \geqq b \geqq c$.

20. Compute K and H for the saddle surface (Example 4.3) by the method given at the end of this section. (*Hint:* Take V and W tangent to the two sets of rulings of M.)

21. *Scherk's minimal surface*, $M: e^z \cos x = \cos y$. Let $\Re$ be the region in the xy plane on which $\cos x \cos y > 0$; $\Re$ is a checkerboard pattern of open squares, with *vertices* $((\pi/2) + \pi m, (\pi/2) + n\pi)$. Show that
    (a) M is a surface.
    (b) For each point (u,v) in $\Re$ there is exactly one point (u, v, w) in M. The only other points of M are entire vertical lines over each of the vertices of $\Re$ (Fig. 5.30).
    (c) M is a minimal surface with $K = -e^{2z}/(e^{2z} \sin^2 x + 1)^2$. (*Hint:*

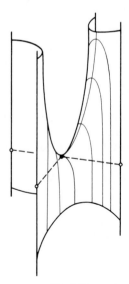

FIG. 5.30

$V = \cos x U_1 + \sin x U_3$ is a tangent vector field.) Further properties of this surface are given in Ex. 11 of VI.8.

22. Let Z be a never-zero normal vector field on M. Show that a tangent vector $\mathbf{v}$ to M at $\mathbf{p}$ is principal if and only if

$$\mathbf{v} \cdot Z(\mathbf{p}) \times \nabla_v Z = 0$$

(see the hint for Exercise 10).

The equation above together with the tangency condition

$$Z(\mathbf{p}) \cdot \mathbf{v} = 0$$

may be solved for the principal directions. Thus umbilics may be located using these equations: $\mathbf{p}$ is umbilic if and only if every tangent vector $\mathbf{v}$ at $\mathbf{p}$ is principal.

23. Consider the ellipsoid $M: \sum x_i^2/a_i^2 = 1$. Show that
(a) A tangent vector $\mathbf{v}$ at the point $\mathbf{p}$ is principal if and only if

$$0 = p_1 v_2 v_3 (a_2^2 - a_3^2) + p_2 v_1 v_3 (a_3^2 - a_1^2) + p_3 v_1 v_2 (a_1^2 - a_2^2).$$

(b) Assuming $a_1 > a_2 > a_3$, show that there are exactly four umbilics on M, with coordinates

$$p_1 = \pm a_1 \left(\frac{a_1^2 - a_2^2}{a_1^2 - a_3^2}\right)^{1/2} \qquad p_2 = 0 \qquad p_3 = \pm a_3 \left(\frac{a_2^2 - a_3^2}{a_1^2 - a_3^2}\right)^{1/2}.$$

5 Special Curves in a Surface

We shall briefly consider three geometrically significant types of curves in a surface $M \subset \mathbf{E}^3$. Neither this section nor the next is really essential for the theory; their purpose is to illustrate some of the ideas already introduced, and supply examples for later work.

5.1 Definition   A regular curve α in $M \subset \mathbf{E}^3$ is a *principal curve* (or *line of curvature*) provided that the velocity α' of α always points in a principal direction.

Thus principal curves always travel in directions in which the bending of M in $\mathbf{E}^3$ takes its extreme values. Neglecting changes of parametrization, there are exactly two principal curves through each nonumbilic point of M—and these necessarily cut orthogonally across each other. (At an umbilic point $\mathbf{p}$, every direction is principal and near $\mathbf{p}$ the pattern of the principal curves can become quite complicated.)

5.2 Lemma   Let α be a regular curve in $M \subset \mathbf{E}^3$, and let U be a unit normal vector field restricted to α. Then

(1) The curve α is principal if and only if U' and α' are collinear at each point.

(2) If α is a principal curve, then the principal curvature of M in the direction of α' is $\alpha'' \cdot U / \alpha' \cdot \alpha'$.

Proof. (1) Exercise 1 from Section 1 shows that $S(\alpha') = -U'$. Thus U' and α' are collinear if and only if $S(\alpha')$ and α' are collinear. But by Theorem 2.5, this amounts to saying that α' always points in a principal direction, or, equivalently, that α is a principal curve.

(2) Since α is a principal curve, the vector field $\alpha'/\|\alpha'\|$ consists entirely of (unit) principal vectors belong to, say, the principal curvature k_i. Thus

$$k_i = k(\alpha'/\|\alpha'\|) = S(\alpha'/\|\alpha'\|) \cdot \alpha'/\|\alpha'\|$$

$$= \frac{S(\alpha') \cdot \alpha'}{\alpha' \cdot \alpha'} = \frac{\alpha'' \cdot U}{\alpha' \cdot \alpha'}$$

where the last equality uses Lemma 2.1.   ∎

In this lemma, (1) is a simple criterion for a curve to be principal, while (2) gives the principal curvature along a curve known to be principal.

5.3 Lemma Let α be a curve cut from a surface $M \subset \mathbf{E}^3$ by a plane P. If the angle between M and P is constant along α, then α is a principal curve of M.

Proof. Let U and V be unit normal vector fields to M and P (respectively) along the curve α, as shown in Fig. 5.31. Since P is a plane, V is parallel, that is, $V' = 0$. Now the constant angle assumption means that $U \cdot V$ is constant; thus

$$0 = (U \cdot V)' = U' \cdot V.$$

Since U is a unit vector, U' is orthogonal to U as well as V. The same is of

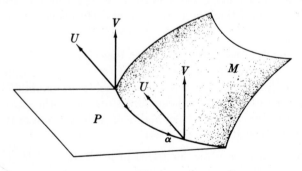

FIG. 5.31

course true of α', since α lies in both M and P. If U and V are linearly independent (as in Fig. 5.31) we conclude that U' and α' are collinear; hence by Lemma 5.2, α is principal.

However, linear independence fails only when $U = \pm V$. But then $U' = 0$, so α is (trivially) principal in this case as well. ∎

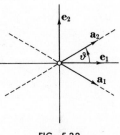

FIG. 5.32

(It is scarcely any harder to prove the generalization given in Exercise 5.) Using this result it is easy to see that *the meridians and parallels of a surface of revolution M are its principal curves.* Indeed, each meridian μ is sliced from M by a plane *through* the axis of revolution and hence orthogonal to M along μ, while each parallel π is sliced from M by a plane *orthogonal* to the axis, and by rotational symmetry such a plane makes a constant angle with M along π.

Directions tangent to $M \subset \mathbf{E}^3$ in which the normal curvature is zero are called *asymptotic directions.* Thus a tangent vector $\mathbf{v}$ is *asymptotic* provided $k(\mathbf{v}) = S(\mathbf{v}) \cdot \mathbf{v} = 0$, so in an asymptotic direction M is (instantaneously, at least) not bending away from its tangent plane.

Using Corollary 2.6 we can get a complete analysis of asymptotic directions in terms of Gaussian curvature.

5.4 Lemma   Let $\mathbf{p}$ be a point of $M \subset \mathbf{E}^3$.

(1) If $K(\mathbf{p}) > 0$, there are no asymptotic directions at $\mathbf{p}$.

(2) If $K(\mathbf{p}) < 0$, then there are exactly two asymptotic directions at $\mathbf{p}$ which are bisected by the principal directions (Fig. 5.32) at angle ϑ such that

$$\tan^2 \vartheta = \frac{-k_1(\mathbf{p})}{k_2(\mathbf{p})}.$$

(3) If $K(\mathbf{p}) = 0$, then *every* direction is asymptotic if $\mathbf{p}$ is a planar point; otherwise there is exactly one asymptotic direction which is also principal.

Proof. These cases all derive from Euler's formula

$$k(\mathbf{u}) = k_1(\mathbf{p}) \cos^2\vartheta + k_2(\mathbf{p}) \sin^2\vartheta$$

in Corollary 2.6.

(1) Since $k_1(\mathbf{p})$ and $k_2(\mathbf{p})$ have the same sign, $k(\mathbf{u})$ is never zero.

(2) Here $k_1(\mathbf{p})$ and $k_2(\mathbf{p})$ have opposite signs and we can solve the equation $0 = k_1(\mathbf{p}) \cos^2\vartheta + k_2(\mathbf{p}) \sin^2\vartheta$ to obtain the two asymptotic directions.

(3) If $\mathbf{p}$ is planar, then

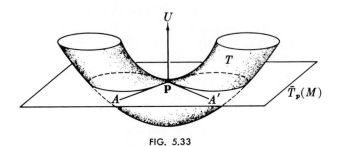

FIG. 5.33

$$k_1(\mathbf{p}) = k_2(\mathbf{p}) = 0;$$

hence $k(\mathbf{u})$ is identically zero. If just

$$k_2(\mathbf{p}) = 0,$$

then

$$k(\mathbf{u}) = k_1(\mathbf{p}) \cos^2\vartheta$$

is zero only when $\cos \vartheta = 0$, that is, in the principal direction $\mathbf{u} = \mathbf{e}_2$. ∎

We can get an approximate idea of the asymptotic directions at a point
$\mathbf{p}$ of a given surface M by picturing the intersection of the tangent plane
$\bar{T}_p(M)$ with M near $\mathbf{p}$. When $K(\mathbf{p})$ is negative, this intersection will consist
of two curves through $\mathbf{p}$ whose tangent lines (at $\mathbf{p}$) are asymptotic directions (Exercise 5 of Section 3).

Figure 5.33 shows the two asymptotic directions A and A' at a point $\mathbf{p}$
on the inner equator of a torus. (The two intersection curves merge into a
single figure-8.)

5.5 Definition   A regular curve α in $M \subset E^3$ is an *asymptotic curve* provided its velocity α' always points in an asymptotic direction.

Thus α is asymptotic if and only if

$$k(\alpha') = S(\alpha') \cdot \alpha' = 0.$$

Since $S(\alpha') = -U'$, this gives a criterion, $U' \cdot \alpha' = 0$, for α to be asymptotic. Asymptotic curves are more sensitive to Gaussian curvature than are
principal curves: Lemma 5.3 shows that there are none in regions where K
is positive, but two cross (at an angle depending on K) at each point of a
region where K is negative.

The simplest criterion for a curve in M to be asymptotic is that *its acceleration α'' always be tangent to M.* In fact, differentiation of $U \cdot \alpha' = 0$
gives

$$U' \cdot \alpha' + U \cdot \alpha'' = 0,$$

so $U' \cdot \alpha' = 0$ (α asymptotic) if and only if $U \cdot \alpha'' = 0$.

The analysis of asymptotic directions in Lemma 5.4 has consequences for both flat and minimal surface. First, *a surface M in $\mathbf{E}^3$ is minimal if and only if there exist two orthogonal asymptotic directions at each of its points.* In fact, $H(\mathbf{p}) = 0$ is equivalent to $k_1(\mathbf{p}) = -k_2(\mathbf{p})$, and an examination of the possibilities in Lemma 5.4 shows that $k_1(\mathbf{p}) = -k_2(\mathbf{p})$ if and only if either (a) $\mathbf{p}$ is planar (so the criterion holds trivially) or (b)

$$K(\mathbf{p}) < 0 \qquad \text{with } \vartheta = \pm \pi/4,$$

which means that the two asymptotic directions are orthogonal.

Thus a surface is minimal if and only if through each point there are two asymptotic curves which cross *orthogonally*. This observation gives geometric meaning to the calculations in Example 4.3 which show that the helicoid is a minimal surface. In fact, the u- and v-parameter curves of the patch $\mathbf{x}$ are orthogonal since $F = 0$, and their accelerations are tangent to the surface since $\ell = U \cdot \mathbf{x}_{uu} = 0$ and $n = U \cdot \mathbf{x}_{vv} = 0$.

Roughly speaking, a *ruled surface M* is one which is swept out by a straight line moving through $\mathbf{E}^3$—the various positions of the line are called the *rulings* of M. Thus M has a parametrization in the *ruled form*

$$\mathbf{x}(u, v) = \beta(u) + v\delta(u), \qquad \text{or} \qquad \beta(v) + u\delta(v)$$

where β and δ are curves in $\mathbf{E}^3$ with δ never 0 (see Exercise 4–9 of Section 2 of Chapter IV). For example, the helicoid is a ruled surface, since the patch in Example 4.3 may be written as

$$\mathbf{x}(u, v) = (0, 0, bv) + u(\cos v, \sin v, 0).$$

This shows how the helicoid is swept out by a line rotating as it rises along the z axis. The saddle surface $M: z = xy$ is *doubly ruled*, since

$$\mathbf{x}(u, v) = (u, v, uv) = \begin{cases} (u, 0, 0) + v(0, 1, u) \\ (0, v, 0) + u(1, 0, v). \end{cases}$$

It is no accident that both these surfaces have K negative, for:

5.6 Lemma   A ruled surface M has Gaussian curvature $K \leqq 0$. Furthermore $K = 0$ if and only if the unit normal U is parallel along each ruling of M (so all points $\mathbf{p}$ on a ruling have the same tangent plane $\bar{T}_p(M)$.)

Proof. A straight line $t \rightarrow \mathbf{p} + t\mathbf{q}$ in any surface is certainly *asymptotic*, since its acceleration is zero, and thus trivially tangent to M. By definition, a ruled surface contains a straight line through each of its points. Hence there is an asymptotic direction at each point, so by Lemma 5.4, $K \leqq 0$.

Now let $\alpha(t) = \mathbf{p} + t\mathbf{q}$ be an arbitrary ruling in M. If U is parallel along α, then $S(\alpha') = U' = 0$. Thus α is a principal curve with principal curvature $k(\alpha') = 0$, and so $K = k_1k_2 = 0$.

Conversely, if $K = 0$, we deduce from the case (3) in Lemma 5.4 that asymptotic directions (and curves) in M are also *principal*. Thus each ruling α is principal $(S(\alpha') = k(\alpha')\alpha')$ as well as asymptotic $(k(\alpha') = 0)$: hence $U' = -S(\alpha') = 0$. ∎

We come now to the last and most important of the three types of curves under discussion.

5.7 Definition   A curve α in $M \subset \mathbf{E}^3$ is a *geodesic* of M provided its acceleration α'' is always normal to M.

Since α'' is normal to M, the inhabitants of M perceive no acceleration at all—for them the geodesic α is a "straight line." A full study of geodesics is given in Chapter VII, where in particular we examine their character as shortest routes of travel. Geodesics are far more plentiful on a surface M than are principal or asymptotic curves; indeed by Theorem 4.2 of Chapter VII there is a geodesic through every point of M in every direction.

Since its acceleration α'' is, in particular, orthogonal to its velocity α', a geodesic α has *constant speed*, for differentiation of

$$\| \alpha' \|^2 = \alpha' \cdot \alpha' \text{ yields } 2\alpha' \cdot \alpha'' = 0.$$

A straight line $\alpha(t) = \mathbf{p} + t\mathbf{q}$ in M is always a geodesic of M, since its acceleration $\alpha'' = 0$ is trivially normal to M. Unlike principal and asymptotic curves, geodesics cannot be defined in terms of the shape operator; however, a (unit speed) geodesic α with positive curvature bears an interesting relation to S, which uses the Frenet apparatus of α. Since the principal normal $N = \alpha''/\kappa$ of α is normal to the surface M, we have

$$-N' = S(\alpha') = S(T).$$

Thus by a Frenet equation, $S(T) = \kappa T - \tau B$.

These remarks are sufficient to derive the geodesics of three rather special surfaces.

5.8 Example   *Geodesics on some surfaces in* $\mathbf{E}^3$.

(1) *Planes.* If α is a geodesic in a plane P orthogonal to $\mathbf{u}$, $\alpha' \cdot \mathbf{u} = 0$, hence $\alpha'' \cdot \mathbf{u} = 0$. But α'' is by definition normal to P; hence $\alpha'' = 0$. Thus α is a straight line, and since every such line is a geodesic, we conclude that *the geodesics of P are all straight lines in P.*

(2) *Spheres.* If α is a (unit speed) geodesic in a sphere Σ of radius r, then, by a remark above, $S(T) = \kappa T - \tau B$. (We saw in Chapter II, Section 3, that any curve in Σ has positive curvature, so the Frenet apparatus

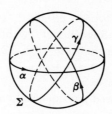

FIG. 5.34

is available.) But Example 1.3 shows that $S(T) = \pm(1/r)T$, depending on which of the two unit normals is used. These two equations for $S(T)$ imply that $\kappa = 1/r$ and $\tau = 0$. Hence by Lemma 3.6 of Chapter II, α lies on a circle C of radius r. This maximum radius r forces C to be a *great circle* of Σ, that is, one sliced from Σ by a plane through its center. Conversely, any constant-speed curve running along a great circle has its acceleration α'' point toward the center of that circle, which is also the center of Σ, so α'' is normal to Σ. We conclude that *the geodesics of Σ are the constant-speed parametrizations of its great circles* (Fig. 5.34).

(3) **Cylinders.** The geodesics of, say, the circular cylinder $M: x^2 + y^2 = r^2$ are all curves of the form

$$\alpha(t) = (r \cos (at + b), r \sin (at + b), ct + d).$$

In fact, any curve in M may be written

$$\alpha(t) = (r \cos \vartheta(t), r \sin \vartheta(t), h(t)),$$

and a vector normal to M has z-coordinate zero. Thus if α is a geodesic, then $h'' = 0$, so $h(t) = ct + d$. The speed $(r^2\vartheta'^2 + h'^2)^{1/2}$ of α is constant, so ϑ' is constant; hence $\vartheta(t) = at + b$.

When the constants a and c are both nonzero, α is a *helix* on M. The extreme case $a = 0$ gives the rulings of M, and $c = 0$ gives the cross-sectional circles.

The essential properties of the three types of curves we have considered may be summarized as follows:

Principal curves	$k(\alpha') = k_1$ or k_2	$S(\alpha')$ collinear α'	
Asymptotic curves	$k(\alpha') = 0$	$S(\alpha')$ orthogonal α'	α'' tangent M
Geodesics			α'' normal M

EXERCISES

1. Prove that a curve α in M is a straight line of $\mathbf{E}^3$ if and only if α is both geodesic and asymptotic.

2. To which of the three types—principal, asymptotic, geodesic—do the following curves belong?
(a) The top circle α on the torus (Fig. 5.35).
(b) The outer equator β of the torus.
(c) The x axis on $M: z = xy$.
(Assume a constant-speed parametrization.)

FIG. 5.35

3. On a surface of revolution, show that all meridians are geodesics, but
that the parallel through a point $\alpha(t)$ of the profile curve is a geodesic
if and only if $\alpha'(t)$ is parallel to the axis of revolution.

4. Let α be an asymptotic curve in $M \subset \mathbf{E}^3$ with curvature $\kappa > 0$.
(a) Prove that the binormal B of α is normal to the surface along α,
and deduce that $S(T) = \tau N$.
(b) Show that along α the surface has Gaussian curvature $K = -\tau^2$.
(c) Use (b) to find the Gaussian curvature of the helicoid (Example
4.3).

5. Suppose that a curve α lies in two surfaces M and $\bar{M}$ which make a
constant angle along α ($U \cdot \bar{U}$ constant). Show that α is principal in M
if and only if principal in $\bar{M}$.

6. If $\mathbf{x}$ is a patch in M, prove that a curve $\alpha(t) = \mathbf{x}(a_1(t), a_2(t))$ is
(a) Principal if and only if

$$\begin{vmatrix} a_2'^2 & -a_1'a_2' & a_1'^2 \\ E & F & G \\ \ell & m & n \end{vmatrix} = 0,$$

(b) Asymptotic if and only if $\ell a_1'^2 + 2m a_1'a_2' + n a_2'^2 = 0$.

7. Let α be a unit-speed curve in $M \subset \mathbf{E}^3$. Instead of the Frenet frame
field on α, consider the frame field T, V, U—where T is the unit tangent
of α, U is the surface normal restricted to α, and $V = U \times T$ (Fig.
5.36).
(a) Show that

$$T' = \quad\quad gV + kU$$
$$V' = -gT \quad\quad + tU$$
$$U' = -kT - tV.$$

where $k = S(T) \cdot T$ is the normal curvature $k(T)$ of M in the
T direction, and $t = S(T) \cdot V$.

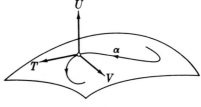

FIG. 5.36

The new function g is called the *geodesic curvature* of α.

(b) Deduce that α is

$$\text{geodesic} \Leftrightarrow g = 0$$

$$\text{asymptotic} \Leftrightarrow k = 0$$

$$\text{principal} \Leftrightarrow t = 0.$$

8. If α is a (unit speed) curve in M, show that
   (a) α is both principal and geodesic if and only if it lies in a plane everywhere orthogonal to M along α.
   (b) α is both principal and asymptotic if and only if it lies in a plane everywhere tangent to M along α.

9. On the monkey saddle M (Ex. 5 of IV.4) find *three* asymptotic curves and *three* principal curves passing through the origin **0**. (This is possible because only because **0** is a planar umbilic point.)

10. Show that the ruled surface $\mathbf{x}(u, v) = \beta(u) + v\delta(u)$ has Gaussian curvature

$$K = \frac{-m^2}{EG - F^2} = \frac{-(\beta' \cdot \delta \times \delta')^2}{W^4}$$

where

$$W = \| \beta' \times \delta + v\delta' \times \delta \|.$$

11. (*Flat ruled surfaces*).
    (a) Show that cones and cylinders are flat (see Exs. 5 and 6 of IV.2)
    (b) If β is a unit-speed curve in $\mathbf{E}^3$ with $\kappa > 0$, the ruled surface

$$\mathbf{x}(u,v) = \beta(u) + vT(u), \qquad v > 0,$$

is called the *tangent surface* of β (Fig. 5.37). Prove that $\mathbf{x}$ is regular, and that the tangent surface is flat.

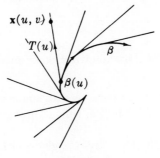

FIG. 5.37

12. Let α be a regular curve in $M \subset \mathbf{E}^3$, and let U be the unit normal of M along α. Show that α is a principal curve of M if and only if the ruled surface $\mathbf{x}(u, v) = \alpha(u) + vU(u)$ is flat.

13. A *closed geodesic* of M is a geodesic which is a periodic function $\alpha \colon \mathbf{R} \to M$. Find all closed geodesics in a sphere, a plane, and a circular cylinder.

14. A ruled surface is *noncylindrical* if its rulings are always changing directions; thus for any director curve, $\delta \times \delta' \neq 0$. Show that
    (a) a noncylindrical ruled surface has a parametrization

$$\mathbf{x}(u, v) = \sigma(u) + v\delta(u)$$

    for which $\| \delta \| = 1$ and $\sigma' \cdot \delta' = 0$.
    (b) for this parametrization,

$$K = \frac{-p^2(u)}{(p^2(u) + v^2)^2} \qquad \text{where } p = \frac{\sigma' \cdot \delta \times \delta'}{\delta' \cdot \delta'}.$$

    The curve σ is called the *striction curve*, and the function p is the *distribution parameter*.
    (*Hint:* For (a), if $\| \delta \| = 1$, find f so that $\sigma = \alpha + f\delta$. For (b), show $\sigma' \times \delta = p\delta'$.)

15. Describe the qualitative behavior of Gaussian curvature K on an arbitrary ruling of a (noncylindrical) ruled surface. Show that the route of the striction curve is independent of the choice of parametrization, and that the distribution parameter is essentially a function on the set of rulings.

16. Show that the striction curve of the helicoid is its central axis, and that its distribution function is constant.

17. Find the striction curve and distribution parameter for
    (a) Both sets of ruling of the saddle surface (Example 4.3)
    (b) Both sets of rulings of the hyperboloid of revolution

$$M \colon \frac{x^2 + y^2}{a^2} - \frac{z^2}{b^2} = 1$$

(Fig. 5.38) (Find a ruled parametrizetion by modifying Ex. 9 of IV.2.)

18. If $\mathbf{x}(u,v) = \alpha(u) + v\delta(u)$ parametrizes a noncylindrical ruled surface, let $L(u)$ be the ruling through $\alpha(u)$. Show that
    (a) If ϑ_ϵ is the (smallest positive) angle from $L(u)$ to $L(u + \epsilon)$, and

FIG. 5.38

d_ε is the orthogonal distance from $L(u)$ to $L(u + \varepsilon)$, then $\lim\limits_{\varepsilon \to 0} (d_\varepsilon/$
$\vartheta_\varepsilon) = \mathbf{p}(u)$. Thus the distribution parameter is the reciprocal of
the rate of turning of L—its sign describes the direction of the
turning.)

(b) There is a unique point p_ε of $L(u)$ which is nearest to $L(u + \varepsilon)$,
and $\lim\limits_{\varepsilon \to 0} p_\varepsilon = \sigma(u)$. (This gives another characterization of the
striction curve.)

(c) The distance from $\sigma(u)$ to $\sigma(u + \varepsilon)$ need not be a good approximation of the distance d_ε from $L(u)$ to $L(u + \varepsilon)$. Give an example.

19. Let $\mathbf{x}(u,v) = \alpha + v\delta$, $\| \delta \| = 1$, parametrize a *flat* ruled surface M.
Show that

(a) If α' is always zero, then M is a cone.†
(b) If δ' is always zero, then M is a cylinder.
(c) If both α' and δ' are *never* zero, then M is the tangent surface of
its striction curve. (Exercise 11b.)

Of course these cases are far from exhausting all the possibilities,
but in a sense they show that an arbitrary flat ruled surface ("de-
velopable surface") is a mixture of the three types in Exercise 11. If
such a surface is closed in $\mathbf{E}^3$ (Ex. 10 of IV.8), it must be a cylinder,
since the closure condition implies that the rulings are *entire* straight
lines.

20. A *right conoid* is a ruled surface whose rulings all pass through a fixed
axis (Fig. 5.39). Taking this axis as the z axis of $\mathbf{E}^3$, we get the parametrization

$$\mathbf{x}(u,v) = (u \cos \vartheta (v), \, u \sin \vartheta (v), \, h(v))$$

(a) Find the Gaussian and mean curvature.
(b) Show that the surface is noncylindrical if
ϑ' is never zero; find its striction curve
and parameter of distribution.

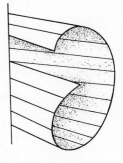

21. Sketch the conoid M parametrized by

$$\mathbf{x}(u, v) = (u \cos v, \, u \sin v, \, \cos 2\,v),$$

and find its Gaussian and mean curvature.
Express M in the form $z = f(x, y)$ (origin
omitted).

22. Prove that a surface which is both *ruled* and   **FIG. 5.39**
minimal is part of either a plane or a helicoid.

† In each case, we assume there are no restriction on v except those necessary
to ensure that $\mathbf{x}$ is regular.

(*Hint:* Flat regions in M are planar; thus arguing as in Theorem 6.2 we may suppose $K < 0$. Use the parametrization in Exercise 14 with the additional feature that δ is a unit-speed curve. Then $H = 0$ leads to three equations. Deduce that δ is a unit circle; we may assume $\delta(u) = (\cos u, \sin u, 0)$.)

6 Surfaces of Revolution

The geometry of a surface of revolution is rather simple, yet these surfaces exhibit a wide variety of geometric behavior; thus they offer a good field for experiment.

We shall apply the methods of Section 4 to study an arbitrary surface of revolution M, parametrized as in Example 2.5 of Chapter IV by

$$\mathbf{x}(u, v) = (g(u), h(u) \cos v, h(u) \sin v).$$

Recall that $h(u)$ is the radius of the parallel of M at distance $g(u)$ along the axis of revolution, as shown in Fig. 4.17. This geometric significance of g and h means that our results do not depend on the particular position of M relative to the coordinate axes of $\mathbf{E}^3$.

Because g and h are functions of u alone, we write

$$\mathbf{x}_u = (g', h' \cos v, h' \sin v) \qquad \begin{aligned} E &= g'^2 + h'^2 \\ F &= 0 \end{aligned}$$
$$\mathbf{x}_v = (0, -h \sin v, h \cos v) \qquad G = h^2.$$

Here E is the square of the speed of the profile curve—and of all meridians (u-parameter curves)—while G is the speed squared of the parallels (v-parameter curves). Next

$$\mathbf{x}_u \times \mathbf{x}_v = (hh', -hg' \cos v, -hg' \sin v)$$
$$\| \mathbf{x}_u \times \mathbf{x}_v \| = \sqrt{EG - F^2} = h\sqrt{g'^2 + h'^2}$$
$$U = (h', -g' \cos v, -g' \sin v)/\sqrt{g'^2 + h'^2}.$$

Taking second derivatives, we obtain

$$\mathbf{x}_{uu} = (g'', h'' \cos v, h'' \sin v)$$
$$\mathbf{x}_{uv} = (0, -h' \sin v, h' \cos v)$$
$$\mathbf{x}_{vv} = (0, -h \cos v, -h \sin v)$$
$$\ell = (-g'h'' + g''h')/\sqrt{g'^2 + h'^2}$$
$$m = 0$$
$$n = g'h/\sqrt{g'^2 + h'^2}$$

Since $F = m = 0$, it is easy to check (Exercise 9 of Section 4) that for the shape operator S derived from U,

$$S(\mathbf{x}_u) = \frac{\ell}{E}\,\mathbf{x}_u \qquad S(\mathbf{x}_v) = \frac{n}{G}\,\mathbf{x}_v.$$

(Thus we have an analytical proof that the *meridians* and *parallels* of M are its principal curves.) Hence if the corresponding principal curvature functions are denoted by k_μ and k_π (instead of k_1 and k_2), we have

$$k_\mu = \frac{\ell}{E} = \frac{-\begin{vmatrix} g' & h' \\ g'' & h'' \end{vmatrix}}{(g'^2 + h'^2)^{3/2}}, \qquad k_\pi = \frac{n}{G} = \frac{g'}{h(g'^2 + h'^2)^{1/2}}. \tag{1}$$

Thus the Gaussian curvature of M is

$$K = k_\mu k_\pi = \frac{-g'\begin{vmatrix} g' & h' \\ g'' & h'' \end{vmatrix}}{h(g'^2 + h'^2)^2}. \tag{2}$$

Note that this formula defines K as a real-valued function on the domain I of the profile curve

$$\alpha(u) = (g(u), h(u), 0).$$

By the conventions in Section 4, $K(u)$ *is the Gaussian curvature* $K(\mathbf{x}(u, v))$ *of every point on the parallel through* $\alpha(u)$. Similarly, with the other functions above—because of the rotational symmetry of M about the axis of revolution, its geometry is "constant on parallels" and completely determined by the profile curve.

In the special case when the profile curve passes at most *once* over each point of the axis, we can usually arrange for the function g to be simply $g(u) = u$ (Exercise 13 of Chapter IV, Section 2). Thus the formulas above reduce to

$$k_\mu = \frac{-h''}{(1 + h'^2)^{3/2}}$$

$$k_\pi = \frac{1}{h(1 + h'^2)^{1/2}} \tag{3}$$

$$K = \frac{-h''}{h(1 + h'^2)^2}.$$

6.1 Example Surfaces of Revolution

(1) *Torus of revolution* T. The parametrization $\mathbf{x}$ in Example 2.6 of Chapter IV has

$$g(u) = r \sin u, \qquad h(u) = R + r \cos u.$$

Although the axis of revolution is now the z axis, formulas (1) and (2) above remain valid and we compute

$$E = r^2 \qquad F = 0 \qquad G = (R + r \cos u)^2$$

$$\ell = r \qquad m = 0 \qquad n = (R + r \cos u) \cos u$$

$$k_\mu = \frac{1}{r} \qquad k_\pi = \frac{\cos u}{R + r \cos u}$$

$$K = \frac{\cos u}{r(R + r \cos u)}.$$

We now have an analytical proof that K is positive on the outer half of the torus and negative on the inner half. In fact K has its maximum value $1/r(R + r)$ on the outer equator $(u = 0)$ and its minimum value

$$-1/r(R - r)$$

on the inner equator $(u = \pi)$, and K is zero on the top and bottom circles $(u = \pm \pi/2)$.

(2) **Catenoid.** The curve $y = c \cosh (x/c)$ is a *catenary;* its shape is that of a chain hanging under the influence of gravity. The surface obtained by revolving this curve about the x axis is called a *catenoid* (Fig. 5.40). From the formulas in (3) we get

$$-k_\mu = k_\pi = \frac{1}{c \cosh^2 (u/c)}$$

Hence

$$K = \frac{-1}{c^2 \cosh^4 (u/c)}.$$

Since its mean curvature H is zero, the catenoid is a minimal surface. Its Gaussian curvature interval is $-1/c^2 \leq K < 0$, with minimum value $K = -1/c^2$ on the circle $u = 0$.

The following result shows that *catenoids are the only complete†* *nonplanar surfaces of revolution which are minimal.* (A plane is trivially a minimal surface, since $k_1 = k_2 = 0$.)

6.2 Theorem   If a surface of revolution M is a minimal surface, then M is contained in either a plane or a catenoid.

---

† We use this word with its dictionary meaning; a mathematical definition is given in Section 4 of Chapter VII.

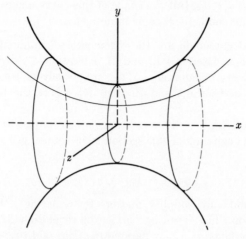

FIG. 5.40

Proof. We use the parametrization

$$\mathbf{x}(u, v) = (g(u), h(u) \cos v, h(u) \sin v)$$

of M, with u in some interval I, and v arbitrary.

Case 1. g' is identically zero. Then g is constant, so M is part of a plane orthogonal to the axis of revolution.

Case 2. g' is never zero. Then by an earlier exercise, M has a parametrization of the form

$$\mathbf{y}(u, v) = (u, f(u) \cos v, f(u) \sin v).$$

The formulas for k_1 and k_2 in (3) on page 235 then show that the minimality condition $k_1 + k_2 = 0$ is equivalent to

$$ff'' = 1 + f'^2.$$

Because u does not appear explicitly in this equation, there is a standard method for solving it, which may be found in any elementary text on differential equations. We merely record that the solution is

$$f(u) = a \cosh\left(\frac{u}{a} + b\right),$$

where a and b are arbitrary constants. Thus M is part of a catenoid.

Case 3. g' is zero at some points, nonzero at others. This cannot occur. For definiteness suppose that $g'(u_0) = 0$, but $g'(u) > 0$ for $u < u_0$. By case 2, the profile curve

$$\alpha(u) = (g(u), h(u), 0)$$

is a catenary for $u < u_0$. But the shape of this curve makes it clear that its slope h'/g' cannot suddenly become infinite at $u = u_0$. ∎

Helicoids and catenoids are the "elementary" minimal surfaces. Two others are given in Exercises 12 and 21 of Section 4. Soap-film models of an immense variety of minimal surfaces may easily be constructed by the methods given in Courant and Robbins [4], where the term *minimal* is explained.

The expression $\sqrt{g'^2 + h'^2}$ which appears so frequently in the general formulas at the beginning of this section is, of course, just the speed of the profile curve

$$\alpha(u) = (g(u), h(u), 0).$$

Thus we can radically simplify matters by replacing α by a *unit-speed* reparametrization. The resulting surface of revolution M is unchanged: We have merely given it a new parametrization, said to be *canonical*.

6.3 Lemma   If **x** is a canonical parametrization of a surface of revolution M $(g'^2 + h'^2 = 1)$, then

$$E = 1, \qquad F = 0, \qquad G = h^2$$

and

$$K = \frac{-h''}{h}.$$

Proof. Since $g'^2 + h'^2 = 1$, these expressions for E, F, and G follow immediately from those on page 234, and K in (2) becomes

$$K = \frac{-g'}{h} \begin{vmatrix} g' & h' \\ g'' & h'' \end{vmatrix} = \frac{-g'^2 h'' + g'g''h'}{h}.$$

Differentiation of $g'^2 = 1 - h'^2$ yields $g'g'' = -h'h''$; hence

$$K = \frac{-(1 - h'^2)h'' - h'^2 h''}{h} = -\frac{h''}{h}. \qquad ∎$$

The effect of canonical parametrization is to shift the emphasis somewhat from measurements in the space outside M (for example, along the axis of revolution) to measurements within M itself. This idea will be developed more fully in Chapters VI and VII.

6.4 Example   Canonical parametrization of the catenoid ($c = 1$). An arc-length function for the catenary

$$\alpha(u) = (u, \cosh u) \quad \text{is} \quad s(u) = \sinh u;$$

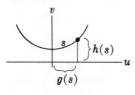

FIG. 5.41

hence a unit-speed reparametrization is

$$\beta(s) = (g(s), h(s)) = (\sinh^{-1}s, \sqrt{1 + s^2})$$

as indicated in Fig. 5.41. The resulting canonical parametrization of the catenoid is then

$$\bar{\mathbf{x}}(s, v) = (\sinh^{-1}s, \sqrt{1 + s^2} \cos v, \sqrt{1 + s^2} \sin v).$$

Thus by the preceding lemma

$$K(s) = -\frac{h''(s)}{h(s)} = \frac{-1}{(1 + s^2)^2}.$$

This formula for Gaussian curvature in terms of $\bar{\mathbf{x}}$ is consistent with the formula $K(u) = -1/\cosh^4 u$ found in Example 6.1 for the parametrization $\mathbf{x}$. In fact, since $s(u) = \sinh u$, we have

$$K(s(u)) = \frac{-1}{(1 + s^2(u))^2} = \frac{-1}{(1 + \sinh^2 u)^2} = \frac{-1}{\cosh^4 u}.$$

The simple formula for K in Lemma 6.3 suggests a way to construct surfaces of revolution with *prescribed* Gaussian curvature. Given a function K on some interval, we first solve the differential equation $h'' + Kh = 0$ for h, subject to initial conditions $h(0) > 0$ and $|h'(0)| < 1$. To get a canonical parametrization we need a function g such that $g'^2 + h'^2 = 1$. Evidently

$$g(u) = \int_0^u \sqrt{1 - h'^2(t)}\ dt$$

will do the job. Thus for any interval around 0 on which the conditions $h > 0$ and $|h'| < 1$ both hold, we can revolve

$$\alpha(u) = (g(u), h(u), 0)$$

around the x axis to obtain a surface of revolution whose Gaussian curvature is, by Lemma 6.3, precisely $-h''/h = K$.

6.5 Example *Surfaces of revolution with constant positive curvature.*

We apply the procedure above to the constant function $K = 1/c^2$. The differential equation $h'' + (1/c^2)h = 0$ has general solution

$$h(u) = a \cos \left(\frac{u}{c} + b \right).$$

The constant b represents merely a translation of coordinates; we may suppose that $b = 0$ and $a > 0$. Thus the functions

$$g(u) = \int_0^u \sqrt{1 - \frac{a^2}{c^2} \sin^2 \frac{t}{c}} \, dt$$

$$h(u) = a \cos \frac{u}{c}$$

give rise to a surface of revolution M_a with constant Gaussian curvature $K = 1/c^2$. The necessary conditions $h > 0$ and $|h'| < 1$ determine the interval I to which u must be restricted. The constant c is fixed throughout the discussion, but the constant a is at our disposal; we consider three cases.

Case 1. $a = c$. Here

$$g(u) = \int_0^u \cos \frac{t}{c} \, dt = c \sin \frac{u}{c},$$

and $h(u) = c \cos (u/c)$. The interval I is thus $-\pi c/2 < u < \pi c/2$. Since this profile curve $\alpha(u) = (g(u), h(u))$ is a semicircle, revolution about the x axis produces the sphere Σ of radius c—except for its two points on the x axis.

Case 2. $0 < a < c$ (Fig. 5.42). Here h is positive on the same interval as above, and $|h'| < 1$ is always true, so g is well-defined. The profile curve $u \to (g(u), h(u))$ still has the same length $\pi c/2$, but it now makes a shallower arch, which rests on the x axis at $\pm a^*$, where

$$a^* = g\left(\frac{\pi c}{2}\right) = \int_0^{\pi c/2} \sqrt{1 - \frac{a^2}{c^2} \sin^2 \frac{t}{c}} \, dt.$$

Approximate values for this elliptic integral may be found in tables, but clearly as a shrinks from c down to 0, a^* increases from c to $\pi c/2$. The

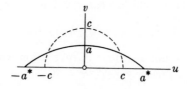

FIG. 5.42

resulting surface of revolution M_a—round for $a = c$—becomes football-shaped at first, and for a very small is a needle of length just less than $\pi c/2$. By contrast with Case 1, the intercepts $(\pm a^*, 0, 0)$ cannot be added to M_a now, since this surface is actually pointed at each end. The differential equation $h'' + (1/c^2)h = 0$ has delicately adjusted the shape of M_a so that its principal curvatures are no longer equal, but still give

$$K = k_\mu k_\pi = 1/c^2.$$

Case 3. $a > c$ (Fig. 5.43). Here the interval I is reduced in size, since the expression under the square root (in the formula for g) becomes zero at t^* such that $\sin(t^*/c) = c/a < 1$.

Thus

$$h(t^*) = a \cos\left(\frac{t^*}{c}\right) = \sqrt{a^2 - c^2}.$$

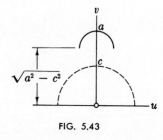

FIG. 5.43

As a increases from $a = c$, the resulting surface of revolution M_a is at first somewhat like the outer half of a torus, but when a is very large, it becomes an enormous circular band whose very short profile curve is sharply curved. (k_μ must be large, since $k_\pi \sim 1/a$ is small and $k_\mu k_\pi = 1/c^2$.)

A similar analysis for constant *negative* curvature leads to an infinite family of surfaces of revolution with $K = -1/c^2$ (Exercise 9). The simplest of these is

6.6 Example The *bugle surface* B. The profile curve of B (in the xy plane) is characterized by this geometric condition: It starts at the point $(0, c)$ and moves so that its tangent line always reaches the x axis after running for distance exactly c. This curve (a tractrix) may thus be described analytically by $\alpha(u) = (u, h(u))$, $u > 0$, where h is the solution of the differential equation $h' = -h/\sqrt{c^2 - h^2}$ such that $h(u) \to c$ as $u \to 0$. (The resulting surface of revolution is shown in Fig. 5.44.)

Using this differential equation, we deduce from the formulas (3) on page 235 that the principal curvatures of B are

$$k_\mu = \frac{-h'}{c} \qquad k_\pi = \frac{1}{ch'}.$$

Thus the bugle surface (or *tractroid*) has *constant negative curvature*

$$K = -1/c^2.$$

This surface cannot be extended across its rim (not part of B) to form a larger surface in $\mathbf{E}^3$, since $k_\mu(u) \to \infty$ as $u \to 0$.

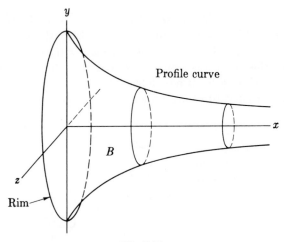

FIG. 5.44

When this surface was first discovered, it seemed to be the analogue, for K constant negative, of the sphere; it was thus called a *pseudosphere*. However as we shall see in Chapter VII the true analogue of the sphere is quite a different surface and cannot be found in $\mathbf{E}^3$.

EXERCISES

1. Find the Gaussian curvature of the surface obtained by revolving the curve $y = e^{-x^2/2}$ around the x axis. Sketch this surface and indicate the regions where $K > 0$ and $K < 0$.

2. (*Sign of Gaussian curvature*). When $y = f(x)$ is revolved about the x axis, and K is expressed in terms of x, show that K has the same sign $(-, 0, +)$ as $-f''$ for each value of x. So K is *positive* on parallels through *convex* intervals on the profile curve, and *negative* for *concave* intervals, as shown Fig. 5.45. (The same result is true for an arbitrary surface of revolution, with convexity and concavity taken relative to the axis of revolution A.)

3. (*Magnitude of Gaussian curvature*). Show that
   $| k_\mu(u) |$ = curvature $\kappa(u)$ of profile curve α at $\alpha(u)$,
   $| k_\pi(u) |$ = $h(u) | \cos \varphi(u) |$, where $\varphi(u)$ is the slope angle of the
          profile curve at $\alpha(u)$,
   hence that
   $$| K | = \kappa h | \cos \varphi |.$$

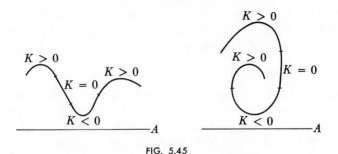

FIG. 5.45

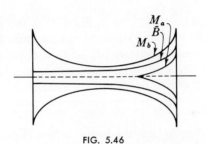

FIG. 5.46

4. An *elliptical torus* M is obtained by revolving an ellipse

$$(x - R)^2/a^2 + y^2/b^2 = 1$$

about the y axis $(R > a)$. Find a parametrization for M and compute its Gaussian curvature. (Check your answer by setting $a = b = r$.)

5. If $r = \sqrt{x^2 + y^2} > 0$ is the usual polar-coordinate function on the xy plane, and f is a differentiable function, show that $M: z = f(r)$ is a surface of revolution, and that its Gaussian curvature (expressed in terms of r) is

$$K = \frac{f'(r)f''(r)}{r(1 + f'(r)^2)^2}.$$

6. Find the Gaussian curvature of the surface $M: z = e^{-r^2/2}$. Sketch this surface, indicating the regions where $K > 0$ and $K < 0$.

7. Prove that a flat surface of revolution is part of a cone or a cylinder.

8. Let M be the surface obtained by revolving one arch $(-\pi < t < \pi)$ of the cycloid $\gamma(t) = (t + \sin t, 1 + \cos t)$ about the x axis.
   (a) Compute K relative to the usual parametrization of M.
   (b) Find the function h giving the height of α in terms of its arc length (measured from the top of the arch). Compute $\bar{K} = -h''/h$. (*Hint:* Use half-angles.)
   (c) Show that the results in (a) and (b) are consistent.

9. (*Surfaces of revolution with constant negative curvature* $K = -1/c^2$).
As in the case $K = 1/c^2$, there is a family of such surfaces, divided into
two subfamilies by a special surface. The solutions of

$$h'' - h/c^2 = 0$$

giving, by canonical parametrization, essentially all these surfaces are:

(a) $h(u) = a \sinh(u/c)$, $0 < a < c$, $u > 0$. Show that the profile curve
$\alpha(u) = (g(u), h(u))$ leaves the origin with slope $a/\sqrt{c^2 - a^2}$ and
rises toward a maximum height of $\sqrt{c^2 - a^2}$. Sketch the resulting
surface of revolution M_a for a small value of the parameter a, and
for a near c.

(b) $h(u) = b \cosh(u/c)$, $b > 0$. Show that the profile curve α rises
symmetrically (for $\pm u$) toward a maximum height of $\sqrt{c^2 + b^2}$.
Sketch the resulting surface M_b for b small and for b large.

(c) $h(u) = ce^{u/c}$, $u < 0$. This surface $\bar{B}$ is merely a mirror image of
the surface B gotten from

$$h(u) = ce^{-u/c}, \qquad u > 0.$$

Show that B is, in fact, the bugle surface (Example 6.6). How does
the surface $\bar{B}$ separate the two subfamilies, that is, for which values
of a and b do M_a and M_b resemble $\bar{B}$? (See Fig. 5.46 where M_a and
M_b have been translated along the axis of revolution.)

7 Summary

The shape operator S of a surface M in $\mathbf{E}^3$ measures the rate of change of
a unit normal U in any direction on M. If we imagine U as the "first
derivative" of M, then S is the "second derivative." But the shape operator is also an algebraic object consisting of linear operators on the tangent
planes of M. And it is by an algebraic analysis of S that we have been led
to the main geometric invariants of a surface in $\mathbf{E}^3$: its principal curvatures
and directions, and its Gaussian and mean curvatures.

CHAPTER VI

Geometry of Surfaces in $\mathbf{E}^3$

Now that we know how to measure the shape of a surface M in $\mathbf{E}^3$, the next step is to see how the shape of M is related to its other properties. Near each point of M, the Gaussian curvature has a strong influence on shape (Remark 3.3 of Chapter V), but we are now interested in the situation *in the large*—over the whole extent of M. For example, what can be said about the shape of M if it is compact, or flat, or both?

Almost 150 years ago Gauss raised a question that led to a new and deeper understanding of what geometry is: How much of the geometry of a surface in $\mathbf{E}^3$ is *independent of* its shape? At first glance this seems a strange question—what can we possibly say about a sphere, for example, if we ignore the fact that it is round? To get some grip on Gauss's question, let us imagine that the surface $M \subset \mathbf{E}^3$ has inhabitants who are unaware of the space outside their surface, and thus have no conception of its shape in $\mathbf{E}^3$. Nevertheless, they will still be able to measure the distance from place to place in M and find the area of regions in M. In this chapter and the next we shall see that in fact they can construct an "intrinsic geometry" for M that is richer and no less interesting than the familiar Euclidean geometry of the plane $\mathbf{E}^2$.

1 The Fundamental Equations

To study the geometry of a surface M in $\mathbf{E}^3$ we shall apply the Cartan methods outlined in Chapter II. As with the Frenet theory of a curve in $\mathbf{E}^3$, this requires that we put frames on M, and examine their rates of change along M. Formally, a *Euclidean frame field* on $M \subset \mathbf{E}^3$ consists of three

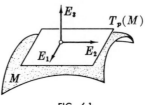

FIG. 6.1

Euclidean vector fields (Definition 3.7, Chapter IV) that are orthonormal at each point. Such a frame field can be fitted to its surface as follows.

1.1 Definition   An *adapted frame field* E_1, E_2, E_3 on a region $\mathcal{O}$ in $M \subset \mathbf{E}^3$ is a Euclidean frame field such that E_3 is always normal to M (hence E_1 and E_2 are tangent to M) (Fig. 6.1).

Thus the normal vector field denoted by U in the preceding chapter now becomes E_3. For brevity we shall refer to an adapted frame field "on M," but the actual domain of definition is in general only some region in M, since an adapted frame field need not exist on *all* of M.

1.2 Lemma   There is an adapted frame field on a region $\mathcal{O}$ in $M \subset \mathbf{E}^3$ if and only if $\mathcal{O}$ is orientable and there exists a nonvanishing tangent vector field on $\mathcal{O}$.

Proof. This condition is certainly necessary, since E_3 orients $\mathcal{O}$, and E_1 and E_2 are unit tangent vector fields. To show that it is sufficient, let $\mathcal{O}$ be oriented by a unit normal vector field U, and let V be a tangent vector field that does not vanish on $\mathcal{O}$. But then it is easy to see that

$$E_1 = \frac{V}{\| V \|}, \qquad E_2 = U \times E_1, \qquad E_3 = U$$

is an adapted frame field on $\mathcal{O}$.   ∎

1.3 Example   Adapted frame fields.
(1) Cylinder $M: x^2 + y^2 = r^2$. The gradient of $g = x^2 + y^2$ leads to the unit normal vector field $E_3 = (xU_1 + yU_2)/r$. Obviously the unit vector field U_3 is tangent to M at each point. Setting $E_2 = U_3 \times E_3$, we then get the adapted frame field

$$E_1 = U_3$$

$$E_2 = \frac{-yU_1 + xU_2}{r}$$

$$E_3 = \frac{xU_1 + yU_2}{r}$$

on the whole cylinder M (Fig. 6.2).

(2) Sphere Σ: $x^2 + y^2 + z^2 = r^2$. The outward
unit normal

$$E_3 = \frac{xU_1 + yU_2 + zU_3}{r}$$

is defined on all of Σ, but as we shall see in
Chapter VII, every tangent vector field on Σ
must vanish somewhere. For example, the "due
east" vector field $V = -yU_1 + xU_2$ is zero at the
the north and south poles $(0, 0, \pm r)$. Thus the
adapted frame field

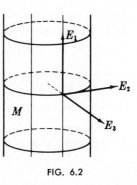

FIG. 6.2

$$E_1 = \frac{V}{\| V \|}$$

$$E_2 = E_3 \times E_1$$

$$E_3 = \frac{xU_1 + yU_2 + zU_3}{r}$$

(Fig. 6.3) is defined on the region $\mathcal{O}$ in Σ gotten by deleting the north and
south poles.

Lemma 1.2 implies in particular that there is an adapted frame field on
the image $\mathbf{x}(D)$ of any patch in M; thus such fields exist *locally* on any
surface in $\mathbf{E}^3$.

Now we shall bring the connection equations (Theorem 7.2 of Chapter
II) to bear on the study of a surface M in $\mathbf{E}^3$. Let E_1, E_2, E_3 be an adapted
frame field on M. By moving each frame $E_1(\mathbf{p})$, $E_2(\mathbf{p})$, $E_3(\mathbf{p})$ over a short

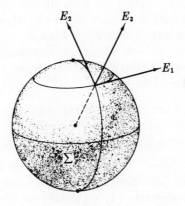

FIG. 6.3

interval on the normal line at each point $\mathbf{p}$, we can extend the given frame field to one defined on an open set in $\mathbf{E}^3$. Thus the connection equations

$$\nabla_v E_i = \sum \omega_{ij}(\mathbf{v}) E_j(\mathbf{p})$$

are available for use. *We shall apply them* only *to vectors* $\mathbf{v}$ *tangent to* M. In particular, *the connection forms* ω_{ij} *become* 1-*forms on* M in the sense of Section 4 of Chapter IV. Thus we have

1.4 Theorem   If E_1, E_2, E_3 is an adapted frame field on $M \subset \mathbf{E}^3$, and $\mathbf{v}$ is tangent to M at $\mathbf{p}$, then

$$\nabla_v E_i = \sum_{j=1}^{3} \omega_{ij}(\mathbf{v}) E_j(\mathbf{p}) \qquad (1 \leqq i \leqq 3).$$

The usual interpretation of the connection forms may be read from these equations, and it bears repetition: $\omega_{ij}(\mathbf{v})$ *is the initial rate at which* E_i *rotates toward* E_j *as* $\mathbf{p}$ *moves in the* $\mathbf{v}$ *direction.* Since E_3 is a unit normal vector field on M, the shape operator of M can be described by connection forms.

1.5 Corollary   Let S be the shape operator gotten from E_3, where E_1, E_2, E_3 is an adapted frame field on $M \subset \mathbf{E}^3$. Then for each tangent vector $\mathbf{v}$ to M at $\mathbf{p}$,

$$S(\mathbf{v}) = \omega_{13}(\mathbf{v}) E_1(\mathbf{p}) + \omega_{23}(\mathbf{v}) E_2(\mathbf{p})$$

Proof.   By definition, $S(\mathbf{v}) = -\nabla_v E_3$. Thus the connection equation for $i = 3$ gives the result, since the connection form $\omega = (\omega_{ij})$ is skew-symmetric: $\omega_{ij} = -\omega_{ji}$.   ∎

In addition to its connection forms, the adapted frame field E_1, E_2, E_3 also has *dual* 1-*forms* θ_1, θ_2, θ_3 (Definition 8.1 of Chapter II) which give the coordinates $\theta_i(\mathbf{v}) = \mathbf{v} \cdot E_i(\mathbf{p})$, of any tangent vector $\mathbf{v}_p$ with respect to the frame $E_1(\mathbf{p})$, $E_2(\mathbf{p})$, $E_3(\mathbf{p})$. As with the connection forms, the dual forms will be applied *only* to vectors tangent to M, so they become forms on M. This restriction is fatal to θ_3, for if $\mathbf{v}$ is tangent to M, it is orthogonal to E_3, so $\theta_3(\mathbf{v}) = \mathbf{v} \cdot E_3(\mathbf{p}) = 0$. Thus θ_3 is identically zero on M.

Because of the skew-symmetry of the connection form, we are left with essentially only five 1-forms:

θ_1, θ_2 provide a dual description of the tangent vector fields E_1, E_2

ω_{12} gives the rate of rotation of E_1, E_2

ω_{13}, ω_{23} describe the shape operator derived from E_3

1.6 Example   *The sphere.*   Consider the adapted frame field E_1, E_2, E_3 defined on the (doubly punctured) sphere Σ in Example 1.3. By extending this frame field to an open set of $\mathbf{E}^3$ we get the *spherical frame field* given in Example 6.2 of Chapter II, provided the indices of the latter are shifted by

$1 \rightarrow 3, 2 \rightarrow 1, 3 \rightarrow 2$. Thus, in terms of the spherical coordinate functions, Example 8.4 of Chapter II gives

$$\theta_1 = r \cos \varphi \, d\vartheta \qquad \theta_2 = r \, d\varphi$$

$$\omega_{12} = \sin \varphi \, d\vartheta \qquad \omega_{13} = -\cos \varphi \, d\vartheta \qquad \omega_{23} = -d\varphi$$

Because all forms (including functions) are now restricted to the surface Σ, the spherical coordinate function ρ has become a constant: the radius r of the sphere.

In general the forms associated with an adapted frame field obey the following remarkable set of equations.

1.7 Theorem   If E_1, E_2, E_3 is an adapted frame field on $M \subset \mathbf{E}^3$, then its dual forms and connection forms on M satisfy:

(1) $\begin{cases} d\theta_1 = \omega_{12} \wedge \theta_2 \\ d\theta_2 = \omega_{21} \wedge \theta_1 \end{cases}$ *First structural equations*

(2) $\omega_{31} \wedge \theta_1 + \omega_{32} \wedge \theta_2 = 0$ *Symmetry equation*

(3) $d\omega_{12} = \omega_{13} \wedge \omega_{32}$ *Gauss equation*

(4) $\begin{cases} d\omega_{13} = \omega_{12} \wedge \omega_{23} \\ d\omega_{23} = \omega_{21} \wedge \omega_{13} \end{cases}$ *Codazzi equations*

Proof. We merely apply the structural equations in Theorem 8.3 of Chapter II. The first structural equation

$$d\theta_i = \sum_j \omega_{ij} \wedge \theta_j$$

yields (1) and (2) above. In fact, for $i = 1, 2$, we get (1), since $\theta_3 = 0$ on the surface M. But $\theta_3 = 0$ implies $d\theta_3 = 0$, so for $i = 3$ we get (2).

Then the second structural equation yields the Gauss (3) and Codazzi (4) equations.   ∎

Because connection forms are skew-symmetric, and a wedge product of 1-forms satisfies $\phi \wedge \psi = -\psi \wedge \phi$, the fundamental equations above can be rewritten in a variety of equivalent ways. However, we shall stick to the index pattern used above, which, on the whole, seems the easiest to remember.

We emphasize that the forms introduced in this section describe, not the surface M directly, but only the particular adapted frame field E_1, E_2, E_3 from which they are derived: A different choice of the frame field will produce different forms. Nevertheless the six fundamental equations in Theorem 1.7 contain a tremendous amount of information about the surface $M \subset \mathbf{E}^3$, and we shall call on each in turn as we come to a geometric situation that it governs. For example, since ω_{13} and ω_{23} describe the shape operator of M, the Codazzi equations (4) express the rate at which the shape of M is *changing* from one point to another.

The first of the following exercises shows how the Cartan approach automatically singles out the three types of curves considered in Chapter V, Section 5.

EXERCISES

1. Let α be a unit-speed curve in $M \subset E^3$. If E_1, E_2, E_3 is an adapted frame field such that E_1 restricted to α is its unit tangent T, show that
   (a) α is a geodesic of M if and only if $\omega_{12}(T) = 0$.
   (b) If $E_3 = E_1 \times E_2$, then

$$g = \omega_{12}(T), \qquad k = \omega_{13}(T), \qquad t = \omega_{23}(T)$$

   where g, k, and t are the functions defined in Ex. 7 of V.5. (*Hint:* If $T = E_1$ along α, then $E_i' = \nabla_{E_1} E_i$ along α.)

2. (*Sphere*). For the frame field in Example 1.6:
   (a) Verify the fundamental equations (Theorem 1.7).
   (b) Deduce from the formulas for θ_1 and θ_2 that

$$E_1[\vartheta] = 1/r \cos \varphi \qquad E_1[\varphi] = 0$$

$$E_2[\vartheta] = 0 \qquad E_2[\varphi] = 1/r$$

   (c) Use Corollary 1.5 to find the shape operator S of the sphere.

3. (*Torus*). Let E_1, E_2, E_3 be the adapted frame field on the torus T (radii $R > r$) such that E_2 is tangent to meridians and E_1 is tangent to parallels, as in Fig. 6.4. Use the toroidal frame field in E^3 to get

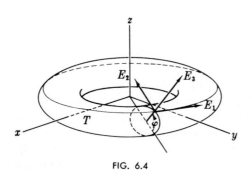

FIG. 6.4

$$\theta_1 = (R + r \cos \varphi) \, d\vartheta$$

$$\theta_2 = r \, d\varphi$$

$$\omega_{12} = \sin \varphi \, d\vartheta$$

$$\omega_{13} = -\cos \varphi \, d\vartheta$$

$$\omega_{23} = -d\varphi$$

Check the fundamental equations for these forms.

4. (*Continuation*). By the methods of this section,

compute $S(E_1)$ and $S(E_2)$ for the frame field above. Deduce that meridians and parallels are principal curves, and find the principal curvature functions. (Compare Example 6.1 of Chapter V, where the unit normal is "inward.")

5. Use the cylindrical frame field in $\mathbf{E}^3$ (Example 6.2 of Chapter II) to compute the shape operator of the cylinder $M: x^2 + y^2 = r^2$.

6. Give a new proof that shape operators are symmetric by using the symmetry equation (Theorem 1.7).

2 Form Computations

From now on, our study of the geometry of surfaces will be carried on mostly in terms of differential forms, so the reader may wish to look back over their general properties in Sections 4 and 5 of Chapter IV. Increasingly we shall tend to compare M with the Euclidean plane $\mathbf{E}^2$. Thus if E_1, E_2, E_3 is an adapted frame field on $M \subset \mathbf{E}^3$, we say that E_1, E_2 constitutes a *tangent frame field* on M. Any tangent vector field V on M may be expressed in terms of E_1 and E_2 by the orthonormal expansion

$$V = V \cdot E_1 \, E_1 + V \cdot E_2 \, E_2$$

To show that two forms are equal, we do not have to check their values on *all* tangent vectors, but only on the "basis" vector fields E_1, E_2. (See the remarks preceding Example 4.7 of Chapter IV). Explicitly: *1-forms ϕ and ψ are equal if and only if*

$$\phi(E_1) = \psi(E_1) \qquad and \qquad \phi(E_2) = \psi(E_2);$$

2-forms μ and ν are equal if and only if

$$\mu(E_1, E_2) = \nu(E_1, E_2).$$

The dual forms θ_1, θ_2 are, as we have emphasized, merely another description of the tangent frame field E_1, E_2; they are completely characterized by the equations

$$\theta_i(E_j) = \delta_{ij} \qquad (1 \leq i, j \leq 2).$$

These forms provided a "basis" for the forms on M (or, strictly speaking, on the region of definition of E_1, E_2).

2.1 Lemma (The Basis Formulas) Let θ_1, θ_2 be the dual 1-forms of E_1, E_2 on M. If ϕ is a 1-form and μ a 2-form, then

(1) $\phi = \phi(E_1) \, \theta_1 + \phi(E_2) \, \theta_2$

(2) $\mu = \mu(E_1, E_2) \, \theta_1 \wedge \theta_2$

Proof. Apply the equality criteria above, observing for (2) that by definition of the wedge product,

$$(\theta_1 \wedge \theta_2)(E_1, E_2) = \theta_1(E_1)\theta_2(E_2) - \theta_1(E_2)\theta_2(E_1)$$
$$= 1 \cdot 1 - 0 \cdot 0 = 1. \qquad \blacksquare$$

Assuming throughout that the forms θ_1, θ_2, ω_{12}, ω_{13}, ω_{23} derive as in Section 1 from an adapted frame field E_1, E_2, E_3 on a region in M, let us see what some of the concepts introduced in Chapter V look like when expressed in terms of forms. We begin with the analogue of Lemma 3.4 of Chapter V.

2.2 Lemma

(1) $\omega_{13} \wedge \omega_{23} = K\theta_1 \wedge \theta_2$.

(2) $\omega_{13} \wedge \theta_2 + \theta_1 \wedge \omega_{23} = 2H\theta_1 \wedge \theta_2$.

Proof. To apply the definitions $K = \det S$, $2H = \text{trace } S$, we shall find the matrix of S with respect to E_1 and E_2. As in Corollary 1.5 the connection equations give

$$S(E_1) = -\nabla_{E_1}E_3 = -\omega_{31}(E_1)E_1 - \omega_{32}(E_1)E_2$$
$$S(E_2) = -\nabla_{E_2}E_3 = -\omega_{31}(E_2)E_1 - \omega_{32}(E_2)E_2.$$

Thus the matrix of S is

$$\begin{pmatrix} \omega_{13}(E_1) & \omega_{23}(E_1) \\ \omega_{13}(E_2) & \omega_{23}(E_2) \end{pmatrix}.$$

Now, by the second formula in Lemma 2.1, what we must show is that

$$(\omega_{13} \wedge \omega_{23})(E_1, E_2) = K \qquad \text{and} \qquad (\omega_{13} \wedge \theta_2 + \theta_1 \wedge \omega_{23})(E_1, E_2), = 2H.$$

But

$$(\omega_{13} \wedge \omega_{23})(E_1, E_2) = \omega_{13}(E_1)\omega_{23}(E_2) - \omega_{13}(E_2)\omega_{23}(E_1)$$
$$= \text{determinant of matrix of } S = \det S = K$$

and a similar computation gives the trace formula. $\qquad \blacksquare$

Comparing the first formula above with the Gauss equation (3) in Theorem 1.7, we get

2.3 Corollary $\quad d\omega_{12} = -K\theta_1 \wedge \theta_2$.

We shall call this the *second structural equation*,† and derive from it a new interpretation of Gaussian curvature: ω_{12} measures the rate of rotation of the tangent frame field E_1, E_2—and since K determines the exterior derivative $d\omega_{12}$, it becomes a kind of "second derivative" of E_1, E_2.

---

† This equation will be shown to be the analogue for M of the second structural equation (Theorem 8.3 of Chapter II) for E³.

For example, on a sphere Σ of radius r, the formulas in Example 1.6 give

$$\theta_1 \wedge \theta_2 = r^2 \cos \varphi \, d\vartheta \, d\varphi = -r^2 \cos \varphi \, d\varphi \, d\vartheta.$$

But

$$d\omega_{12} = d(\sin \varphi \, d\vartheta) = d(\sin \varphi) \wedge d\vartheta = \cos \varphi \, d\varphi d\vartheta.$$

Thus the second structural equation gives the expected result, $K = 1/r^2$.

This new description of curvature may be rewritten in still another way.

2.4 Corollary $K = E_2[\omega_{12}(E_1)] - E_1[\omega_{12}(E_2)] - \omega_{12}(E_1)^2 - \omega_{12}(E_2)^2.$

Proof. By Lemma 2.1, we have

$$\omega_{12} = f_1 \theta_1 + f_2 \theta_2,$$

where

$$f_i = \omega_{12}(E_i) \qquad \text{for } i = 1, 2.$$

Then

$$
\begin{aligned}
d\omega_{12} &= df_1 \wedge \theta_1 + df_2 \wedge \theta_2 + f_1 \, d\theta_1 + f_2 \, d\theta_2 \\
&= df_1 \wedge \theta_1 + df_2 \wedge \theta_2 + f_1 \omega_{12} \wedge \theta_2 + f_2 \omega_{21} \wedge \theta_1
\end{aligned}
$$

where we have used the first structural equations (Theorem 1.7). Now apply this formula to E_1, E_2. Since $\theta_i(E_j) = \delta_{ij}$, we get

$$d\omega_{12}(E_1, E_2) = -df_1(E_2) + df_2(E_1) + f_1 \omega_{12}(E_1) - f_2 \omega_{21}(E_2)$$

Hence, using the previous corollary,

$$-K = -E_2[f_1] + E_1[f_2] + f_1 \omega_{12}(E_1) + f_2 \omega_{12}(E_2)$$

which, by definition of f_1 and f_2, is the required result. ∎

For instance, from Example 1.6 we readily compute that

$$\omega_{12}(E_1) = \frac{1}{r} \tan \varphi \qquad \text{and} \qquad \omega_{12}(E_2) = 0.$$

Thus for the sphere, the formula above yields

$$K = E_2 \left[\frac{1}{r} \tan \varphi \right] - \left(\frac{1}{r} \tan \varphi \right)^2 = \frac{\sec^2 \varphi - \tan^2 \varphi}{r^2} = \frac{1}{r^2},$$

since by Exercise 2 of Section 1 we have

$$E_2 [\tan \varphi] = \sec^2 \varphi \, E_2[\varphi] = \frac{\sec^2 \varphi}{r}.$$

We emphasized in Section 1 that, in general, adapted frame fields on $M \subset \mathbf{E}^3$ give only *indirect* information about M. If such a frame field is to give *direct* geometric information, it must be derived in some natural way from the geometry of M itself, as was the case with the Frenet frame field of a curve. There is a way to do this:

2.5 Definition A *principal frame field* on $M \subset E^3$ is an adapted frame field E_1, E_2, E_3 such that at each point E_1 and E_2 are principal vectors of M.

So long as its domain of definition contains no umbilics, a principal frame field is uniquely determined—except for changes of sign—by the two principal directions at each point.

Occasionally it may be possible to get a principal frame field on an entire surface. For example, on a surface of revolution, we can take E_1 tangent to meridians, E_2 tangent to parallels. In general, however, about the best we can do is as follows.

2.6 Lemma If $\mathbf{p}$ is a nonumbilic point of $M \subset E^3$, then there exists a principle frame field on some neighborhood of $\mathbf{p}$ in M.

Proof. Let F_1, F_2, F_3 be an arbitrary adapted frame field on a neighborhood $\mathfrak{N}$ of $\mathbf{p}$. Since $\mathbf{p}$ is not umbilic, we can assume (by rotating F_1, F_2 if necessary) that $F_1(\mathbf{p})$ and $F_2(\mathbf{p})$ are not principal vectors at $\mathbf{p}$. By hypothesis $k_1(\mathbf{p}) \neq k_2(\mathbf{p})$; hence by continuity k_1 and k_2 remain distinct near $\mathbf{p}$. On a small enough neighborhood $\mathfrak{N}$ of $\mathbf{p}$, all these conditions are thus in force.

Let S_{ij} be the matrix of S with respect to F_1, F_2. It is now just a standard problem in linear algebra to compute—*simultaneously at all points of $\mathfrak{N}$*—characteristic vectors of S, that is, principal vectors of M. In fact, at each point the tangent vector fields

$$V_1 = S_{12}F_1 + (k_1 - S_{11})F_2$$

$$V_2 = (k_2 - S_{22})F_1 + S_{12}F_2$$

give characteristic vectors of S. (This can be checked by a direct computation if one does not care to appeal to linear algebra.) Furthermore, the function $S_{12} = S(F_1) \cdot F_2$ is never zero on our selected neighborhood $\mathfrak{N}$, so $\| V_1 \|$ and $\| V_2 \|$ are never zero. Thus the vector fields

$$E_1 = \frac{V_1}{\| V_1 \|} \qquad E_2 = \frac{V_2}{\| V_2 \|}$$

consist only of principal vectors, so E_1, E_2, $E_3 = F_3$ is a principal frame field on $\mathfrak{N}$. ∎

If E_1, E_2, E_3 is a principal frame field on M, then the vector fields E_1 and E_2 consist of characteristic vectors of the shape operator derived from E_3. Thus we can label the principal curvature functions so that $S(E_1) = k_1 E_1$ and $S(E_2) = k_2 E_2$. Comparison with Corollary 1.5 then yields

$$\omega_{13}(E_1) = k_1 \qquad \omega_{13}(E_2) = 0$$
$$\omega_{23}(E_1) = 0 \qquad \omega_{23}(E_2) = k_2.$$

Thus the basis formula (1) in Lemma 2.1 gives

$$\omega_{13} = k_1\theta_1 \qquad \omega_{23} = k_2\theta_2 \qquad\qquad (^*)$$

This leads to an interesting version of the Codazzi equations.

2.7 Theorem If E_1, E_2, E_3 is a principal frame field on $M \subset \mathbf{E}^3$, then

$$E_1[k_2] = (k_1 - k_2)\,\omega_{12}(E_2)$$
$$E_2[k_1] = (k_1 - k_2)\,\omega_{12}(E_1)$$

Proof. The Codazzi equations (Theorem 1.7) read

$$d\omega_{13} = \omega_{12} \wedge \omega_{23} \qquad d\omega_{23} = \omega_{21} \wedge \omega_{13}.$$

The proof is now an exercise in the calculus of forms as discussed in Chapter IV, Section 4. Substituting from (*), above, in the first of these equations, we get

$$d(k_1\theta_1) = \omega_{12} \wedge k_2\theta_2;$$

hence

$$dk_1 \wedge \theta_1 + k_1\,d\theta_1 = k_2\omega_{12} \wedge \theta_2$$

If we substitute the structural equation $d\theta_1 = \omega_{12} \wedge \theta_2$, this becomes

$$dk_1 \wedge \theta_1 = (k_2 - k_1)\,\omega_{12} \wedge \theta_2$$

Now apply these 2-forms to the pair of vector fields E_1, E_2 to obtain

$$0 - dk_1(E_2) = (k_2 - k_1)\,\omega_{12}(E_1) - 0,$$

hence

$$E_2[k_1] = dk_1(E_2) = (k_1 - k_2)\,\omega_{12}(E_1).$$

The other required equation derives in the same way from the Codazzi equation $d\omega_{23} = \omega_{21} \wedge \omega_{13}$. ∎

Note that for a principal frame field, $\omega_{12}(\mathbf{v})$ tells how the principal directions are changing in the $\mathbf{v}$ direction.

EXERCISES

1. Verify the Codazzi equations (Theorem 2.7) for the principal frame field on the torus given in Exercise 3 of Section 1. (*Hint:*

$$V\left[\frac{f}{g}\right] = \frac{gV[f] - fV[g]}{g^2};$$

the derivatives of ϑ and φ with respect to E_1 and E_2 may be found from the formulas for θ_1 and θ_2 in this exercise.)

2. If E_1, E_2, E_3 is an adapted frame field on M with $E_1 \cdot E_2 \times E_3 = 1$, let

$$h_i(\mathbf{p}) = \mathbf{p} \cdot E_i(\mathbf{p}) \qquad \text{for } i = 1, 2, 3.$$

In particular, h_3 is the support function h of M defined on page 218. Show that

$$dh_1 = \theta_1 + \omega_{12}h_2 + \omega_{13}h_3$$
$$dh_2 = \theta_2 + \omega_{21}h_1 + \omega_{23}h_3$$

(*Hint:* Strictly speaking, $h_i = X \cdot E_i$, where X is the remarkable vector field such that $\nabla_V X = V$ used also on page 218.)

There is a rough rule for computing exterior derivatives of 1-forms in terms of an adapted frame field: Express the form in terms of θ_1 and θ_2 (or perhaps ω_{ij}); then apply d and use the fundamental equations. The proof of Theorem 2.7 is one example; another is as follows.

3. (*Continuation*).
  (a) If ψ is the 1-form such that $\psi(\mathbf{v}) = \mathbf{p} \cdot \mathbf{v} \times E_3(\mathbf{p})$, show that

$$d\psi = 2(1 + hH)\,\theta_1 \wedge \theta_2,$$

  where h is the support function.
  (b) If ζ is the 1-form such that $\zeta(\mathbf{v}) = \psi(S(\mathbf{v}))$, show that

$$d\zeta = 2(H + hK)\,\theta_1 \wedge \theta_2.$$

(*Hint:* $\psi = h_1\,\theta_2 - h_2\,\theta_1$, and ζ has an analogous expression.)

3 Some Global Theorems

We have claimed all along that the shape operator S is the analogue for a surface M of the curvature and torsion of a curve in $\mathbf{E}^3$. Simple hypotheses on κ and τ singled out some important special types of curves. Let us now see what can be done with S in the case of surfaces. (Recall that we are dealing exclusively with *connected* surfaces.)

3.1 Theorem If its shape operator is identically zero, then M is (part of) a plane in $\mathbf{E}^3$.

Proof. The scheme used is analogous to that of Corollary 3.5 of Chapter II. By definition of the shape operator, $S = 0$ means that any unit normal vector field E_3 on M is Euclidean parallel; thus it may be identified with a point of $\mathbf{E}^3$. Fix a point $\mathbf{p}$ of M. We shall show that M lies in the plane

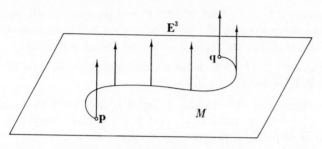

FIG. 6.5

through $\mathbf{p}$ orthogonal to E_3. If $\mathbf{q}$ is an arbitrary point of M, then since M is connected, there is a curve α in M from $\alpha(0) = \mathbf{p}$ to $\alpha(1) = \mathbf{q}$. Consider the function

$$f(t) = (\alpha(t) - \mathbf{p}) \cdot E_3.$$

Now

$$\frac{df}{dt} = \alpha' \cdot E_3 = 0 \qquad \text{and} \qquad f(0) = 0;$$

hence f is identically zero. In particular,

$$f(1) = (\mathbf{q} - \mathbf{p}) \cdot E_3 = 0,$$

so every point $\mathbf{q}$ of M is in the required plane (Fig. 6.5). ∎

We saw in Chapter V, Section 3 that requiring a single point $\mathbf{p}$ of $M \subset \mathbf{E}^3$ to be planar ($k_1 = k_2 = 0$, or equivalently $S = 0$) produces no significant effect on the shape of M near $\mathbf{p}$. But the result above shows that if *every* point is planar, then M is, in fact, part of a plane.

Perhaps the next simplest hypothesis on a surface M in $\mathbf{E}^3$ is that at each point $\mathbf{p}$, the shape operator is merely scalar multiplication by some number—which *a priori* may depend on $\mathbf{p}$. This means that M is *all-umbilic*, that is, consists entirely of umbilic points.

3.2 Lemma If M is an all-umbilic surface in $\mathbf{E}^3$, then M has constant Gaussian curvature $K \geqq 0$.

Proof. Let E_1, E_2, E_3 be an adapted frame field on some region $\mathcal{O}$ in M. Since M is all-umbilic, the principal curvature functions on $\mathcal{O}$ are equal, $k_1 = k_2 = k$, and furthermore E_1, E_2, E_3 is actually a principal frame field (since every direction on M is principal). Thus we can apply Theorem 2.7 to conclude that $E_1[k] = E_2[k] = 0$. Alternatively we may write

$$dk(E_1) = dk(E_2) = 0,$$

so by Lemma 2.1, $dk = 0$ on $\odot$. But $K = k_1 k_2 = k^2$, so $dK = 2k\, dk = 0$ on $\odot$. Since every point of M is in such a region $\odot$, we conclude that $dK = 0$ on all of M. It follows by an earlier exercise that K is constant. ∎

3.3 Theorem If $M \subset \mathbf{E}^3$ is all-umbilic and $K > 0$, then M is part of a sphere in $\mathbf{E}^3$ of radius $1/\sqrt{K}$.

Proof. (This time the scheme of proof is analogous to Lemma 3.6 of Chapter II.) Pick at random a point $\mathbf{p}$ in M and a unit normal vector $E_3(\mathbf{p})$ to M at $\mathbf{p}$. *We shall prove that the point*

$$\mathbf{c} = \mathbf{p} + \frac{1}{k(\mathbf{p})}\, E_3(\mathbf{p})$$

is equidistant from every point of M. (Here $k(\mathbf{p}) = k_1(\mathbf{p}) = k_2(\mathbf{p})$ is the principal curvature corresponding to $E_3(\mathbf{p})$.)

Now let $\mathbf{q}$ be any point of M, and let α be a curve segment in M from $\alpha(0) = \mathbf{p}$ to $\alpha(1) = \mathbf{q}$. Extend $E_3(\mathbf{p})$ to a unit normal vector field E_3 on α, as shown in Fig. 6.6, and consider the curve

$$\gamma = \alpha + \frac{1}{k}\, E_3 \qquad \text{in } \mathbf{E}^3.$$

Here we understand that the principal curvature function k derives from E_3, thus k is continuous. But $K = k^2$ and by the preceding lemma, K is constant, so k is constant. Thus

$$\gamma' = \alpha' + \frac{1}{k}\, E_3{}'.$$

But

$$E_3{}' = -S(\alpha') = -k\alpha',$$

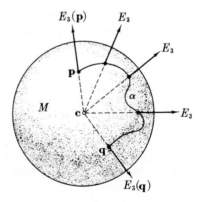

FIG. 6.6

since by the all-umbilic hypothesis, S is scalar multiplication by k. Thus

$$\gamma' = \alpha' + \frac{1}{k} (-k\alpha') = 0,$$

so the curve γ must be constant. In particular

$$c = \gamma(0) = \gamma(1) = q + \frac{1}{k} E_3(q)$$

so $d(c, q) = 1/\,|\,k\,|$ for every point q of M. Since $K = k_1 k_2 = k^2$, we have shown that M is contained in the sphere of center c and radius $1/\sqrt{K}$.   ■

Using all three of the preceding results, we conclude that *a surface M in E^3 is all-umbilic if and only if M is part of a plane or a sphere.*

3.4 Corollary   A compact all-umbilic surface M in E^3 is an entire sphere.

Proof. By the preceding remark, we deduce from Theorem 7.6, Chapter IV, that M must be an *entire* plane or sphere. The former is impossible, since M is—by hypothesis—compact, but planes are not.   ■

Gaussian curvature was used in the preceding results mostly because it is well-defined and differentiable on all of M, and is thus easier to work with than principal curvatures.

We now turn to a more serious examination of the Gaussian curvature K of a surface $M \subset E^3$.

3.5 Theorem   On every compact surface M in E^3, there is a point at which the Gaussian curvature K is strictly positive.

Proof. Consider the real-valued function f on M such that $f(p) = \|\,p\,\|^2$. Thus in terms of the natural coordinates of E^3, $f = \sum x_i^2$. Now f is differentiable, hence continuous, and M is compact. Thus by Lemma 7.3 of Chapter IV, f takes on its maximum at some point m of M. Since f measures the square of the distance to the origin, m is simply a point of M at maximum distance $r = \|\,m\,\| > 0$ from the origin. Intuitively it is clear that M is tangent at p to the sphere Σ of radius r—and that M lies inside Σ, hence is more curved than Σ (Fig. 6.7). Thus we would expect that $K(m) \geq 1/r^2 > 0$. Let us *prove* this inequality.

Given any unit tangent vector u to M at the maximum point m, pick a unit-speed curve α in M such that $\alpha(0) = m$, $\alpha'(0) = u$. It follows from the derivation of m that the composite function $f(\alpha)$ also has a maximum at $t = 0$. Thus

$$\frac{d}{dt} (f\alpha)(0) = 0, \qquad \frac{d^2}{dt^2} (f\alpha)(0) \leq 0. \qquad (1)$$

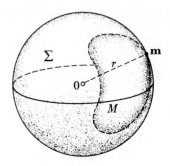

FIG. 6.7

But $f(\alpha) = \alpha \cdot \alpha$, so $\dfrac{d}{dt}(f\alpha) = 2\alpha \cdot \alpha'$. Evaluating at $t = 0$, we find

$$0 = \frac{d(f\alpha)}{dt}(0) = 2\alpha(0) \cdot \alpha'(0) = 2\mathbf{m} \cdot \mathbf{u}.$$

Since $\mathbf{u}$ was any unit tangent vector to M at $\mathbf{m}$, this means that $\mathbf{m}$ (considered as a vector) is normal to M at $\mathbf{m}$.

Differentiating again, we get

$$\frac{d^2(f\alpha)}{dt^2} = 2\alpha' \cdot \alpha' + 2\alpha \cdot \alpha''.$$

By (1), at $t = 0$ this yields

$$\begin{aligned} 0 &\geqq \mathbf{u} \cdot \mathbf{u} + \mathbf{m} \cdot \alpha''(0) \\ &= 1 + \mathbf{m} \cdot \alpha''(0). \end{aligned} \tag{2}$$

The discussion above shows that $\mathbf{m}/r$ may be considered as a unit normal vector to M at $\mathbf{m}$ as shown in Fig. 6.8. Thus $(\mathbf{m}/r) \cdot \alpha''$ is precisely the normal curvature $k(\mathbf{u})$ of M in the $\mathbf{u}$ direction, and it follows from (2) that $k(\mathbf{u}) \leqq -1/r$. In particular, both principal curvatures satisfy this inequality, so

$$K(\mathbf{m}) \geqq \frac{1}{r^2} > 0. \qquad \blacksquare$$

Thus there are no compact surfaces in $\mathbf{E}^3$ *with* $K \leqq 0$.

Maintaining the hypothesis of compactness, we consider the effect of requiring that Gaussian curvature be constant. Theorem 3.5 shows that the only possibility is $K > 0$. Spheres are obvious examples of compact surfaces in $\mathbf{E}^3$ with constant positive Gaussian curvature. It is one of the most remarkable facts of surface theory that they are the *only* such surfaces. To prove this we need a rather deep preliminary result.

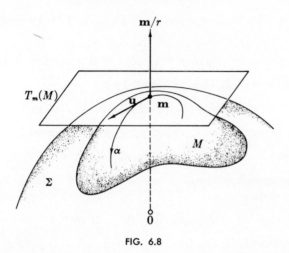

FIG. 6.8

3.6 Lemma (Hilbert)   Let **m** be a point of $M \subset \mathbf{E}^3$ such that
(1) k_1 has a local maximum at **m**.
(2) k_2 has a local minimum at **m**.
(3) $k_1(\mathbf{m}) > k_2(\mathbf{m})$.
Then $K(\mathbf{m}) \leqq 0$.

For example, it is easy to see that these hypotheses hold at any point on the inner equator of a torus or on the minimal circle ($x = 0$) of the catenoid. And K is, in fact, negative in both these examples.

To convert hypotheses (1) and (2) into usable form in the proof that follows, we recall some facts about maxima and minima. If f is a (differentiable) function on a surface M and V is a tangent vector field, then the *first* derivative $V[f]$ is again a function on M. Thus we can apply V again to obtain the *second* derivative $V[V[f]] = VV[f]$. A straightforward computation shows that if f has a local maximum at a point **m**, then the analogues of the usual conditions in elementary calculus hold, namely

$$V[f] = 0, \quad VV[f] \leqq 0 \qquad \text{at } \mathbf{m}.$$

For a local minimum, of course, the inequality is reversed.

Proof. Since $k_1(\mathbf{m}) > k_2(\mathbf{m})$, **m** is not umbilic, hence by Lemma 2.6 there exists a principal frame field E_1, E_2, E_3 on a neighborhood of **m** in M. By the remark above, the hypotheses of minimality and maximality at **m** imply in particular

$$E_1[k_2] = E_2[k_1] = 0 \qquad \text{at } \mathbf{m} \tag{1}$$

and

$$E_1 E_1[k_2] \geqq 0 \qquad \text{and} \qquad E_2 E_2[k_1] \leqq 0 \qquad \text{at } \mathbf{m}. \tag{2}$$

Now we use the Codazzi equations (Theorem 2.7). From (1) it follows that

$$\omega_{12}(E_1) \;=\; \omega_{12}(E_2) \;=\; 0 \qquad \text{at } \mathbf{m}$$

since $k_1 - k_2 \neq 0$ at $\mathbf{m}$, thus by Corollary 2.4

$$K \;=\; E_2[\omega_{12}(E_1)] \;-\; E_1[\omega_{12}(E_2)] \qquad \text{at } \mathbf{m}. \tag{3}$$

Applying E_1 to the first Codazzi equation in Theorem 2.7 yields

$$E_1 E_1[k_2] \;=\; (E_1[k_1] - E_1[k_2])\,\omega_{12}(E_2) \;+\; (k_1 - k_2)\,E_1[\omega_{12}(E_2)].$$

But at the special point $\mathbf{m}$, we have $\omega_{12}(E_2) = 0$ and $k_1 - k_2 > 0$; hence from (2) we deduce

$$E_1[\omega_{12}(E_2)] \geqq 0 \qquad \text{at } \mathbf{m}. \tag{4}$$

A similar argument starting from the second Codazzi equation gives

$$E_2[\omega_{12}(E_1)] \leqq 0 \qquad \text{at } \mathbf{m}. \tag{5}$$

Using (4) and (5) in the expression (3) for the Gaussian curvature at $\mathbf{m}$, we conclude that $K(\mathbf{m}) \leqq 0$. ∎

3.7 Theorem (*Liebmann*)   If M is a compact surface in $\mathbf{E}^3$ with constant Gaussian curvature K, then M is a sphere of radius $1/\sqrt{K}$. (Theorem 3.5 implies K is positive.)

Proof. We do not know that M is orientable, so principal curvature functions are not available on all of M. Nevertheless, the function

$$H^2 - K \;=\; (k_1 - k_2)^2/4$$

is well-defined and continuous on all of M, since squaring eliminates ambiguity of sign. Because M is compact, the function $H^2 - K \geqq 0$ has a maximum point $\mathbf{m}$. Now *if $H^2 - K$ is zero at* $\mathbf{m}$, it is identically zero; thus M is all-umbilic and Corollary 3.4 gives the required result.

So what we must show is that $H^2 - K$ cannot be positive at $\mathbf{m}$. Assume *it is*; then $\mathbf{m}$ is not umbilic and by suitably orienting a neighborhood $\mathfrak{N}$ of $\mathbf{m}$ we can arrange that $k_1 > k_2 > 0$ on $\mathfrak{N}$ (since $K > 0$). Then $k_1 - k_2$ has a maximum at $\mathbf{m}$, since $(k_1 - k_2)^2$ does. Since $K = k_1 k_2$ is constant, it follows that k_1 has a local maximum at $\mathbf{m}$ and k_2 has a local minimum. But now we can apply Hilbert's lemma to obtain the contradiction

$$K(\mathbf{m}) \leqq 0. ∎$$

Liebmann's theorem is false if the compactness hypothesis is omitted, for we saw in Chapter V, Section 6, that there are many nonspherical surfaces in $\mathbf{E}^3$ with constant (positive) curvature. Both Theorem 3.5 and Liebmann's theorem depend on the fundamental topological fact (Lemma

7.3 of Chapter IV) that a continuous real-valued function on a compact surface has a maximum. More advanced topological methods are required for a full investigation of the influence of Gaussian curvature on the shape of surfaces in $\mathbf{E}^3$. For example, one might ask what the situation is for constant curvature surfaces when compactness is weakened to *closure in* $\mathbf{E}^3$.† Here are the answers:

A closed surface $M \subset \mathbf{E}^3$ with K constant positive is compact—hence by Liebmann's theorem it is a sphere.

A closed surface $M \subset \mathbf{E}^3$ with $K = 0$ is a generalized cylinder (Massey).

There are no closed surfaces in $\mathbf{E}^3$ with K constant negative (Hilbert).

We shall prove the first result in Chapter VII. Proofs of the last two may be found in Hicks [5] and Willmore [3], respectively.

EXERCISES

1. If M is a flat minimal surface, prove that M is part of a plane.

2. *Flat surfaces in* $\mathbf{E}^3$ *can be bent only along straight lines:* If $k_1 = 0$, but k_2 is never zero, show that the principal curves of k_1 are straight-line segments in $\mathbf{E}^3$. (*Hint:* Use Ex. 1 of VI.1.)

   This is the starting point for the proof of Massey's theorem.

3. Let $M \subset \mathbf{E}^3$ be a compact orientable surface with $K > 0$. If M has constant mean curvature, show that M is a sphere.

4. Prove that in a region free of umbilics there are exactly two principal curves (ignoring different parametrizations) through each point, these crossing orthogonally. (Hint: Use Ex. 7 of IV.8.)

5. If the principal curvatures of a surface $M \subset \mathbf{E}^3$ are constant, show that M is part of either a plane, a sphere, or a circular cylinder. (*Hint:* In the nontrivial case $k_1 \neq k_2$, assume there is an adapted frame field on all of M, and show that, say, $k_1 = 0$.)

4   Isometries and Local Isometries

We referred earlier to properties of a surface M in $\mathbf{E}^3$ that could be discovered by inhabitants of M unaware of the space outside their surface. We asserted that the inhabitants of M could determine the distance *in* M

---

† This condition is defined in Exercise 10 of Chapter IV, Section 7. Roughly speaking, it means that M has no edges or rims. For surfaces in $\mathbf{E}^3$ it is equivalent to the intrinsic property of completeness (Definition 4.4, of Chapter VII).

between any two points, just as the distance along the surface of the earth is found by its inhabitants. The mathematical formulation is as follows.

4.1 Definition   If $\mathbf{p}$ and $\mathbf{q}$ are points of $M \subset \mathbf{E}^3$, consider the collection of all curve segments α *in* M from $\mathbf{p}$ to $\mathbf{q}$. The *intrinsic distance* $\rho(\mathbf{p}, \mathbf{q})$ from $\mathbf{p}$ to $\mathbf{q}$ in M is the greatest lower bound of the lengths $L(\alpha)$ of these curve segments.

There need not be a curve α whose length is exactly $\rho(\mathbf{p}, \mathbf{q})$ (see Exercise 3). The intrinsic distance $\rho(\mathbf{p}, \mathbf{q})$ will generally be greater than the straight-line Euclidean distance $d(\mathbf{p}, \mathbf{q})$, since the curves α are obliged to stay in M (Fig. 6.9).

On the surface of the earth (sphere of radius ca. 4000 miles) it is, of course, intrinsic distance that is of practical interest. One says, for example, that it is 12,500 miles from the north pole to the south pole—the Euclidean distance through the center of the earth is only 8000.

We saw in Chapter III how Euclidean geometry is based on the notion of isometry, a distance-preserving mapping. For surfaces in M we shall *prove* the distance-preserving property and use its infinitesimal form (Corollary 2.2 of Chapter III) as definition.

4.2 Definition   An *isometry* $F \colon M \to \bar{M}$ of surfaces in $\mathbf{E}^3$ is a one-to-one mapping of M onto $\bar{M}$ that preserves dot products of tangent vectors. Explicitly, if F_* is the derivative map of F, then

$$F_*(\mathbf{v}) \cdot F_*(\mathbf{w}) = \mathbf{v} \cdot \mathbf{w}$$

for any pair of tangent vectors $\mathbf{v}$, $\mathbf{w}$ to M.

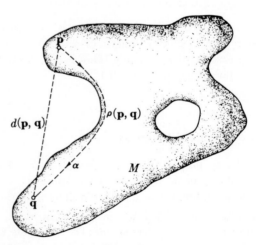

FIG. 6.9

If F_* preserves dot products, then it also preserves lengths of tangent vectors. It follows that an isometry is a regular mapping (Chapter IV, Section 5), for if $F_*(\mathbf{v}) = 0$, then

$$\| \mathbf{v} \| = \| F_*(\mathbf{v}) \| = 0,$$

hence $\mathbf{v} = 0$. Thus by the remarks following Theorem 5.4 of Chapter IV, an isometry $F: M \to \bar{M}$ is in particular a diffeomorphism, that is, has an inverse mapping $F^{-1}: \bar{M} \to M$.

4.3 Theorem Isometries preserve intrinsic distance: if $F: M \to \bar{M}$ is an isometry of surfaces in $\mathbf{E}^3$, then

$$\rho(\mathbf{p}, \mathbf{q}) = \bar{\rho}(F(\mathbf{p}), F(\mathbf{q}))$$

for any two points $\mathbf{p}$, $\mathbf{q}$ in M.

(Here ρ and $\bar{\rho}$ are the intrinsic distance functions of M and $\bar{M}$ respectively.)

Proof. First note that isometries preserve the speed and length of curves. The proof is just like the Euclidean case: If α is a curve segment in M, then $\bar{\alpha} = F(\alpha)$ is a curve segment in $\bar{M}$ with velocity $\bar{\alpha}' = F_*(\alpha')$. Since F_* preserves dot products, it preserves norms, so $\| \alpha' \| = \| F_*(\alpha') \| = \| F(\alpha)' \| = \| \bar{\alpha}' \|$. Hence

$$L(\alpha) = \int_a^b \| \alpha'(t) \| \, dt = \int_a^b \| \bar{\alpha}'(t) \| \, dt = L(\bar{\alpha}).$$

Now if α runs from $\mathbf{p}$ to $\mathbf{q}$ in M, its image $\bar{\alpha} = F(\alpha)$ runs from $F(\mathbf{p})$ to $F(\mathbf{q})$ in $\bar{M}$. Reciprocally, if β is a curve segment in $\bar{M}$ from $F(\mathbf{p})$ to $F(\mathbf{q})$ in $\bar{M}$, then $F^{-1}(\beta)$ runs from $\mathbf{p}$ to $\mathbf{q}$ in M. We have in fact established a one-to-one correspondence between the collection of curve segments used to define $\rho(\mathbf{p}, \mathbf{q})$ and those used for $\bar{\rho}(F(\mathbf{p}), F(\mathbf{q}))$. But as shown above, corresponding curves have the same length, so it follows immediately that $\rho(\mathbf{p}, \mathbf{q}) = \bar{\rho}(F(\mathbf{p}), F(\mathbf{q}))$. ∎

Thus we may think of an isometry as bending a surface into a different shape without changing the intrinsic distance between any of its points. *Consequently the inhabitants of the surface are not aware of any change at all, for their geometric measurements all remain exactly the same.*

If there exists an isometry from M to $\bar{M}$, then these two surfaces are said to be *isometric*. For example, if a piece of paper is bent into various shapes without creasing or stretching, the resulting surfaces are all isometric (Fig. 6.10).

To study isometries it is convenient to separate the geometric condition of preservation of dot products from the one-to-one and onto requirements.

4.4 Definition A *local isometry* $F: M \to N$ of surfaces is a mapping that preserves dot products of tangent vectors (that is, F_* does).

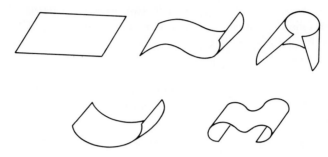

FIG. 6.10

Thus an isometry is a local isometry that is one-to-one and onto.

If F is a local isometry, the earlier argument still shows that F is a regular mapping. Then for each point $\mathbf{p}$ of M the inverse function theorem (5.4 of Chapter IV) asserts that there is a neighborhood $\mathfrak{U}$ of $\mathbf{p}$ in M that F carries diffeomorphically onto a neighborhood $\mathfrak{V}$ of $F(\mathbf{p})$ in N. Now $\mathfrak{U}$ and $\mathfrak{V}$ are themselves surfaces in $\mathbf{E}^3$, and thus the mapping $F \mid \mathfrak{U} \colon \mathfrak{U} \to \mathfrak{V}$ is an isometry. In this sense a local isometry is, indeed, *locally an isometry*.

There is a simple patch criterion for local isometries using the functions E, F, and G defined in Section 4 of Chapter V.

4.5 Lemma   Let $F \colon M \to N$ be a mapping. For each patch $\mathbf{x} \colon D \to M$, consider the composite mapping

$$\bar{\mathbf{x}} = F(\mathbf{x}) \colon D \to N.$$

Then F is a local isometry if and only if for each patch $\mathbf{x}$ we have

$$E = \bar{E} \qquad F = \bar{F} \qquad G = \bar{G}$$

(Here $\bar{\mathbf{x}}$ need not be a patch, but $\bar{E}$, $\bar{F}$, and $\bar{G}$ are defined for it as usual.)

Proof. Suppose the criterion holds—and only for enough patches to cover all of M. Then by one of the equivalences in Exercise 1, to show that F_* preserves dot products we need only prove that

$$\| \mathbf{x}_u \| = \| F_*(\mathbf{x}_u) \|, \qquad \mathbf{x}_u \cdot \mathbf{x}_v = F_*(\mathbf{x}_u) \cdot F_*(\mathbf{x}_v), \qquad \| \mathbf{x}_v \| = \| F_*(\mathbf{x}_v) \|$$

But as we saw in Chapter IV, it follows immediately from the definition of F_* that $F_*(\mathbf{x}_u) = \bar{\mathbf{x}}_u$ and $F_*(\mathbf{x}_v) = \bar{\mathbf{x}}_v$. Thus the equations above follow from the hypotheses $E = \bar{E}$, $F = \bar{F}$, $G = \bar{G}$. Hence F is a local isometry.

Reversing the argument, we deduce the converse assertion. ∎

This result can sometimes be used to *construct* local isometries. In the simplest case, suppose that M is the image of a single patch $\mathbf{x} \colon D \to M$.

Then if $\mathbf{y}\colon D \to N$ is a patch in another surface, define a mapping $F\colon M \to N$ by

$$F(\mathbf{x}(u,v)) = \mathbf{y}(u,v) \qquad \text{for } (u,v) \text{ in } D.$$

If

$$E = \bar{E},\, F = \bar{F},\, G = \bar{G},$$

then by the above criterion, F is a local isometry.

4.6 Example

(1) *Local isometry of a plane onto a cylinder.* The plane $\mathbf{E}^2$ may be considered as a surface, with natural frame field U_1, U_2. If $\mathbf{x}\colon \mathbf{E}^2 \to M$ is a parametrization of some surface, then Exercise 1 shows that $\mathbf{x}$ is a local isometry if

$$\mathbf{x}_*(U_i)\boldsymbol{\cdot}\mathbf{x}_*(U_j) = U_i\boldsymbol{\cdot}U_j,$$

Since $\mathbf{x}_*(U_1) = \mathbf{x}_u$, $\mathbf{x}_*(U_2) = \mathbf{x}_v$, and $U_i\boldsymbol{\cdot}U_j = \delta_{ij}$, this is the same as requiring $E = 1$, $F = 0$, $G = 1$.

To take a concrete case, the parametrization

$$\mathbf{x}(u,\,v) = \left(r\,\cos\left(\frac{u}{r}\right),\, r\,\sin\left(\frac{u}{r}\right),\, v \right)$$

of the cylinder $M\colon x^2 + y^2 = r^2$ has $E = 1$, $F = 0$, $G = 1$. Thus $\mathbf{x}$ is a local isometry which wraps the plane $\mathbf{E}^2$ around the cylinder, with horizontal lines going around cross-sectional circles, and vertical lines to rulings of the cylinder.

(2) *Local isometry of a helicoid onto a catenoid.* Let H be the helicoid which is the image of the patch

$$\mathbf{x}(u,v) = (u\,\cos v,\, u\,\sin v,\, v).$$

Furnish the catenoid C with the canonical parametrization $\mathbf{y}\colon \mathbf{E}^2 \to C$ discussed in Example 6.4 of Chapter V. Thus

$$\mathbf{y}(u,v) = (g(u),\, h(u)\,\cos v,\, h(u)\,\sin v)$$

$$g(u) = \sinh^{-1}u \qquad h(u) = \sqrt{1 + u^2}.$$

Let $F\colon H \to C$ be the mapping such that

$$F(\mathbf{x}(u,v)) = \mathbf{y}(u,v).$$

To prove that F is a local isometry, it suffices to check that

$$E = 1 = \bar{E}, \qquad F = 0 = \bar{F}, \qquad G = 1 + u^2 = h^2 = \bar{G}.$$

F carries the *rulings* (v constant) of H onto *meridians* of the surface of revolution C, and wraps the *helices* (u constant) of H around the *parallels* of C. In particular, the central axis of H (z axis) is wrapped around the minimal circle $x = 0$ of C.

Figure 6.11 shows how a sample strip of H is carried over to C.

Suppose that the helicoid H (or at least a finite region of it) has been stamped like an automobile fender out of a flexible sheet of steel—the patch $\mathbf{x}$ does this. Then H may be wrapped into the shape of a catenoid with no further distortion of the metal (Exercise 5 of Section 5).

A similar experiment may be performed by cutting a hole in a ping-pong ball representing a sphere in $\mathbf{E}^3$. Mild pressure will then deform the ball into various nonround shapes, all of which are isometric. For arbitrary isometric surfaces M and $\bar{M}$ in $\mathbf{E}^3$, however, it is generally not possible to bend M (through a whole family of isometric surfaces) so as to produce $\bar{M}$.

There are special types of mappings other than (local) isometries that are of interest in geometry.

4.7 Definition A mapping of surfaces $F: M \to N$ is *conformal* provided there exists a real-valued function $\lambda > 0$ on M such that

$$\| F_*(\mathbf{v}_p) \| = \lambda(\mathbf{p}) \| \mathbf{v}_p \|$$

for all tangent vectors to M. The function λ is called the *scale factor* of F.

Note that if F is a conformal mapping for which λ has constant value 1, F is a local isometry. Thus a conformal mapping is a generalized isometry

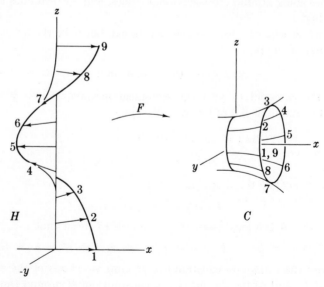

FIG. 6.11

for which lengths of tangent vectors need not be preserved—but at each point $\mathbf{p}$ of M the tangent vectors at $\mathbf{p}$ all have their lengths stretched by the same factor.

The criteria in Lemma 4.5 and in Exercise 1 may easily be adapted from isometries to conformal mappings by introducing the scale factor (or its square). In Lemma 4.5, for example, replace $E = \bar{E}$ by $E\lambda^2(\mathbf{x}) = \bar{E}$, and similarly for the other two equations.

An essential property of conformal mappings is discussed in Exercise 8.

EXERCISES

1. If $F: M \to N$ is a mapping, show that the following conditions on its derivative map at one point $\mathbf{p}$ are logically equivalent:

   (a) F_* preserves inner products.

   (b) F_* preserves lengths of tangent vectors, that is, $\| F_*(\mathbf{v}) \| = \| \mathbf{v} \|$ for all $\mathbf{v}$ at $\mathbf{p}$.

   (c) F_* preserves frames: If $\mathbf{e}_1$, $\mathbf{e}_2$ is a tangent frame at $\mathbf{p}$, then

   $$F_*(\mathbf{e}_1),\ F_*(\mathbf{e}_2)$$

   is a tangent frame at $F(\mathbf{p})$.

   (d) For some one pair of linearly independent tangent vectors $\mathbf{v}$ and $\mathbf{w}$ at $\mathbf{p}$

   $$\| F_*(\mathbf{v}) \| = \| \mathbf{v} \|,\quad \| F_*(\mathbf{w}) \| = \| \mathbf{w} \|,\qquad \text{and}\qquad F_*(\mathbf{v}) \cdot F_*(\mathbf{w}) = \mathbf{v} \cdot \mathbf{w}$$

   [Hint: It suffices, for example, to prove that $(a) \Rightarrow (c) \Rightarrow (d) \Rightarrow (b) \Rightarrow (a)$.]

   These are general facts from linear algebra; in this context they provide useful criteria for F to be a local isometry.

2. Show that each of the following conditions is necessary and sufficient for $F: M \to N$ to be a local isometry.

   (a) F preserves the speeds of curves: $\| F(\alpha)' \| = \| \alpha' \|$ for all curves α in M.

   (b) F preserves lengths of curves: $L(F(\alpha)) = L(\alpha)$ for all curve segments α in M.

3. Let M be the xy plane in $\mathbf{E}^3$ with the origin removed. Show that the intrinsic distance from $(-1, 0, 0)$ to $(1, 0, 0)$ in M is 2, but that there is no curve segment in M which joins these points and has length 2. (*Hint:* Ex. 11 of II. 2.)

4. Formulate precisely and prove: Local isometries can shrink but not increase intrinsic distance.

5. Let $\alpha, \beta \colon I \to \mathbf{E}^3$ be unit-speed curves with the same curvature function $\kappa > 0$, and assume that the ruled parametrization

$$\mathbf{x}(u, v) = \alpha(u) + vT(u)$$

of the tangent surface of α is actually a patch. Find a local isometry from the tangent surface of α to:
(a) The tangent surface of β.
(b) A region D in the plane.
(*Hint:* Ex. 9 of III.5.)

6. Show that the preceding exercise applies to the tangent surface of a helix and find the image region D in the plane.

7. Modify the conditions in Exercise 1 so that they provided criteria for F to be a conformal mapping. Then prove that a patch $\mathbf{x} \colon D \to M$ is a conformal mapping if and only if $E = G$ and $F = 0$.

8. Show that a conformal mapping $F \colon M \to N$ *preserves angles* in this sense: If ϑ is an angle $(0 \leqq \vartheta \leqq \pi)$ between $\mathbf{v}$ and $\mathbf{w}$ at $\mathbf{p}$, then ϑ is also an angle between $F_*(\mathbf{v})$ and $F_*(\mathbf{w})$ at $F(\mathbf{p})$.

9. If $F \colon M \to \bar{M}$ is an isometry, prove that the inverse mapping $F^{-1} \colon \bar{M} \to M$ is also an isometry. If $F \colon M \to N$ and $G \colon N \to P$ are (local) isometries, prove that the composite mapping $GF \colon M \to P$ is a (local) isometry.

10. Let $\mathbf{x}$ be a parametrization of all of M, $\bar{\mathbf{x}}$ a parametrization in N. If $F \colon M \to N$ is a mapping such that $F(\mathbf{x}(u, v)) = \bar{\mathbf{x}}(f(u), g(v))$, then
(a) Describe the effect of F on the parameter curves of $\mathbf{x}$.
(b) Show that F is a local isometry if and only if

$$E = \bar{E}(f, g)\left(\frac{df}{du}\right)^2 \qquad F = \bar{F}(f, g)\frac{df}{du}\frac{dg}{dv} \qquad G = \bar{G}(f, g)\left(\frac{dg}{dv}\right)^2$$

(In the general case, f and g are functions of both u and v, and this criterion becomes more complicated.)
(c) Find analogous conditions for F to be a conformal mapping.

11. Let M be a surface of revolution, and let $F \colon H \to M$ be a local isometry of the helicoid which (as in Example 4.6) carries rulings to meridians, and helices to parallels. Show that M must be a catenoid. (*Hint:* Use Exercise 10.)

12. Let M be the image of a patch $\mathbf{x}$ with $E = 1$, $F = 0$, and G a function of u only ($G_v = 0$). If the derivative $d(\sqrt{G})/du$ is bounded, show that there is a local isometry of M into a surface of revolution.

Thus any small enough region in M is isometric to a region in a surface of revolution.

13. Let **x** be the geographical patch in the sphere Σ of radius r (Example 2.2 of Chapter IV). Stretch **x** in the north-south direction to produce a conformal mapping. Explicitly, let

$$\mathbf{y}(u,v) = \mathbf{x}(u, g(v)) \qquad \text{with } g(0) = 0,$$

and determine g so that **y** is a conformal. Find the scale factor of **y** and the domain D such that $\mathbf{y}(D)$ omits only a semicircle of Σ. (Mercator's map of the earth derives from **y**: its inverse is *Mercator's projection*.)

14. Show that stereographic projection $P\colon \Sigma_0 \to E^2$ (Example 5.2 of Chapter IV) is conformal, with scale factor

$$\lambda(\mathbf{p}) = 1 + \frac{\| P(\mathbf{p}) \|^2}{4}.$$

15. Let M be a surface of revolution whose profile curve is not closed, hence has a one-to-one parametrization. Find a conformal mapping $F\colon M \to E^2$ such that meridians go to lines through the origin and parallels go to circles centered at the origin.

5 Intrinsic Geometry of Surfaces in E^3

In Chapter III we defined Euclidean geometry to consist of those concepts preserved by Euclidean isometries. The same definition applies to surfaces: The *intrinsic geometry* of $M \subset E^3$ consists of those concepts—called *isometric invariants*—that are preserved by all isometries $F\colon M \to \bar{M}$. For example, Theorem 4.3 shows that intrinsic distance is an isometric invariant. We can now state Gauss's question (mentioned in the start of the chapter) more precisely: *Which of the properties of a surface M in E^3 belong to its intrinsic geometry?* The definition of isometry (Definition 4.2) suggests that isometric invariants must depend only on the dot product as applied to *tangent* vectors to M. But the shape operator derives from a *normal* vector field, and the examples in Section 4 show that isometric surfaces in E^3 can have quite different shapes. In fact, these examples provide a formal proof that shape operators, principal directions, principal curvatures, and mean curvature definitely do not belong to the intrinsic geometry of $M \subset E^3$.

To build a systematic theory of intrinsic geometry, we must look back at Section 1 and see how much of our work there is intrinsic to M. Using the dot product only on tangent vectors to M, we can still define a tangent frame field E_1, E_2 on M. Thus from an adapted frame field we can salvage the two tangent vector fields E_1, E_2—and hence also their dual 1-forms

θ_1, θ_2. It is somewhat surprising to find that these completely determine the connection form ω_{12}.

5.1 Lemma The connection form $\omega_{12} = -\omega_{21}$ is the only 1-form that satisfies the first structural equations

$$d\theta_1 = \omega_{12} \wedge \theta_2 \qquad d\theta_2 = \omega_{21} \wedge \theta_1.$$

Proof. Apply these equations to the tangent vector fields E_1, E_2. Since $\theta_i(E_j) = \delta_{ij}$, the definition of wedge product (Definition 4.3 of Chapter IV) gives

$$\omega_{12}(E_1) = d\theta_1(E_1, E_2)$$

$$\omega_{12}(E_2) = -\omega_{21}(E_2) = d\theta_2(E_1, E_2).$$

Thus by Lemma 2.1, $\omega_{12} = -\omega_{21}$ is uniquely determined by θ_1, θ_2. ∎

5.2 Remark In fact, this proof shows how to *construct* $\omega_{12} = -\omega_{21}$ without the use of Euclidean covariant derivatives (as in Section 1). Given E_1, E_2 and thus θ_1, θ_2; take the equations in the above proof as the *definition* of ω_{12} on E_1 and E_2. Then the usual linearity condition

$$\omega_{12}(V) = \omega_{12}(v_1 E_1 + v_2 E_2) = v_1 \omega_{12}(E_1) + v_2 \omega_{12}(E_2)$$

makes ω_{12} a 1-form on M, and one can easily check (by reversing the argument above) that $\omega_{12} = -\omega_{21}$ satisfies the first structural equations.

If $F: M \to \bar{M}$ is an isometry, then we can transfer a tangent frame field E_1, E_2 on M to a tangent frame field $\bar{E}_1$, $\bar{E}_2$ on $\bar{M}$: For each point $\mathbf{q}$ in $\bar{M}$ there is a unique point $\mathbf{p}$ in M such that $F(\mathbf{p}) = \mathbf{q}$. Then define

$$\bar{E}_1(\mathbf{q}) = F_*(E_1(\mathbf{p}))$$

$$\bar{E}_2(\mathbf{q}) = F_*(E_2(\mathbf{p})).$$

(Fig. 6.12).

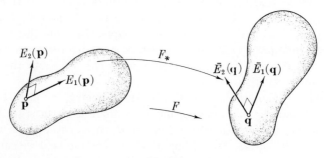

FIG. 6.12

In practice we shall abbreviate these formulas somewhat carelessly to

$$\bar{E}_1 = F_*(E_1), \quad \bar{E}_2 = F_*(E_2).$$

Because F_* preserves dot products, $\bar{E}_1$, $\bar{E}_2$ is a frame field on $\bar{M}$, since

$$\bar{E}_i \cdot \bar{E}_j = F_*(E_i) \cdot F_*(E_j) = E_i \cdot E_j = \delta_{ij}.$$

5.3 Lemma Let $F: M \to \bar{M}$ be an isometry, and let E_1, E_2 be a tangent frame field on M. If $\bar{E}_1$, $\bar{E}_2$ is the transferred frame field on $\bar{M}$, then
(1) $\theta_1 = F^*(\bar{\theta}_1), \quad \theta_2 = F^*(\bar{\theta}_2)$
(2) $\omega_{12} = F^*(\bar{\omega}_{12}).$

Proof. (1) It suffices to prove that θ_i and $F^*(\bar{\theta}_i)$ have the same value on E_1 and E_2. But for $1 \leq i, j \leq 2$ we have

$$F^*(\bar{\theta}_i)(E_j) = \bar{\theta}_i(F_* E_j) = \bar{\theta}_i(\bar{E}_j) = \delta_{ij} = \theta_i(E_j).$$

(2) Consider the structural equation $d\bar{\theta}_1 = \bar{\omega}_{12} \wedge \bar{\theta}_2$ on $\bar{M}$. If we apply F^*, then by the results in Chapter IV, Section 5, we get

$$d(F^*\bar{\theta}_1) = F^*(d\bar{\theta}_1) = F^*(\bar{\omega}_{12}) \wedge F^*(\bar{\theta}_2).$$

Hence, by (1), we have

$$d\theta_1 = F^*(\bar{\omega}_{12}) \wedge \theta_2.$$

The other structural equation

$$d\bar{\theta}_2 = \bar{\omega}_{21} \wedge \bar{\theta}_1$$

gives a corresponding equation, so

$$d\theta_1 = F^*(\bar{\omega}_{12}) \wedge \theta_2$$
$$d\theta_2 = F^*(\bar{\omega}_{21}) \wedge \theta_1.$$

But now (2) is an immediate consequence of the uniqueness property (Lemma 5.1), since

$$F^*(\bar{\omega}_{21}) = F^*(-\bar{\omega}_{12}) = -F^*(\bar{\omega}_{12}). \qquad \blacksquare$$

From this rather routine lemma we easily derive a proof of the celebrated *theorema egregium* of Gauss.

5.4 Theorem Gaussian curvature is an isometric invariant; that is if $F: M \to \bar{M}$ is an isometry, then

$$K(\mathbf{p}) = \bar{K}(F(\mathbf{p}))$$

for every point $\mathbf{p}$ in M.

Proof. For an arbitrary point $\mathbf{p}$ of M, pick a tangent frame field E_1, E_2

on some neighborhood of $\mathbf{p}$ and transfer via F_* to $\bar{E}_1$, $\bar{E}_2$ on $\bar{M}$. By the previous lemma, $F^*(\bar{\omega}_{12}) = \omega_{12}$. According to Corollary 2.3, we have

$$d\bar{\omega}_{12} = -\bar{K}\bar{\theta}_1 \wedge \bar{\theta}_2.$$

Apply F^* to this equation. By the results in Chapter IV, Section 5, we get

$$d(F^*\bar{\omega}_{12}) = F^*(d\bar{\omega}_{12}) = -F^*(\bar{K})F^*(\bar{\theta}_1) \wedge F^*(\bar{\theta}_2)$$

where $F^*(\bar{K})$ is simply the composite function $\bar{K}(F)$. Thus by the preceding lemma,

$$d\omega_{12} = -\bar{K}(F)\theta_1 \wedge \theta_2.$$

Comparison with $d\omega_{12} = -K\theta_1 \wedge \theta_2$ yields $K = \bar{K}(F)$; hence, in particular, $K(\mathbf{p}) = \bar{K}(F(\mathbf{p}))$.   ∎

Gauss's theorem is one of the great discoveries of nineteenth-century mathematics, and we shall see in the next chapter that its implications are far-reaching. The essential step in the proof is the second structural equation

$$d\omega_{12} = -K\theta_1 \wedge \theta_2.$$

Once we prove Lemma 5.1, all the ingredients of this equation, except K, are known to derive from M alone—thus K must also. This means that the inhabitants of $M \subset \mathbf{E}^3$ can determine the Gaussian curvature of their surface even though they cannot generally find S and have no conception of the shape of M in $\mathbf{E}^3$.

The machinery of differential forms puts this heuristic reasoning beyond doubt by supplying the formal proof of isometric invariance in Theorem 5.4. This remarkable situation is perhaps best illustrated by the formula $K = k_1 k_2$: An isometry need not preserve the principal curvatures, nor their sum, but *it must preserve their product*. Thus the shapes which isometric surfaces may have—although possibly quite different—are by no means unrelated.

A local isometry is, as we have shown, an isometry on all sufficiently small neighborhoods. Thus it follows from Theorem 5.4 that *local isometries preserve Gaussian curvature*. For example, in Example 4.6 the plane and the cylinder both have $K = 0$. (This is why we did not hesitate to call the curved cylinder "flat". Intrinsically it is as flat as a plane.) In the second part of Example 4.6, at corresponding points

$$\mathbf{x}(u,v) \quad \text{and} \quad F(\mathbf{x}(u,v)) = \mathbf{y}(u,v),$$

the helicoid and catenoid have exactly the same Gaussian curvature: $-1/(1 + u^2)^2$ (see Examples 4.3 and 6.4 of Chapter V.)

Gauss's *theorema egregium* can obviously be used to show that given surfaces are *not* isometric. For example, there can be no isometry of the sphere Σ (or even a very small region of it) onto part of the plane, since their Gaussian curvatures are different. This is the dilemma of the map maker: The intrinsic geometry of the earth's surface is misrepresented by any flat map.

The next section is computational; Section 7 will provide more isometric invariants.

EXERCISES

1. Geodesics belong to intrinsic geometry: If α is a geodesic in M, and $F: M \to N$ is a (local) isometry, then $F(\alpha)$ is a geodesic of N. (*Hint:* See Ex. 1 of VI.1.)

2. Use Exercise 1 to derive the geodesics of the circular cylinder (Example 5.8 of Chapter V). Generalize to an arbitrary cylinder.

3. For a (connected) surface, the values of its Gaussian curvature fill an interval. If there exists a local isometry of M onto N (in particular if M and N are isometric), show that M and N have the same curvature interval. Give an example to show that the converse is false.

4. Prove that no two of the following surfaces are isometric: sphere, torus, helicoid, circular cylinder, saddle surface.

5. *Bending of the helicoid into the catenoid* (4.6). For each number t in the interval $0 \leq t \leq \pi/2$, let $\mathbf{x}_t: E^2 \to E^3$ be the mapping such that

$$\mathbf{x}_t(u,v) = \cos t\,(Sc,\ Ss,\ v) + \sin t\,(-Cs,\ Cc,\ u)$$

where $C = \cosh u$, $S = \sinh u$, $c = \cos v$, and $s = \sin v$.

Now $\mathbf{x}_0$ is a patch covering the helicoid, and $\mathbf{x}_{\pi/2}$ is a parametrization of the catenoid—these are mild variants of our usual parametrizations, and the catenoid now has the z axis as its axis of rotation. If we imagine t to be time, then $\mathbf{x}_t$ for $0 \leq t \leq \pi/2$ describes a *bending* of the helicoid M_0 in space which carries it onto the catenoid $M_{\pi/2}$ through a whole family of intermediate surfaces $M_t = \mathbf{x}_t(E^2)$. Prove

(a) M_t *is a surface.* (Show merely that $\mathbf{x}_t$ is regular.)

(b) M_t *is isometric to the helicoid* M_0 *if* $t < \pi/2$. (Show that $F_t: M_0 \to M_t$ is an isometry, where

$$F_t(\mathbf{x}_0(u,v)) = \mathbf{x}_t(u,v).$$

Also show that for $t = \pi/2$, $F_{\pi/2}$ is a local isometry.)

(c) *Each* M_t *is a minimal surface.* (Compute $\mathbf{x}_{uu} + \mathbf{x}_{vv} = 0$.)

(d) *Unit normals are parallel on orbits:* Along the curve $t \to \mathbf{x}_t(u, v)$ by which the point $\mathbf{x}_0(u, v)$ of M_0 moves to $M_{\pi/2}$, the unit normals U_t of successive surfaces are *parallel*.

(e) *Gaussian curvature is constant on orbits.* (Find $K_t(\mathbf{x}_t(u, v))$, where K_t is the Gaussian curvature of M_t).

A brilliant series of illustrations of this bending is given in Struik [6].

6. Show that *every* local isometry of the helicoid H to the catenoid C must carry the axis of H to the minimal circle of C, and the rulings of H to the meridians of C, as in Example 4.6. (Compare Ex 11 of VI.4.)

6 Orthogonal Coordinates

We have seen that the intrinsic geometry of a surface $M \subset \mathbf{E}^3$ may be expressed in terms of the dual forms θ_1, θ_2, and connection form ω_{12} derived from a tangent frame field E_1, E_2. These forms satisfy
the first structural equations:

$$d\theta_1 = \omega_{12} \wedge \theta_2$$

$$d\theta_2 = \omega_{21} \wedge \theta_1$$

the second structural equation:

$$d\omega_{12} = -K\theta_1 \wedge \theta_2$$

In this section we develop a practical way to compute these forms—and hence a new way to find the Gaussian curvature of M.

The starting point is an *orthogonal* coordinate patch $\mathbf{x}: D \to M$, one for which $F = \mathbf{x}_u \cdot \mathbf{x}_v = 0$. Since $\mathbf{x}_u$ and $\mathbf{x}_v$ are orthogonal, dividing by their lengths $\| \mathbf{x}_u \| = \sqrt{E}$ and $\| \mathbf{x}_v \| = \sqrt{G}$ will produce frames.

6.1 Definition The *associated frame field* E_1, E_2 of an orthogonal patch $\mathbf{x}: D \to M$ consists of the orthogonal unit vector fields E_1 and E_2 whose values at each point $\mathbf{x}(u, v)$ of $\mathbf{x}(D)$ are

$$\mathbf{x}_u(u, v)/\sqrt{E}(u, v) \quad \text{and} \quad \mathbf{x}_v(u, v)/\sqrt{G}(u, v).$$

In Exercise 9 of Section 4 of Chapter IV, we associated with each patch $\mathbf{x}$ the coordinate functions $\tilde{u}$ and $\tilde{v}$, which assign to each point $\mathbf{x}(u, v)$ the numbers u and v, respectively. For example, for the geographical patch $\mathbf{x}$ of Example 2.2 of Chapter IV, the coordinate functions are the longitude and latitude functions on the sphere Σ. In the extreme case when $\mathbf{x}$ is the identity map of $\mathbf{E}^2$, the coordinate functions are just the natural coordinate functions $(u, v) \to u$, $(u, v) \to v$ on $\mathbf{E}^2$.

For an orthogonal patch $\mathbf{x}$ with associated frame field E_1, E_2, we shall express θ_1, θ_2, and ω_{12} in terms of the coordinate functions $\tilde{u}$, $\tilde{v}$. Since $\mathbf{x}$ is fixed throughout the discussion, we shall run the risk of omitting the inverse mapping $\mathbf{x}^{-1}$ from the notation. With this convention, the coordinate functions $\tilde{u} = u(\mathbf{x}^{-1})$ and $\tilde{v} = v(\mathbf{x}^{-1})$ are written simply u and v, and similarly $\mathbf{x}_u$ and $\mathbf{x}_v$ now become tangent vector fields on M itself. Thus the associated frame field of $\mathbf{x}$ has the concise expression

$$E_1 = \mathbf{x}_u/\sqrt{E} \qquad E_2 = \mathbf{x}_v/\sqrt{G}. \tag{1}$$

Now the dual forms θ_1, θ_2 are characterized by $\theta_i(E_j) = \delta_{ij}$, and in the exercise referred to above it is shown that

$$du(\mathbf{x}_u) = 1 \qquad dv(\mathbf{x}_u) = 0$$

$$du(\mathbf{x}_v) = 0 \qquad dv(\mathbf{x}_v) = 1.$$

Thus we deduce from (1) that

$$\theta_1 = \sqrt{E}\, du \qquad \theta_2 = \sqrt{G}\, dv. \tag{2}$$

By using the structural equations we shall find analogous formulas for ω_{12} and K. Recall that for a function f, $df = f_u\, du + f_v\, dv$, where the subscripts indicate partial derivatives. Hence from (2) we get

$$d\theta_1 = d(\sqrt{E}) \wedge du = (\sqrt{E})_v\, dv\, du = \frac{-(\sqrt{E})_v}{\sqrt{G}}\, du \wedge \theta_2$$

$$d\theta_2 = d(\sqrt{G}) \wedge dv = (\sqrt{G})_u\, du\, dv = \frac{-(\sqrt{G})_u}{\sqrt{E}}\, dv \wedge \theta_1$$

where we have used the alternation rule for wedge products and substituted $dv = \theta_2/\sqrt{G}$ and $du = \theta_1/\sqrt{E}$ from (2). Comparison with the first structural equations $d\theta_1 = \omega_{12} \wedge \theta_2$ and $d\theta_2 = -\omega_{12} \wedge \theta_1$ shows that

$$\omega_{12} = \frac{-(\sqrt{E})_v}{\sqrt{G}}\, du + \frac{(\sqrt{G})_u}{\sqrt{E}}\, dv \tag{3}$$

The logic is simple: By the computations above, this form satisfies the first structural equations; hence by uniqueness (Lemma 5.1), *it must be* ω_{12}.

6.2 Example *Geographical coordinates on the sphere.* For the geographical patch $\mathbf{x}$ in the sphere Σ (Example 2.2 of Chapter IV), we have computed $E = r^2 \cos^2 v$, $F = 0$, $G = r^2$. Thus by formula (2) above,

$$\theta_1 = r \cos v\, du \qquad \theta_2 = r\, dv.$$

Now $(\sqrt{E})_v = -r \sin v$ and $(\sqrt{G})_u = 0$; hence by (3),

$$\omega_{12} = \sin v\, du.$$

The associated frame field of this patch is the same one we got in Example 1.6 from the spherical frame field in $\mathbf{E}^3$. With the notational shift $u \to \vartheta$, $v \to \varphi$, the forms above are (necessarily) also the same. But now we have a simple way to compute them *directly in terms of the surface* with no appeal to the geometry of $\mathbf{E}^3$.

Finally we derive a new expression for the Gaussian curvature. In this context, exterior differentiation of ω_{12} as given in (3) yields

$$d\omega_{12} = -((\sqrt{E})_v/\sqrt{G})_v \, dv \, du + ((\sqrt{G})_u/\sqrt{E})_u \, du \, dv.$$

From (2) we get

$$\theta_1 \wedge \theta_2 = \sqrt{EG} \, du \, dv;$$

hence

$$-dv \, du = du \, dv = \frac{1}{\sqrt{EG}} \, \theta_1 \wedge \theta_2.$$

Thus the formula above becomes

$$d\omega_{12} = \frac{1}{\sqrt{EG}} \left\{ \left(\frac{(\sqrt{G})_u}{\sqrt{E}} \right)_u + \left(\frac{(\sqrt{E})_v}{\sqrt{G}} \right)_v \right\} \theta_1 \wedge \theta_2$$

We now compare this with the second structural equation,

$$d\omega_{12} = -K\theta_1 \wedge \theta_2.$$

6.3 Lemma If $\mathbf{x} \colon D \to M$ is an orthogonal patch, then the Gaussian curvature K is given in terms of $\mathbf{x}$ by

$$K = \frac{-1}{\sqrt{EG}} \left\{ \left(\frac{(\sqrt{G})_u}{\sqrt{E}} \right)_u + \left(\frac{(\sqrt{E})_v}{\sqrt{G}} \right)_v \right\}$$

By contrast with the formula for K in Corollary 4.1 of Chapter V, the functions l, m, and n (which describe the shape operator) no longer appear. Indeed since K is now expressed solely in terms of E, F, and G, we have, using Lemma 4.5, another proof of the isometric invariance of Gaussian curvature.

In applications it is generally easier to repeat the derivation of Lemma 6.3 in each case, rather than look it up and substitute in it. For example, consider the polar parametrization $\mathbf{x}(u, v) = (u \cos v, u \sin v)$ of the Euclidean plane $\mathbf{E}^2$. Here $E = 1$, $F = 0$ (so $\mathbf{x}$ is orthogonal), and $G = u^2$. Thus by (2), $\theta_1 = du$ and $\theta_2 = u \, dv$. Since

$$d\theta_1 = 0 \qquad \text{and} \qquad d\theta_2 = du\ dv = -dv \wedge \theta_1,$$

we find $\omega_{12} = dv$. But then $d\omega_{12} = 0$, which shows again that $\mathbf{E}^2$ is flat.

6.4 Example   The natural frame field of a surface of revolution.   For a canonical parametrization

$$\mathbf{x}(u, v) = (g(u), h(u)\cos v, h(u)\sin v)$$

of a surface of revolution, the associated frame field has E_1 in the direction of the meridians and E_2 in the direction of the parallels (Fig. 6.13). Since $\mathbf{x}$ is orthogonal, with $E = 1$, $G = h^2$, we get $\theta_1 = du$ and $\theta_2 = h\ dv$. Here h is a function of u alone, so $h_v = 0$, and h_u is the ordinary derivative h'. From (3)—or by direct computation—$\omega_{12} = h'\ dv$, so

$$d\omega_{12} = h''\ du\ dv = \frac{h''}{h}\ \theta_1 \wedge \theta_2$$

We conclude that the Gaussian curvature is $K = -h''/h$, in agreement with the results of Lemma 6.3 of Chapter V.

EXERCISES

1. Compute the dual forms, connection form, and Gaussian curvature for the associated frame field of the following orthogonal patches:
   (a) $\mathbf{x}(u, v) = (u\cos v, u\sin v, bv)$, helicoid
   (b) $\mathbf{x}(u, v) = (u\cos v, u\sin v, u^2/2)$, paraboloid of revolution.
   (c) $\mathbf{x}(u, v) = (u\cos v, u\sin v, au)$, cone

2. Let the patch $\mathbf{x}\colon D \to M$ be a conformal mapping. (The associated coordinate system is said to be *isothermal*.) Prove:
   (a) $K = -\Delta(\log E)/2E$, where Δ is the *Laplacian:* $\Delta f = f_{uu} + f_{vv}$.

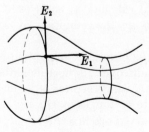

FIG. 6.13

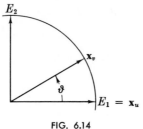

FIG. 6.14

(b) The mean curvature H is zero if and only if $\mathbf{x}_{uu} + \mathbf{x}_{vv} = 0$. (*Hint:* Use Ex. 7 of VI.4.)

3. For a patch with $E = G = 1$, show that $K = -\vartheta_{uv}/\sin \vartheta$, where ϑ is the coordinate angle. (*Hint:* For the frame field with $E_1 = \mathbf{x}_u$ as in Fig. 6.14, show that $\theta_1 = du + \cos \vartheta \, dv$, $\theta_2 = \sin \vartheta \, dv$.)

4. If $\mathbf{x}$ is a principal patch (Ex. 9 of V.4), prove

(a) $\omega_{13} = \dfrac{\ell}{\sqrt{E}} \, du$ 　　　　　 (b) $\ell_v = HE_v$

　　 $\omega_{23} = \dfrac{n}{\sqrt{G}} \, dv.$ 　　　　　　 $n_u = HG_u.$

5. By refining the argument in the text, show that equations (2) and (3) are *literally* true provided θ_1, θ_2, and ω_{12} are replaced by their coordinate expressions

$$\mathbf{x}^*(\theta_1), \ \mathbf{x}^*(\theta_2), \ \mathbf{x}^*(\omega_{12}).$$

7　Integration and Orientation

The main aim of this section is to define the integral of a 2-form over a compact oriented surface. This notion does not involve geometry at all; it belongs to the integral calculus on surfaces (Chapter IV, Section 6). However, we shall motivate the definition by considering some related geometric problems.

Perhaps the simplest use of double integration in geometry is in finding the area of a surface. To discover a proper definition of area, we start with a patch $\mathbf{x}: D \to M$ and ask what the area of its image $\mathbf{x}(D)$ should be. Let ΔR be a small coordinate rectangle in D with sides Δu and Δv. Now $\mathbf{x}$ distorts ΔR into a small curved region $\mathbf{x}(\Delta R)$ in M, marked off in an obvious way by four segments of parameter curves, as shown in Fig. 6.15.

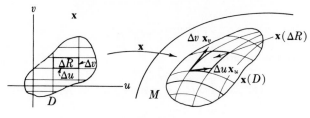

FIG. 6.15

We have seen that the segment from $\mathbf{x}(u, v)$ to $\mathbf{x}(u + \Delta u, v)$ is linearly approximated by $\Delta u \; \mathbf{x}_u$ (evaluated at (u, v)), and the one from $\mathbf{x}(u, v)$ to $\mathbf{x}(u, v + \Delta v)$ by $\Delta v \; \mathbf{x}_v$. Thus the region $\mathbf{x}(\Delta R)$ is approximated by the parallelogram in $T_{\mathbf{x}(u,v)}(M)$ with these vectors as sides. From Chapter II, Section 1, we know that this parallelogram has area

$$\| \Delta u \mathbf{x}_u \times \Delta v \mathbf{x}_v \| = \| \mathbf{x}_u \times \mathbf{x}_v \| \Delta u \Delta v = \sqrt{EG - F^2} \; \Delta u \Delta v$$

We conclude that the area of $\mathbf{x}(\Delta R)$ should be approximately $\sqrt{EG - F^2}$ times the area $\Delta u \Delta v$ of ΔR. At each point (u, v) the familiar expression $\sqrt{EG - F^2}$ gives *the rate at which* $\mathbf{x}$ *is expanding area* at (u, v). Thus it is natural to define the *area* of the whole region $\mathbf{x}(D)$ to be

$$\iint_D \sqrt{EG - F^2} \; du \; dv$$

Such integrals, of course, may well be improper; we shall avoid this difficulty by modifying the notion of patch.

7.1 Definition　The interior R° of a rectangle R: $a \leqq u \leqq b, c \leqq v \leqq d$ is the open set $a < u < b, c < v < d$. A 2-segment $\mathbf{x}: R \to M$ is *patchlike* provided the restricted mapping $\mathbf{x}: R^\circ \to M$ is a patch in M.

The remark preceding Lemma 7.3 of Chapter IV shows that the area of $\mathbf{x}(R)$ is finite, since it implies that $\sqrt{EG - F^2} \geqq 0$ is bounded on R.

A patch-like 2-segment $\mathbf{x}: R \to M$ need not be one-to-one on the boundary of R, so its image may not be very rectangular. In fact we shall now see that the area of an entire compact surface may often be computed by covering it with a *single* 2-segment.

7.2 Example　Areas of surfaces

(1) *The sphere* Σ *of radius* r. If the formula defining the geographical patch is applied to the rectangle R: $-\pi \leq u \leq \pi, -\pi/2 \leq v \leq \pi/2$, we obtain a 2-segment that covers the whole sphere. Now

$$E = r^2 \cos^2 v, \qquad F = 0, \qquad \text{and } G = r^2,$$

so

$$\sqrt{EG - F^2} = r^2 \cos v,$$

and the area of the sphere is thus

$$A(\Sigma) = \int_{-\pi}^{\pi} \int_{-\pi/2}^{\pi/2} r^2 \cos v \, du \, dv = 4\pi r^2.$$

(2) *Torus T of radii* $R > r > 0$. From Example 2.6 of Chapter IV we can derive a patchlike 2-segment covering the torus. Here

$$\sqrt{EG - F^2} = r(R + r \cos u),$$

so the area is

$$A(T) = \int_{-\pi}^{\pi} \int_{-\pi}^{\pi} r(R + r \cos u) \, du \, dv = 4\pi^2 Rr$$

(3) *The bugle surface* (Example 6.6 of Chapter V). Every surface of revolution M has a canonical parametrization with $E = 1, F = 0, G = h^2$. On a rectangle $R: a \leqq u \leqq b, 0 \leqq v \leqq 2\pi$, $\mathbf{x}$ is a patchlike 2-segment whose image is the region of M between the parallels $u = a$ and $u = b$ (Fig. 6.16). Thus the area of this region is

$$A_{ab} = \int_{a}^{b} \int_{0}^{2\pi} h \, du \, dv = 2\pi \int_{a}^{b} h \, du.$$

For the bugle surface, we saw in Chapter V that $h(u) = ce^{-u/c}$; hence

$$A_{ab} = 2\pi c \int_{a}^{b} e^{-u/c} \, du$$

$$= 2\pi c^2 (e^{-a/c} - e^{-b/c}).$$

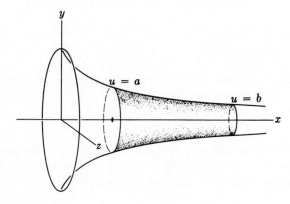

FIG. 6.16

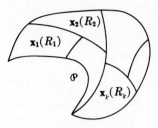

FIG. 6.17

To find the area of the whole bugle—a noncompact surface—we expand this region, letting $a \to 0$ and $b \to \infty$.

Thus (see Remark 7.6) the bugle has finite area

$$\lim_{\substack{a \to 0 \\ b \to \infty}} A_{ab} = 2\pi c^2.$$

To define the area of a complicated region we shall not try to fit a single patchlike 2-segment onto it. Instead we follow the usual scheme of elementary calculus and break the region into simple pieces, then add their areas.

7.3 Definition   A *paving* of a region $\mathcal{P}$ in M consists of a finite number of patchlike 2-segments $x_1, \cdots, x_k$ whose images fill $\mathcal{P}$ in such a way that each point of $\mathcal{P}$ is in at most one set $x_i(R_i^{\circ})$.

In short, the images of the x_i's cover $\mathcal{P}$ exactly and overlap only on their boundaries (Fig. 6.17).

Not every region is pavable; since pavings are finite, compactness is certainly necessary (Definition 7.2 of Chapter IV). It is safe to assume that a compact region is pavable if its boundary consists of a finite number of regular curve segments. In particular, *an entire compact surface is always pavable.*† The area of a pavable region $\mathcal{P}$ is defined to be the sum of the areas of $x_1(R_1), \cdots, x_k(R_k)$ for a paving of $\mathcal{P}$. (The consistency problem here will be discussed following the analogous definition, 7.5.)

The preceding exposition shows that computation of area does not demand differential forms, but integration of 2-forms (Definition 6.3 of Chapter IV) will give area—and much more besides. The first question is this: *Which 2-form should we integrate over a patchlike 2-segment* x *to get the area of its image?* By definition,

$$\iint_{x} \mu = \iint_{R} \mu(x_u, x_v) \, du \, dv.$$

† See remarks and reference following Theorem 8.5 of Chapter VII.

Thus we want a 2-form whose value on $\mathbf{x}_u, \mathbf{x}_v$ is

$$\| \mathbf{x}_u \times \mathbf{x}_v \| = \sqrt{EG - F^2}.$$

In general, a 2-form μ such that

$$\mu(\mathbf{v}, \mathbf{w}) = \pm \| \mathbf{v} \times \mathbf{w} \| \qquad \text{for all } \mathbf{v}, \mathbf{w}$$

is called an *area form*. Such a form assigns to every pair of tangent vectors $\mathbf{v}$, $\mathbf{w}$ either plus or minus the area of the parallelogram with sides $\mathbf{v}$ and $\mathbf{w}$.

7.4 Lemma   A surface M has an area form if and only if it is orientable. On a (connected) orientable surface M there are exactly two area forms, which are negatives of each other. (We denote them by dM and $-dM$.)

Proof. If $\mathbf{v}$ and $\mathbf{w}$ are linearly independent, then $\| \mathbf{v} \times \mathbf{w} \| > 0$; thus area forms are nonvanishing. Hence, by Definition 7.4 of Chapter IV, a nonorientable surface cannot have an area form.

Now suppose that M is an orientable surface in $\mathbf{E}^3$. The proof of Theorem 7.5 of Chapter IV actually establishes a one-to-one correspondence between normal vector fields on M and 2-forms on M. If U is a *unit* normal, then the associated 2-form dM is an area form, since

$$dM(\mathbf{v}, \mathbf{w}) = U(\mathbf{p}) \cdot \mathbf{v} \times \mathbf{w} = \pm \| \mathbf{v} \times \mathbf{w} \|.$$

(In Fig. 6.18 this number is positive, since $\mathbf{v} \times \mathbf{w}$ points in the same direction as $U(\mathbf{p})$, but if $\mathbf{v}$ and $\mathbf{w}$ were reversed, we would get the negative sign.)

Thus the two unit normal vector fields on M determine the two area forms dM and $-dM$ on M.   ▮

To *orient* an orientable surface is to pick one of its two area forms, since that amounts to the same thing as picking one of its unit normals.

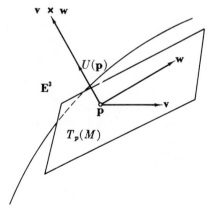

FIG. 6.18

Finding area is not a really typical integration problem, since area is always positive. Thus to find area by integrating an area form we must be careful about signs. Suppose $\mathbf{x}$ is a patchlike 2-segment in a surface oriented by an area form dM. By definition,

$$\iint_{\mathbf{x}} dM = \iint_{R} dM(\mathbf{x}_u, \mathbf{x}_v) \, du \, dv.$$

Now there are two cases:

(1) If $dM(\mathbf{x}_u, \mathbf{x}_v) > 0$, we say that $\mathbf{x}$ is *positively oriented*. Then by definition of area form,

$$dM(\mathbf{x}_u, \mathbf{x}_v) = \| \mathbf{x}_u \times \mathbf{x}_v \|;$$

hence $\iint_{\mathbf{x}} dM$ is the area of $\mathbf{x}(R)$.

(2) If $dM(\mathbf{x}_u, \mathbf{x}_v) < 0$, we say that $\mathbf{x}$ is *negatively oriented*. Then

$$dM(\mathbf{x}_u, \mathbf{x}_v) = - \| \mathbf{x}_u \times \mathbf{x}_v \|,$$

hence $\iint_{\mathbf{x}} dM$ is *minus* the area of $\mathbf{x}(R)$.

Thus to find the area of a pavable oriented region by integrating its selected area form, we cannot use an arbitrary paving; the paving must be *positively oriented*, that is, consist only of positively oriented patchlike 2-segments. Then

$$A(\mathcal{P}) = \sum_i A(\mathbf{x}(R_i)) = \sum_i \iint_{\mathbf{x}_i} dM.$$

Now we replace the area form by an arbitrary 2-form to get the definition we are looking for.

7.5 Definition   Let ν be a 2-form defined on a pavable oriented region $\mathcal{P}$ in a surface. The integral of ν over $\mathcal{P}$ is

$$\iint_{\mathcal{P}} \nu = \sum_{i=1}^{k} \iint_{\mathbf{x}_i} \nu$$

where $\mathbf{x}_1, \cdots, \mathbf{x}_k$ is a positively oriented paving of $\mathcal{P}$.

There is a consistency problem in this definition: We must know that two *different* positively oriented pavings of $\mathcal{P}$ give the *same* value for the sum on the right. A detailed proof would be somewhat long; the general scheme is given on page 103 of Hicks [5].

Since compact surfaces are pavable, the definition above gives in particular the integral of a 2-form over a compact oriented surface.

7.6 Remark   Improper integrals.   We have defined area and the integration of forms for *compact* surfaces; however, the notion of area can easily be extended to a *noncompact* surface N. We define the area of N to be the least upper bound of the set of all areas of pavable regions $\mathcal{P}$ in N:

$$A(N) = \text{l.u.b.} \ A(\mathcal{P}).$$

Thus $A(N) = +\infty$ if no finite upper bound exists.

By contrast, it is generally impossible to assign a value—finite or infinite—to the improper integral $\iint_N f \, dN$. The special case $f \geq 0$, however, may be handled in the same way as area; we set

$$\iint_N f \, dN = \text{l.u.b.} \iint_{\mathcal{P}} f \, dN \qquad (\mathcal{P} \text{ pavable in } N)$$

For $f \leq 0$, switch to greatest lower bound. Thus values $+\infty$ and $-\infty$ are possible in these two cases. Now let $\mathcal{P}_1, \mathcal{P}_2, \cdots$ be a sequence of pavable regions in N such that $\mathcal{P}_i$ is contained in $\mathcal{P}_{i+1}$, and every pavable region in N is contained in some $\mathcal{P}_i$. It then follows that

$$\lim_{i \to \infty} \iint_{\mathcal{P}_i} f \, dN = \iint_N f \, dN.$$

(We used the corresponding fact for area in (3) of Example 7.2.)

If $\mathcal{P}$ is a pavable region in a surface M oriented by dM, we have seen that $\iint_{\mathcal{P}} dM$ is the area of $\mathcal{P}$. More generally, $\iint_{\mathcal{P}} f \, dM$ gives the integral of a function f over $\mathcal{P}$—an obvious analogue of the usual integral $\int_a^b f \, dx$ from elementary calculus. We turn now to an important geometric application of this idea.

7.7 Definition Let K be the Gaussian curvature of a surface M, and let $\mathcal{P}$ be a pavable region in M oriented by dM. Then the number

$$\iint_{\mathcal{P}} K \, dM$$

is called the *total Gaussian curvature* of $\mathcal{P}$.

When $\mathcal{P}$ is an entire compact oriented surface M, we get the total Gaussian curvature of M. The total is an algebraic one: Negative curvature at one place may cancel positive curvature at another.

To compute the total curvature, Definition 7.5 shows that it suffices to know how to integrate the 2-form $K \, dM$ over patchlike 2-segments. But

$$\iint_{\mathbf{x}} K \, dM = \iint_R \mathbf{x}^*(K \, dM)$$

$$= \iint_R K(\mathbf{x}) \mathbf{x}^*(dM) = \int_a^b \int_c^d K(\mathbf{x}) \sqrt{EG - F^2} \, du \, dv$$

with the usual notation for $\mathbf{x}: R \to M$. Then $K(\mathbf{x})$ may be computed explicitly by Corollary 4.1 of Chapter V or by Lemma 6.3. Luckily the orientation problems here solve themselves automatically; see Exercise 4(c).

7.8 Example   Total curvature of some surfaces

(1) *Constant curvature.* If the Gaussian curvature of M is constant, then its total curvature is

$$\iint_M K \, dM = K \iint_M dM = K \, A(M).$$

Thus a sphere of radius r has total curvature 4π (since $K A(M)$ becomes $(1/r^2)(4\pi r^2)$), and the bugle surface has total curvature -2π (since $K A(M)$ becomes $(-1/c^2)(2\pi c^2)$).

(2) *Torus.* Let $\mathbf{x}$ be the 2-segment used on the torus T in Example 7.2. By this example the area form dT has the coordinate expression

$$\mathbf{x}^*(dT) = \sqrt{EG - F^2} \, du \, dv = r(R + r \cos u) \, du \, dv$$

But in Example 6.1 of Chapter V we computed

$$K(\mathbf{x}) = (\cos u)/r(R + r \cos u)$$

for this same $\mathbf{x}$. Hence, the torus has total curvature

$$\iint_T K \, dT = \int_{-\pi}^{\pi} \int_{-\pi}^{\pi} \cos u \, du \, dv = 0.$$

Thus on the torus the negative curvature of its *inner* half exactly balances the positive curvature on its *outer* half, giving total curvature zero.

(3) *Catenoid.* This surface is not compact, and its area is infinite; nevertheless its total curvature—treated by the remark above as an improper integral—is finite. On the rectangle R: $-a \leq u \leq a$, $0 \leq v \leq 2\pi$, the parametrization in Example 6.1 of Chapter V becomes a patchlike 2-segment covering the region between the parallels $u = -a$ and $u = a$. (Fig. 6.19). From Example 6.1 of Chapter V, we get

$$K(\mathbf{x}) = \frac{-1}{c^2 \cosh^4 (u/c)}$$

and

$$\mathbf{x}^*(dM) = \sqrt{EG} \, du \, dv = c \cosh^2 (u/c).$$

Hence the region has total curvature

$$\iint_{\mathbf{x}} K \, dM = - \int_{-a}^{a} \int_0^{2\pi} \frac{du \, dv}{c \cosh^2(u/c)} = -4\pi \tanh \left(\frac{a}{c}\right).$$

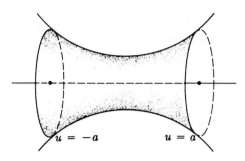

$$u = -a \qquad u = a$$

FIG. 6.19

As $a \to \infty$, this region expands to fill the whole surface; thus the total curvature of the catenoid is

$$\iint_M K\, dM = -4\pi \lim_{a \to \infty} \tanh\left(\frac{a}{c}\right) = -4\pi.$$

The total curvatures computed above are 4π, -2π, 0, and -4π—a rather special set of numbers. Furthermore, none depends on the particular "size" (radius r, constant c, $\cdots$) of its surface. A partial explanation is provided by Corollary 7.10; a rather deeper one comes in Chapter VII, Section 8.

If $F: M \to N$ is a mapping of oriented surfaces, then the *Jacobian* J of F is the real-valued function on M such that

$$F^*(dN) = J\, dM.$$

We can get an idea of the geometric meaning of J by arguing as in the special case at the beginning of this section. If $\mathbf{v}$ and $\mathbf{w}$ are very small tangent vectors at a point $\mathbf{p}$ of M, they span a parallelopiped in $T_p(M)$ which approximates a small region ΔM in M. The character of the derivative map F_* is such that $F_*(\mathbf{v})$ and $F_*(\mathbf{w})$ are the sides of a parallelogram in $T_{F(p)}(N)$ approximating the image region $F(\Delta M)$, as shown in Fig. 6.20. By the definition of Jacobian we have

$$J(\mathbf{p})\, dM(\mathbf{v}, \mathbf{w}) = (F^*\, dN)(\mathbf{v}, \mathbf{w}) = dN(F_*\mathbf{v}, F_*\mathbf{w}). \qquad (*)$$

Now $\| \mathbf{v} \times \mathbf{w} \|$ is approximately the area of ΔM (and similarly for $F(\Delta M)$). Hence by taking absolute values we get

$$|J(\mathbf{p})|\ (\text{area of } \Delta M) \sim \text{area of } F(\Delta M).$$

Thus $|J(\mathbf{p})|$ *gives the rate at which F is expanding area at* $\mathbf{p}$. Furthermore if ΔM is positively oriented, that is, $dM(\mathbf{v}, \mathbf{w}) > 0$, then $(*)$ shows that the sign of $dN(F_*\mathbf{v}, F_*\mathbf{w})$ is the same as that of $J(\mathbf{p})$. Thus *the sign of*

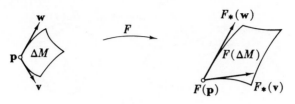

FIG. 6.20

$J(\mathbf{p})$ *tells whether F preserves or reverses the orientation of ΔM.*
In this context we call the number

$$\iint_M J \, dM = \iint_M F^*(dN)$$

the *algebraic area* of $F(M)$. The discussion above shows that, roughly speaking, each small region ΔM in M contributes to this total the *algebraic* area of its image $F(\Delta M)$:

(1) Positive, if the orientation of $F(\Delta M)$ agrees with that of N;

(2) Negative, if these orientations disagree (so F has turned ΔM over);

(3) Zero, if F collapses ΔM to a curve or a point.

Let us consider what this means in the case of the Gauss mapping (Exercise 4 of Section 1 of Chapter V).

7.9 Theorem   The Gaussian curvature K of an oriented surface $M \subset \mathbf{E}^3$ is the Jacobian of its Gauss mapping $G \colon M \to \Sigma$.

(Here Σ is the unit sphere, oriented by the outward normal $\bar{U}$ or the corresponding area form $d\Sigma$.)

Proof. If $U = \Sigma \, g_i U_i$ is the unit normal orienting M, then the corresponding Gauss mapping is $G = (g_1, g_2, g_3)$. Notice that if S is the shape operator of M given by U, then

$$-S(\mathbf{v}) = \nabla_v U = \Sigma \, \mathbf{v}[g_i] U_i(\mathbf{p})$$

and by Theorem 7.5 of Chapter I,

$$G_*(\mathbf{v}) = \Sigma \, \mathbf{v}[g_i] \, U_i(G(\mathbf{p})).$$

Hence $G_*(\mathbf{v})$ and $-S(\mathbf{v})$ are parallel for any tangent vector $\mathbf{v}$ to M, as shown in Fig. 6.21.

To prove the theorem we must show that

$$K \, dM = G^*(d\Sigma),$$

so we evaluate these 2-forms on an arbitrary pair of tangent vectors to M. Using Lemma 3.4 of Chapter V,

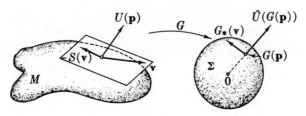

FIG. 6.21

$$(K\,dM)(\mathbf{v},\mathbf{w}) = K(\mathbf{p})\,dM(\mathbf{v},\mathbf{w}) = K(\mathbf{p})\,U(\mathbf{p})\cdot\mathbf{v}\times\mathbf{w}$$

$$= U(\mathbf{p})\cdot S(\mathbf{v})\times S(\mathbf{w}).$$

On the other hand,

$$(G^*\,d\Sigma)(\mathbf{v},\mathbf{w}) = d\Sigma(G_*\mathbf{v},\,G_*\mathbf{w}) = \bar{U}(G(\mathbf{p}))\cdot G_*\mathbf{v}\times G_*\mathbf{w}$$

Now a triple scalar product depends only on the Euclidean coordinates of its vectors, so $G_*(\mathbf{v})$ and $G_*(\mathbf{w})$ may be replaced by the parallel vectors $-S(\mathbf{v})$ and $-S(\mathbf{w})$. Furthermore, by the definition of G and the special character of the unit sphere Σ, the vectors $U(\mathbf{p})$ and $\bar{U}(G(\mathbf{p}))$ are also parallel (Fig. 6.21). Thus the two triple scalar products above are the same—and the proof is complete. ∎

7.10 Corollary The total Gaussian curvature of an oriented surface $M \subset E^3$ is the algebraic area of the image $G(M)$ of its Gauss mapping $G\colon M \to \Sigma$.

To prove this it suffices to integrate the form

$$K\,dM = G^*(d\Sigma)$$

over M.

Algebraic area can be tricky when the mapping $F\colon M \to N$ folds M many times over the same regions in N. Thus for practical purposes, the following special case of Theorem 7.10 is simpler, since it involves only ordinary area.

7.11 Corollary If $\Re$ is an oriented region in $M \subset E^3$ on which (1) the Gauss map G is one-to-one (U is not parallel at different points of $\Re$), and (2) either $K \geq 0$ or $K \leq 0$, then the total curvature of $\Re$ is plus or minus the area of $G(\Re)$, where the sign is that of K. Furthermore this area does not exceed 4π.

(The proof uses improper integrals.) For example, consider the Gauss mapping of an oriented torus. Now G collapses the top and bottom circles of T (where $K = 0$) to the north and south poles of Σ. If, as in Fig. 5.21, $\mathcal{O}$ and $\mathcal{I}$ are the outer and inner halves of T, then G maps $\mathcal{O}$ (where $K \geq 0$)

in one-to-one fashion onto the whole sphere Σ—and does the same for $\mathcal{G}$ (where $K \leq 0$). Thus T has total curvature $+ A(\Sigma) - A(\Sigma) = 0$, as we found in Example 7.8 by an explicit integration.

As another example we find that the entire bugle surface B satisfies the hypotheses in Corollary 7.11. In fact, as Fig. 6.22 suggests, the Gauss mapping carries its profile curve in one-to-one fashion onto a quarter of a great circle in Σ. Thus by moving U around the parallels of B, we see that G is one-to-one from B onto an open hemisphere (minus its central point, since the rim is not part of B). Thus the total curvature of the bugle is $- (\tfrac{1}{2})A(\Sigma) = -2\pi$. Furthermore, since the bugle has constant curvature, we can find its area without explicit integration: Total curvature -2π divided by (constant) curvature $K = -1/c^2$ gives the area $2\pi c^2$, as found in Example 7.2.

On an oriented surface, the ambiguity in the measurement of angles mentioned in Chapter II, Section 1, can be reduced. If the surface is oriented by an unit normal U, then for every tangent vector $\mathbf{v}$ to M, $U \times \mathbf{v}$ is a tangent vector orthogonal to $\mathbf{v}$. We shall think of $U \times \mathbf{v}$ as $\mathbf{v}$ rotated through $+90°$. Then if $\mathbf{v}$ and $\mathbf{w}$ are unit tangent vectors at a point of M, a number ϑ is defined to be an *oriented angle from* $\mathbf{v}$ *to* $\mathbf{w}$ provided

$$\mathbf{w} = \cos \vartheta \, \mathbf{v} + \sin \vartheta \, (U \times \mathbf{v}).$$

(Fig. 6.23). All oriented angles from $\mathbf{v}$ to $\mathbf{w}$ thus have the form $\vartheta + 2\pi n$, where n is an arbitrary integer. (The same scheme applies to any pair of nonzero tangent vectors: It suffices to divide by their norms to get unit vectors.)

Consistency is the essence of orientability—in studying an oriented surface oriented by an area form dM we shall always use positively oriented patches, $dM(\mathbf{x}_u, \mathbf{x}_v) > 0$, and *positively oriented frame fields*, for which

$$dM(E_1, E_2) = +1.$$

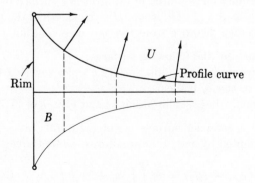

FIG. 6.22

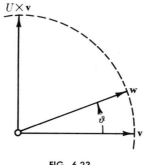

FIG. 6.23

(Note that by definition of area form the only possible values of dM on a frame are ± 1.) For a positively oriented frame field, we can now give geometric meaning to the wedge product of its dual forms:

$$dM = \theta_1 \wedge \theta_2$$

on the domain of the frame field. To prove this useful fact, it suffices to note that both sides have the same value, $+1$, on E_1, E_2. The second structural equation (Corollary 2.3) then becomes $d\omega_{12} = -K \, dM$.

EXERCISES

1. For a Monge patch

$$\mathbf{x}(u,v) = (u, v, f(u,v)),$$

show that the area of $\mathbf{x}(D)$ is given by the usual formula from elementary calculus. Deduce that $A(\mathbf{x}(D)) \geq A(D)$.

2. Find a formula for the area of an arbitrary surface of revolution, and interpret it as $A = 2\pi L\bar{h}$, where L is the length of the profile curve and $\bar{h}$ is its average distance from the axis of revolution (Pappus).

3. Find the area of the following surfaces:
   (a) The region in the saddle surface $z = xy$ covering the disc $x^2 + y^2 \leq c^2$ in the xy plane.
   (b) Catenoid   (c) The Möbius band (Ex. 7 of IV.7).

4. Let M be a compact surface oriented by dM; let $-M$ be the same surface oriented by the other area form $-dM$. Prove

   (a) $\displaystyle\iint_M (c_1\nu_1 + c_2\nu_2) = c_1 \iint_M \nu_1 + c_2 \iint_M \nu_2$ $(c_1, c_2$ constant$)$.

(b) $\iint_{-M} \nu = -\iint_{M} \nu.$

(*Hint:* If $\tilde{\mathbf{x}}(u,v) = \mathbf{x}(v,u)$, show that $\tilde{\mathbf{x}}$ and $\mathbf{x}$ have opposite orientations; then look ahead to Exercise 21.)

(c) $\iint_{M} f \, dM = \iint_{-M} f(-dM).$

(d) If $f \leq g$, then $\iint_{M} f \, dM \leq \iint_{M} g \, dM$

(Note the effect of $f = 0$ or $g = 0$.)

Property (c) shows how to define the integral of a *function* over a compact surface that is merely orient*able*; either choice of area form leads to the same result. In particular, the total curvature of a compact orientable surface is now well-defined.

5. *Total curvature of surfaces of revolution.* On a surface of revolution M with profile curve α, let Z_{ab} be the region ("zone") between the parallels through $\alpha(a)$ and $\alpha(b)$.

(a) Show that the total curvature of Z_{ab} is $2\pi(\sin \varphi_a - \sin \varphi_b)$, where these are the slope angles of α at $\alpha(a)$ and $\alpha(b)$—measured relative to the axis of revolution (Fig. 6.24).

(b) Deduce that every surface of revolution whose profile curve is closed has total curvature zero.

If the profile curve α is not closed, then it is one-to-one on some open interval $A < t < B$. Define the total curvature in this case to be

$$2\pi(\lim_{a \to A} \sin \varphi_a - \lim_{b \to B} \sin \varphi_b)$$

provided both limits exist.

(c) Test this formula on the bugle and catenoid.

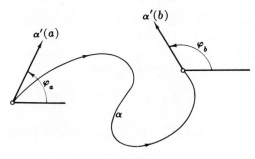

FIG. 6.24

6. Show that the Gauss mapping of a surface $M \subset \mathbf{E}^3$ is conformal if and only if M is part of either a sphere or a minimal surface.

7. Find the total curvatures of the constant curvature surfaces of revolution (Chapter V, Section 6, and Exercises); deduce their areas.

8. The area forms of $\mathbf{E}^2$ are, as expected, $\pm\, du\, dv$ (since du and dv are the dual forms of a frame field). The *natural orientation* of $\mathbf{E}^2$ is by $du\, dv$; this orientation is assumed unless the contrary is explicitly stated.
   (a) Using the general definition in the text, show that the Jacobian of a mapping

$$F = (f, g): \mathbf{E}^2 \to \mathbf{E}^2$$

   is given by the usual formula

$$J = f_u g_v - f_v g_u.$$

   (b) Show that the Jacobian of a patch $\mathbf{x}: D \to M$ in an oriented surface (D connected) is $\pm\, \sqrt{EG - F^2}$, where the sign depends on whether $\mathbf{x}$ is positively or negatively oriented.

9. Let M be a ruled surface whose rulings are entire straight lines, and assume that $K < 0$.
   (a) Show that the total curvature of M is $-2L(\delta)$, where δ is the director curve, with $\|\delta\| = 1$.
   (b) Compute the total curvature of the saddle surface $M\colon z = xy$ by this method, and check the result using Corollary 7.11.

10. *Total curvature of quadric surfaces*
   (a) Find the total curvature of an arbitrary hyperbolic paraboloid, elliptic paraboloid, and ellipsoid.
   (b) Show that the total curvature of the hyperboloid of revolution $M\colon (x^2 + y^2)/a^2 - z^2/c^2 = 1$ is $-4\pi a/\sqrt{a^2 + c^2}$.

   Thus the total curvature of an elliptic hyperboloid is dependent on the particular "dimensions" of the surface—and the same is true for an elliptic hyperboloid of two sheets.

11. A *simple region* S in M is a region that can be expressed as the image $F(D)$ of the disc $u^2 + v^2 \leq 1$ in $\mathbf{E}^2$ under a one-to-one regular mapping F. Show that S can be paved by a single patchlike 2-segment $\mathbf{x}$ in such a way that for any 1-form ϕ we have

$$\iint_S d\phi = \int_\alpha \phi,$$

   where α is an edge curve of $\mathbf{x}$. (*Hint:* See Ex. 12 of IV.6.)

12. (*Continuation*)

(a) If $\mathcal{S}$ is a simple region in $\mathbf{E}^2$, show that the area of $\mathcal{S}$ is

$$\frac{1}{2}\int_\alpha (u\, dv - v\, du),$$

where α is the "boundary curve" of $\mathcal{S}$.

(b) Find the area of the region in $\mathbf{E}^2$ enclosed by the ellipse

$$u^2/a^2 + v^2/b^2 = 1.$$

13. Exercise 7 of Section 8, Chapter VII, will show that if ϕ is any 1-form on a compact oriented surface, then

$$\iint_M d\phi = 0.$$

Combine this result with Exercise 3 of Section 2 to show that if h is the support function of $M \subset \mathbf{E}^3$, then

$$A(M) + \iint_M hH\, dM = 0 \qquad \iint_M H\, dM + \iint_M hK\, dM = 0.$$

Check these formulas on a sphere of radius r, oriented by the outward unit normal.

14. Write $\vartheta = \measuredangle\,(\mathbf{u}, \mathbf{v})$ if ϑ is an oriented angle from $\mathbf{u}$ to $\mathbf{v}$. Show that
    (a) If $\vartheta = \measuredangle\,(\mathbf{u}, \mathbf{v})$ and $\varphi = \measuredangle\,(\mathbf{v}, \mathbf{w})$, then $\vartheta + \varphi = \measuredangle\,(\mathbf{u}, \mathbf{w})$.
    (b) If $\vartheta = \measuredangle\,(\mathbf{u}, \mathbf{v})$, then $-\vartheta = \measuredangle\,(\mathbf{v}, \mathbf{u})$.

15. Let $\alpha: I \to M$ be a curve in an oriented surface M. If V and W are nonzero tangent vector fields on α, show there is a differentiable function ϑ on I such that $\vartheta(t)$ is an oriented angle from $V(t)$ to $W(t)$ for each t in I. We call ϑ an *angle function from* V to W. Note that any two differ by an integer multiple of 2π. (*Hint:* Reduce to Ex. 12 of II.1.)

16. A mapping $F: M \to N$ is *area-preserving* provided the area of any pavable region $\mathcal{R}$ in M is the same as the area of its image $F(\mathcal{R})$ in N. (Note that such a mapping must be one-to-one.) Show that:
    (a) A one-to-one mapping $F: M \to N$ is area-preserving if

$$EG - F^2 = \bar{E}\bar{G} - \bar{F}^2$$

    for every patch $\mathbf{x}$ in M, with $\bar{\mathbf{x}} = F(\mathbf{x})$ in N. (*Hint:* Show that F carries pavings to pavings.)
    (b) Isometries are area-preserving; isometric surfaces have the same area. Include the noncompact case.
    (c) The mapping (1) in Example 5.2 of Chapter IV is area-preserving but not an isometry. Deduce the standard formula for the area

of a zone in the sphere. (It suffices in (a) to consider a single parametrization $\mathbf{x}$ if it covers all of M.)

17. If $F: M \to N$ is a mapping of oriented surfaces with Jacobian J, show that:
    (a) F is regular if and only if J is never zero.
    (b) F is area-preserving if F is one-to-one and $J = \pm 1$. (The converse is true also.)
    (c) If F is an isometry, then $J = \pm 1$, but the converse is false.

Since we are dealing only with connected surfaces, (a) shows that all such mappings F separate into two classes: *orientation-preserving* if $J > 0$, *orientation-reversing* if $J < 0$. Except for patches (Exercise 8), we use this notion mainly in the easy case where F is an isometry.

18. If $F: M \to \bar{M}$ is an isometry of oriented surfaces, show that F is orientation-preserving if and only if any one of the following conditions hold:
    (a) $F^*(d\bar{M}) = dM$.
    (b) $F_*(U(\mathbf{p}) \times \mathbf{v}) = \bar{U}(F(\mathbf{p})) \times F_*(\mathbf{v})$ for all tangent vectors $\mathbf{v}$ to M at $\mathbf{p}$ (U and $\bar{U}$ the unit normals orienting M and $\bar{M}$).
    (c) $U(\mathbf{p}) \cdot \mathbf{v} \times \mathbf{w} = \bar{U}(F(\mathbf{p})) \cdot F_*(\mathbf{v}) \times F_*(\mathbf{w})$ for all pairs of tangent vectors.
    (d) For any positively oriented frame field E_1, E_2 on M, $F_*(E_1), F_*(E_2)$ is a positively oriented frame field on M.

19. If $F: M \to N$ is an orientation-preserving diffeomorphism of compact oriented surfaces, show that
    (a) $$\iint_M F^*(\nu) = \iint_N \nu,$$
    for any 2-form on N. (*Hint:* Ex. 8 of IV.6.)
    (b) Deduce that total curvature is an isometric invariant for compact orientable surfaces.
    (c) Extend (b) to the noncompact case, assuming either $K \geq 0$ or $K \leq 0$.

20. *Gauss mapping G of some minimal surfaces.* Prove:
    (a) *Catenoid.* G is one-to-one and its image covers all the sphere except two points.
    (b) *Helicoid.* The image of G omits exactly two points of the sphere, and each point of the image is hit by an infinite number of points of the helicoid.
    (c) *Scherk's surface* (Ex. 21 of V.4). Same as the helicoid except that exactly four points are omitted. (*Hint:* Consider $Z = \nabla g$ on one of the vertical lines.)
    What are the total curvatures of these surfaces?

21. In an oriented surface M let $\mathbf{x}$ and $\mathbf{y}$ be patchlike 2-segments with the same image $\mathbf{x}(D) = \mathbf{y}(E)$. For any 2-form ν show that

$$\iint_{\mathbf{y}} \nu = \pm \iint_{\mathbf{x}} \nu$$

with plus sign if $\mathbf{x}$ and $\mathbf{y}$ have the *same* orientation (positive or negative), minus sign if *opposite* orientation. (Hint: examine the sign of the Jacobian of $\mathbf{y}^{-1}\mathbf{x}$ and use the change of variables formula for double integrals.)

8 Congruence of Surfaces

Two surfaces M and $\bar{M}$ are *congruent* provided there is an isometry $\mathbf{F}$ of $\mathbf{E}^3$ that carries M exactly onto $\bar{M}$. Thus congruent surfaces have the same shape—only their positions in $\mathbf{E}^3$ can be different. For example, any two spheres of the same radius are congruent (use the translation carrying one center to the other), and the surfaces

$$M: z = xy \qquad \text{and} \qquad M: z = \frac{x^2 + y^2}{2}$$

are congruent under a $45°$ rotation about the z axis.

To simplify the exposition, we shall assume that the surfaces we deal with in this section are *orientable* as well as connected.

8.1 Theorem   If $\mathbf{F}$ is a Euclidean isometry such that $\mathbf{F}(M) = \bar{M}$, then the restriction of $\mathbf{F}$ to M is an isometry $F = \mathbf{F} \mid M: M \to \bar{M}$ of surfaces. Furthermore, if M and $\bar{M}$ are suitably oriented, then F preserves shape operators; that is,

$$F_*(S(\mathbf{v})) = \bar{S}(F_*(\mathbf{v}))$$

for all tangent vectors $\mathbf{v}$ to M.

In short, *congruent surfaces are isometric* and have essentially the same shape operators. We emphasize, however, that *isometric surfaces need not be congruent*, since, as we have seen, they may have quite different shapes in $\mathbf{E}^3$.

Proof. We know from Chapter IV, Section 5, that the restriction $F: M \to \bar{M}$ is a mapping. Furthermore the derivative maps of F and $\mathbf{F}$ agree on tangent vectors to M. In fact, if $\mathbf{v}$ is tangent to M, then $\mathbf{v}$ is the initial velocity of some curve α *in* M—and since $F = \mathbf{F} \mid M$, we have

$$F(\alpha) = \mathbf{F}(\alpha).$$

Thus

$$F_*(\mathbf{v}) = F(\alpha)'(0) = \mathbf{F}(\alpha)'(0) = \mathbf{F}_*(\mathbf{v}).$$

It follows immediately that F_* preserves dot products of tangent vectors to M, for F_* has this property for *all* pairs of tangent vectors (Corollary 2.2, Chapter III). Also, $F: M \to \bar{M}$ is one-to-one (since F is) and onto (by hypothesis); hence F is an isometry of surfaces.

Finally, we show that F preserves shape operators. If M is oriented by the unit normal U, then since F_* preserves dot products (and agrees with F_* on M), it follows that $F_*(U)$ has unit length and is everywhere normal to $F(M) = \bar{M}$. Thus one of the unit normals on M, say $\bar{U}$, has the property that

$$\mathbf{F}_*(U(\mathbf{p}) = \bar{U}(\bar{\mathbf{p}}) \qquad \text{where } \bar{\mathbf{p}} = F(\mathbf{p}).$$

If S and $\bar{S}$ are the shape operators on M and $\bar{M}$ derived from U and $\bar{U}$, respectively, we shall show that

$$\mathbf{F}_*(S(\mathbf{v})) = \bar{S}(\mathbf{F}_*(\mathbf{v})).$$

Again let α be a curve in M with initial velocity $\mathbf{v}$. Thus $\mathbf{F}(\alpha)$ is a curve in $\bar{M}$ with initial velocity $\mathbf{F}_*(\mathbf{v})$. If U is restricted to α, and $\bar{U}$ is restricted to $\mathbf{F}(\alpha)$, then $\mathbf{F}_*(U) = \bar{U}$ (Fig. 6.25). Since $\mathbf{F}_*$ preserves derivatives of vector fields, we get

$$\mathbf{F}_*(S(\mathbf{v})) = -\mathbf{F}_*(U'(0)) = -\bar{U}'(0) = \bar{S}(\mathbf{F}_*(\mathbf{v})).$$

But $\mathbf{v}$ and $S(\mathbf{v})$ are tangent to M; hence $\mathbf{F}_*$ may be replaced by F_*. ∎

Our goal now is the converse of the preceding theorem, that is: If M and $\bar{M}$ are isometric and have the same shape operators, then M and $\bar{M}$ are congruent. This is the analogue of the basic result (Theorem 5.3 of Chapter III) for curves. The condition M and $\bar{M}$ isometric corresponds to the hypothesis that α and β are unit-speed curves defined on the same

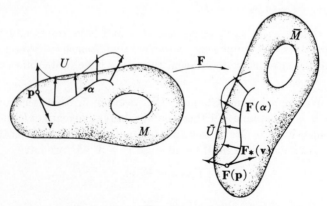

FIG. 6.25

interval, and, of course, "same shape operators" corresponds to

$$\kappa = \bar{\kappa}, \qquad \tau = \pm \bar{\tau}.$$

8.2 Lemma Let $F: M \rightarrow \bar{M}$ be an isometry of oriented surfaces in $\mathbf{E}^3$ that preserves shape operators (as in Theorem 8.1). Let E_1, E_2 be a tangent frame field on M, with $\bar{E}_1$, $\bar{E}_2$ the transferred frame field on $\bar{M}$. If E_3 and $\bar{E}_3$ are the unit normals orienting M and $\bar{M}$, then E_1, E_2, E_3 and $\bar{E}_1$, $\bar{E}_2$, $\bar{E}_3$ are adapted frame fields on M and $\bar{M}$. For the connection forms of these frame fields, we have

$$F^*(\bar{\omega}_{ij}) = \omega_{ij} \qquad (1 \leqq i,j \leqq 3).$$

Proof. We already know from Lemma 5.3 that $F^*(\bar{\omega}_{12}) = \omega_{12}$. It remains to prove that

$$F^*(\bar{\omega}_{i3}) = \omega_{i3}, \qquad \text{for } i = 1, 2.$$

But Corollary 1.5 shows that this merely expresses the preservation of shape operators in terms of connection forms. In fact, for $j = 1, 2$,

$$F^*(\bar{\omega}_{i3})(E_j) = \bar{\omega}_{i3}(F_* E_j) = \bar{S}(F_* E_j) \cdot \bar{E}_i = F_*(S(E_j)) \cdot F_*(E_i)$$

$$= SE_j \cdot E_i = \omega_{i3}(E_j).$$

Hence the forms $F^*(\bar{\omega}_{i3})$ and ω_{i3} are equal. ∎

8.3 Theorem Let $F: M \rightarrow N$ be an isometry of oriented surfaces that preserves shape operators; that is,

$$F_*(S(\mathbf{v})) = \bar{S}(F_*(\mathbf{v}))$$

for all tangent vectors $\mathbf{v}$ to M. Then M and $\bar{M}$ are congruent; in fact there is an isometry $\mathbf{F}$ of $\mathbf{E}^3$ such that $F = \mathbf{F} \mid M$.

(If it should happen that

$$F_*(S(\mathbf{v})) = -\bar{S}(F_*(\mathbf{v}))$$

for all tangent vectors, then it suffices to reverse the orientation of either M or $\bar{M}$ to get the hypothesis as stated.)

Proof. Fix a point $\mathbf{p}$ of M, and let E_3 and $\bar{E}_3$ be the unit normals orienting M and $\bar{M}$. By using Corollary 2.3 of Chapter III, it is easy to show that there is a unique isometry $\mathbf{F}$ of $\mathbf{E}^3$ which agrees with F at the selected point $\mathbf{p}$ in the sense that

$$\mathbf{F}(\mathbf{p}) = F(\mathbf{p})$$
$$\mathbf{F}_*(\mathbf{v}) = F_*(\mathbf{v}) \qquad \text{for every tangent vector } \mathbf{v} \text{ to } M \text{ at } \mathbf{p}$$
$$\mathbf{F}_*(E_3(\mathbf{p})) = \bar{E}_3(F(\mathbf{p})).$$

We shall show that $\mathbf{F}$ is the required Euclidean isometry, in other words,

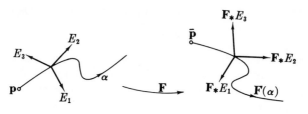

FIG. 6.26

that $\mathbf{F}(\mathbf{q}) = F(\mathbf{q})$ for an arbitrary point $\mathbf{q}$ of M. Thus if α is a curve in M from $\mathbf{p}$ to $\mathbf{q}$, *it suffices to prove that* $\mathbf{F}(\alpha) = F(\alpha)$.

There is no loss of generality in assuming that α lies in the domain of an adapted frame field E_1, E_2, E_3 on M. (If not, we could break α up into segments for which this is true, and repeat essentially the following proof for each segment in turn.) Our plan is to use the general criterion Theorem 5.7 of Chapter III to show that the curves $\mathbf{F}(\alpha)$ and $F(\alpha)$ are identical.

A. The curve $\mathbf{F}(\alpha)$. Restrict the frame field E_1, E_2, E_3 to the curve α (Fig. 6.26). Then by the connection equations,

$$E_i' = \nabla_{\alpha'} E_i = \sum \omega_{ij}(\alpha')E_j \qquad (1 \le i \le 3).$$

Now apply $\mathbf{F}_*$ to this equation; since $\mathbf{F}_*$ is linear and preserves derivatives, we get

$$(\mathbf{F}_* E_i)' = \sum \omega_{ij}(\alpha') \, \mathbf{F}_*(E_j) \qquad (1 \le i \le 3). \tag{A1}$$

Also $\mathbf{F}_*$ preserves dot products; hence $\mathbf{F}_* E_1$, $\mathbf{F}_* E_2$, $\mathbf{F}_* E_3$ is the frame field on the image curve $\mathbf{F}(\alpha)$. Furthermore,

$$\mathbf{F}(\alpha)' \cdot \mathbf{F}_* E_i = \alpha' \cdot E_i \tag{A2}$$

since

$$\mathbf{F}(\alpha)' = \mathbf{F}_*(\alpha').$$

B. The curve $F(\alpha) = \bar{\alpha}$. Use the isometry F to transfer the tangent frame field E_1, E_2 to a tangent frame field $\bar{E}_1$, $\bar{E}_2$ on $\bar{M}$. With the unit normal vector field $\bar{E}_3$, we now have an adapted frame field $\bar{E}_1$, $\bar{E}_2$, $\bar{E}_3$ on $\bar{M}$. We restrict it to the image curve $F(\alpha) = \bar{\alpha}$, and use the connection equations as above to get

$$\bar{E}_i' = \sum \bar{\omega}_{ij}(\bar{\alpha}')\bar{E}_j \qquad (1 \le i \le 3). \tag{B1}$$

Furthermore, we assert that

$$\bar{\alpha}' \cdot \bar{E}_i = \alpha' \cdot E_i \qquad (1 \le i \le 3). \tag{B2}$$

For $i = 1, 2$ this follows immediately from the definition of $\bar{E}_1$, $\bar{E}_2$, since

F_* is an isometry and

$$\bar{\alpha}' = F(\alpha)' = F_*(\alpha').$$

For $i = 3$, both sides are zero, since α and $\bar{\alpha}$ are curves in M and $\bar{M}$, respectively.

C. *Comparison of* $\mathbf{F}(\alpha)$ *and* $F(\alpha) = \bar{\alpha}$. The construction above and the assumption that F preserves shape operators has exactly reproduced the hypotheses of the preceding lemma; hence $F^*(\bar{\omega}_{ij}) = \omega_{ij}$ for $1 \leq i,j \leq 3$. Thus

$$\bar{\omega}_{ij}(\bar{\alpha}') = \bar{\omega}_{ij}(F_*(\alpha')) = (F^*\bar{\omega}_{ij})(\alpha') = \omega_{ij}(\alpha').$$

Using this fact we deduce from (A1) and (B1) that

$$(\mathbf{F}_*E_i)' \cdot \mathbf{F}_*E_j = \bar{E}_i' \cdot \bar{E}_j \qquad (1 \leq i,j \leq 3). \tag{C1}$$

Comparing (A2) and (B2), we get

$$\mathbf{F}(\alpha)' \cdot \mathbf{F}_*E_i = \bar{\alpha}' \cdot \bar{E}_i \qquad (1 \leq i \leq 3). \tag{C2}$$

And by our initial construction we have

$$\mathbf{F}_*E_i = \bar{E}_i \text{ at the point } \bar{\mathbf{p}} = \bar{\alpha}(0) \tag{C3}$$
$$= F(\alpha(0)) \qquad (1 \leq i \leq 3).$$

Refering to equations ($\ddagger$) in Theorem 5.7 of Chapter III, we observe that the three equations (C1), (C2), (C3) are precisely what is needed to conclude that $\mathbf{F}(\alpha) = \bar{\alpha}$; that is,

$$\mathbf{F}(\alpha) = F(\alpha). \qquad \blacksquare$$

This theorem gives a formal proof that the shape operators of a surface M in $\mathbf{E}^3$ do, in fact, completely describe its shape.

EXERCISES

1. A surface $M \subset \mathbf{E}^3$ is *rigid* provided every surface isometric to M is congruent to M. Deduce from Liebmann's theorem that spheres are rigid.

2. If $\alpha, \beta \colon I \to \mathbf{E}^3$ are unit-speed curves with $\kappa_\alpha = \kappa_\beta > 0$ and $\tau_\alpha = \tau_\beta$, show that their tangent surfaces are congruent. (Compare Ex. 5 of VI.4.)

3. If M and N are congruent surfaces in $\mathbf{E}^3$, and $\mathbf{F}$ is a Euclidean isometry such that $\mathbf{F}(M) = N$, prove that $\mathbf{F} \mid M$ preserves Gaussian and mean

curvature, principal curvatures, principal directions, umbilics, asymptotic and principal curves, and geodesics. Which of these are preserved by arbitrary isometries $F: M \to N$? (*Hint:* Orient locally, and ignore ambiguity of signs for H, k_1, and k_2.)

4. If $F: \Sigma \to \Sigma'$ is an isometry of spheres, show that there is a Euclidean isometry $\mathbf{F}$ such that $F = \mathbf{F} \mid \Sigma$.

5. Let M be the saddle surface $(z = xy)$. A rotation of $90°$ followed by a reflection in the xy plane yields an orthogonal transformation C of $\mathbf{E}^3$ whose matrix is

$$\begin{pmatrix} 0 & -1 & 0 \\ 1 & 0 & 0 \\ 0 & 0 & -1 \end{pmatrix}$$

(With our conventions, the columns of the matrix are $C(\mathbf{u}_1)$, $C(\mathbf{u}_2)$, $C(\mathbf{u}_3)$, where $\mathbf{u}_i$ is the ith unit point.)
 (a) Prove that $C(M) = M$.
 (b) Let $F = C \mid M: M \to M$. Orient M (as domain of F) by the unit normal U such that $U(\mathbf{0}) = \mathbf{u}_3$. For which orientation of M (as image of F) does F preserve shape operators?

6. In the general description of a surface of revolution on page 129 (M obtained by revolving C around A), let A be the line through $\mathbf{p}$ in the direction of a unit vector $\mathbf{e}_1$, and let

$$\alpha(u) = \mathbf{p} + g(u)\,\mathbf{e}_1 + h(u)\,\mathbf{e}_2$$

be a parametrization of C, where $\mathbf{e}_2$ is a unit vector orthogonal to $\mathbf{e}_1$.
 (a) Find a regular mapping $\mathbf{x}: D \to \mathbf{E}^3$ whose image is the set M. Then prove:
 (b) M is congruent to a surface of revolution in the special position given in Example 2.5 of Chapter IV.
 (c) M is a surface in $\mathbf{E}^3$.
 (d) Two surfaces of revolution are congruent if and only if they can be described in this way by the same pair of functions g, h.

7. If M is a surface in $\mathbf{E}^3$, a Euclidean isometry $\mathbf{F}$ such that $\mathbf{F}(M) = M$ is called a *Euclidean symmetry* of M. Show that
 (a) The set of all Euclidean symmetries of M forms a subgroup $\mathcal{S}(M)$ of the group $\mathcal{E}$ of all isometries of $\mathbf{E}^3$ (Ex. 7 of III.1). $\mathcal{S}(M)$ is called the *Euclidean symmetry group* of M.
 (b) The Euclidean symmetry groups of congruent surfaces are isomorphic.

8. Show that the Euclidean symmetry group of any sphere is isomorphic to the group of all 3×3 orthogonal matrices.

9. Find all eight Euclidean symmetries of the saddle surface M: $z = xy$.
   Show that they are orthogonal transformations and give their matrices.

10. Find all Euclidean symmetries of the ellipsoid $x^2/a^2 + y^2/b^2 + z^2/c^2 = 1$,
    where $a > b > c$. (*Hint:* Use the fact that Gaussian curvature is
    preserved.)

11. If M is Scherk's surface (Ex. 21 of V.4) and D is the open square
    $-\pi/2 \leqq u, v \leqq \pi/2$, show that:
    (a) The image $\mathbf{x}(D)$ of $\mathbf{x}$ (Ex. 4 of V.4) lies in M.
    (b) The portion of M over any open square (Ex. 21 of V.4) is con-
        gruent to $\mathbf{x}(D)$.
    (c) The curvature formulas given in the abovementioned exercises
        are consistent.

9 Summary

The geometrical study of a surface M in $\mathbf{E}^3$ separates into three distinct
categories:
   (1) The intrinsic geometry of M.
   (2) The shape of M in $\mathbf{E}^3$.
   (3) The Euclidean geometry of $\mathbf{E}^3$.
We saw in Chapters II and III that the geometry of $\mathbf{E}^3$ is based on the dot
product and consists of all concepts preserved by isometries of $\mathbf{E}^3$. Similarly,
we have now found that the intrinsic geometry of M is based on the dot
product—applied only to vectors tangent to M—and that it consists of
all concepts preserved by isometries of M.

The shape of M in $\mathbf{E}^3$ is, in a sense, a link between these two geometries.
For example, Gaussian curvature K is an essential feature of the intrinsic
geometry of M, and the shape operator S dominates category (2)—thus
the equation

$$K = \det S$$

shows that the geometries (1) and (3) can be harmonized only by means
of restrictions on (2). Stated bluntly: *Only certain shapes are possible in*
$\mathbf{E}^3$ *for a surface* M *with prescribed Gaussian curvature.* A strong result of
this character is Liebmann's theorem, which asserts that a compact surface
in $\mathbf{E}^3$ with K constant has only *one* possible shape—spherical.

In the last two chapters, the computation of explicit examples has been
mostly in terms of coordinate patches (Gauss), but the theory itself has
been expressed in terms of frame fields and forms (Cartan). Historically,
coordinates were used for the theory as well, but by now the Cartan approach has largely won the day. We have seen in Section 6 that the two
approaches are not so far apart when the coordinate patch is orthogonal.

Riemannian Geometry

In studying the geometry of a surface in $\mathbf{E}^3$ we found that some of its most important geometric properties belong to the surface itself and not the surrounding Euclidean space. Gaussian curvature is a prime example; although defined in terms of shape operators, it belongs to this intrinsic geometry, since it passes the test of isometric invariance. As this situation gradually became clear to the mathematicians of the 19th century, Riemann drew the correct conclusion: There must exist a geometrical theory of surfaces *completely independent of* $\mathbf{E}^3$, a geometry built from the start solely of isometric invariants. In this chapter we shall give an outline of the resulting theory, concentrating on its dominant features: Gaussian curvature and geodesics. Our constant guides will be the two special cases which led to its discovery: the intrinsic geometry of surfaces in $\mathbf{E}^3$, and Euclidean geometry—particularly that of the plane $\mathbf{E}^2$.

1 Geometric Surfaces

The evidence from earlier work on the intrinsic geometry of surfaces in $\mathbf{E}^3$ (and on Euclidean geometry as well) suggests that we will need the dot product on tangent vectors to do geometry on a surface.

But to free ourselves of confinement to $\mathbf{E}^3$, we must begin with an abstract surface M (Chapter IV, Section 8). Since M need not be in $\mathbf{E}^3$ there is no dot product—and hence no geometry. The dot product, however, is but one instance of the general notion of *inner product*, and Riemann's idea was to *replace the dot product by a quite arbitrary inner product on each tangent plane of* M.

1.1 Definition An *inner product* on a vector space V is a function which assigns to each pair of vectors $\mathbf{v}$, $\mathbf{w}$ in V a number $\mathbf{v} \circ \mathbf{w}$ so that these rules hold:

(1) Bilinearity:

$$(a_1\mathbf{v}_1 + a_2\mathbf{v}_2) \circ \mathbf{w} = a_1\mathbf{v}_1 \circ \mathbf{w} + a_2\mathbf{v}_2 \circ \mathbf{w}$$

$$\mathbf{v} \circ (b_1\mathbf{w}_1 + b_2\mathbf{w}_2) = b_1\mathbf{v} \circ \mathbf{w}_1 + b_2\mathbf{v} \circ \mathbf{w}_2.$$

(2) Symmetry: $\qquad\qquad \mathbf{v} \circ \mathbf{w} = \mathbf{w} \circ \mathbf{v}.$

(3) Positive definiteness:

$$\mathbf{v} \circ \mathbf{v} \geqq 0; \quad \text{and} \quad \mathbf{v} \circ \mathbf{v} = 0 \text{ if and only if } \mathbf{v} = 0.$$

On the vector space $\mathbf{E}^2$ the dot product

$$\mathbf{v} \cdot \mathbf{w} = v_1w_1 + v_2w_2$$

is, of course, an inner product, but there are infinitely many others, such as, for example, $\mathbf{v} \circ \mathbf{w} = 2v_1w_1 + 3v_2w_2$. (See Exercise 8 of Section 2.)

Shifting then from surfaces in $\mathbf{E}^3$ to abstract surfaces and, from the dot product to arbitrary inner products, we get the following definition.

1.2 Definition A *geometric surface* is an abstract surface M furnished with an inner product, $\circ$, on each of its tangent planes. This inner product is required to be *differentiable* in the sense that if V and W are (differentiable) vector fields on M, then $V \circ W$ is a differentiable real-valued function on M.

We emphasize that *each* tangent plane $T_p(M)$ of M has its own inner product, and they are unrelated save for the differentiability condition— an obvious necessity for a theory founded on the calculus. In this definition $V \circ W$ has its usual pointwise meaning: It is the function on M whose value at each point $\mathbf{p}$ is the number $V(\mathbf{p}) \circ W(\mathbf{p})$. An assignment of inner products to tangent planes as in Definition 1.2 is called a *geometric structure* (or *metric tensor* or "ds^2") on M.

In short:

$$\text{Surface} + \text{geometric structure} = \text{geometric surface}$$

and we emphasize that the same surface furnished with two different geometric structures gives rise to two different geometric surfaces.

1.3 Example *Some geometric surfaces.*

(1) The plane $\mathbf{E}^2$, furnished with the usual dot product on tangent vectors, is the best-known geometric surface. Its geometry is two-dimensional *Euclidean geometry*.

(2) A simple way to get new geometric structures is to distort old ones.

For example, if $g > 0$ is any differentiable function on the plane, and $\cdot$ is the usual dot product, define

$$\mathbf{v} \circ \mathbf{w} = \frac{\mathbf{v} \cdot \mathbf{w}}{g^2(\mathbf{p})}$$

for tangent vectors $\mathbf{v}$ and $\mathbf{w}$ to $\mathbf{E}^2$ at $\mathbf{p}$. This is a new geometric structure on the plane, said to be *conformal* to that of the dot product (Exercise 1). We shall see that (unless g is special) the resulting geometric surface has properties quite different from the Euclidean plane (1).

(3) If M is a surface in $\mathbf{E}^3$, then the dot product from $\mathbf{E}^3$ applied to tangent vectors on M furnishes an inner product making M a geometric surface. This, of course, is just what was done in Chapters V and VI. Unless some other inner product is explicitly mentioned, it is always assumed that a surface in $\mathbf{E}^3$ is made geometric in this way.

Here a word about terminology is in order. Euclid's name carries geometric implications. Hence in Chapter I, where geometry did not appear, we should have called $\mathbf{E}^2$ the *Cartesian plane*, reserving the term *Euclidean plane* for the geometric surface (1) above.

From the simple beginning in Definition 1.2, it is rather surprising what a rich geometric theory can be built. But as mentioned earlier, examples (1) and (3) certainly indicate that the theory is there to be explored, and their common features even suggest the kind of results we may expect to find.

The definitions in Chapter VI that are clearly intrinsic in character will be used here without further discussion. In particular, an isometry F: $M \to \bar{M}$ of arbitrary geometric surfaces is exactly as defined in Definition 4.2 of Chapter VI and the geometry of M consists by definition of its isometric invariants. A *frame field* on an arbitrary geometric surface M consists, as usual, of two orthogonal unit vector fields E_1, E_2 defined on some open set of M. The orthonormality equations

$$E_i \circ E_j = \delta_{ij} \qquad (1 \leqq i, j \leqq 2)$$

are expressed, of course, in terms of the inner product of M. As before, we derive the dual 1-forms θ_1, θ_2, characterized by $\theta_i(E_j) = \delta_{ij}$, and then the connection form $\omega_{12} = -\omega_{21}$, characterized by the first structural equations,

$$d\theta_1 = \omega_{12} \wedge \theta_2, \qquad d\theta_2 = \omega_{21} \wedge \theta_1.$$

We emphasize once more that these forms θ_1, θ_2, ω_{12} are not invariantly attached to the geometric surface M; a different choice of frame field $\bar{E}_1$, $\bar{E}_2$ will produce different forms $\bar{\theta}_1$, $\bar{\theta}_2$, $\bar{\omega}_{12}$. Before going any further, we had better see how two such sets of forms are related.

On a small enough neighborhood of a point **p**, careful use of the inverse function $\cos^{-1}$ (or $\sin^{-1}$) will yield a differentiable function φ such that

$$\bar{E}_1 = \cos \varphi \, E_1 + \sin \varphi \, E_2.$$

We call φ an *angle function from* E_1, E_2 *to* $\bar{E}_1, \bar{E}_2$. As shown in Fig. 7.1, there are now two possibilities for $\bar{E}_2$. Either

$$\bar{E}_2 = -\sin \varphi \, E_1 + \cos \varphi \, E_2$$

in which case we say that $\bar{E}_1, \bar{E}_2$ and E_1, E_2 have the *same orientation*, or

$$\bar{E}_2 = \sin \varphi \, E_1 - \cos \varphi \, E_2,$$

which is *opposite orientation*.

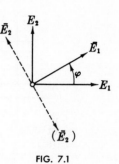

FIG. 7.1

1.4 Lemma   Let E_1, E_2 and $\bar{E}_1, \bar{E}_2$ be frame fields on the same region in M. If these frame fields have

(1) The same orientation, then

$$\bar{\omega}_{12} = \omega_{12} + d\varphi, \qquad \text{and} \qquad \bar{\theta}_1 \wedge \bar{\theta}_2 = \theta_1 \wedge \theta_2.$$

(2) Opposite orientation, then

$$\bar{\omega}_{12} = -(\omega_{12} + d\varphi), \qquad \text{and} \qquad \bar{\theta}_1 \wedge \bar{\theta}_2 = -\theta_1 \wedge \theta_2.$$

Proof. We discuss only the first case, since the second is obtained from it by merely changing signs. By the basis formulas (Lemma 2.1 of Chapter VI), the equations

$$\bar{E}_1 = \cos \varphi \, E_1 + \sin \varphi \, E_2, \qquad \bar{E}_2 = -\sin \varphi \, E_1 + \cos \varphi \, E_2$$

yield

$$\theta_1 = \cos \varphi \, \bar{\theta}_1 - \sin \varphi \, \bar{\theta}_2, \qquad \theta_2 = \sin \varphi \, \bar{\theta}_1 + \cos \varphi \, \bar{\theta}_2. \qquad (^*)$$

Applying the exterior derivative to the first of these, we get

$$d\theta_1 = -\sin \varphi \, d\varphi \wedge \bar{\theta}_1 + \cos \varphi \, d\bar{\theta}_1 - \cos \varphi \, d\varphi \wedge \bar{\theta}_2 - \sin \varphi \, d\bar{\theta}_2.$$

Now we substitute the first structural equations for $d\bar{\theta}_1, d\bar{\theta}_2$ to obtain

$$d\theta_1 = (\bar{\omega}_{12} - d\varphi) \wedge (\sin \varphi \, \bar{\theta}_1 + \cos \varphi \, \bar{\theta}_2)$$
$$= (\bar{\omega}_{12} - d\varphi) \wedge \theta_2.$$

In the same way we get

$$d\theta_2 = -(\bar{\omega}_{12} - d\varphi) \wedge \theta_1.$$

Because the form $\omega_{12} = -\omega_{21}$ *uniquely* satisfies the first structural equations, we conclude from the last two equations that $\omega_{12} = \bar{\omega}_{12} - d\varphi$, as

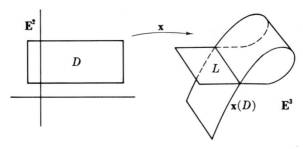

FIG. 7.2

required. Direct computation of $\theta_1 \wedge \theta_2$ using $(^*)$ above shows that this 2-form equals $\bar{\theta}_1 \wedge \bar{\theta}_2$.   ∎

The concept of geometric surface can be used to fill a gap in our earlier work. Occasionally we met regular mappings $\mathbf{x}: D \rightarrow \mathbf{E}^3$ which were not parametrizations of any surface in $\mathbf{E}^3$. For example, the image $\mathbf{x}(D)$ of D might fold back through itself as indicated in Fig. 7.2, so that the definition of surface in $\mathbf{E}^3$ fails along the crossing line L.

This technical difficulty can be eliminated by assigning D (which is a surface) not the usual dot product, but instead the *induced inner product*

$$\mathbf{v} \circ \mathbf{w} = \mathbf{x}_* (\mathbf{v}) \cdot \mathbf{x}_* (\mathbf{w}).$$

Thus D becomes a geometric surface, and *if* $\mathbf{x}(D)$ *were a surface in* $\mathbf{E}^3$, $\mathbf{x}$ *would evidently be an isometry.* In short, D has exactly the intrinsic geometry we might intuitively expect $\mathbf{x}(D)$ to have.

In this chapter, as earlier, the restriction to low dimensions is not essential. A surface is the two-dimensional case of the general notion of manifold (Chapter IV, Section 8). A manifold M of arbitrary dimension furnished with a (differentiable) inner product on each of its tangent spaces is called a *Riemannian manifold*, and the resulting geometry is *Riemannian geometry*. (Euclidean geometry, as discussed in Chapter III, is the special case of Riemannian geometry obtained on the Euclidean space $\mathbf{E}^n$, with its usual dot product.) *A geometric surface is thus the same thing as a two-dimensional Riemannian manifold*, and the subject of this chapter is two-dimensional Riemannian geometry.†

EXERCISES

1. For a conformal geometric structure on the plane (Example 1.3), show that

---

† We would prefer to call a geometric surface a *Riemann(ian) surface* but this term has a firmly established and distinctly different meaning.

(a) The formula

$$\mathbf{v} \circ \mathbf{w} = \| \mathbf{v} \| \; \| \mathbf{w} \| \cos \vartheta \qquad (\text{where } \| \mathbf{v} \| = \sqrt{\mathbf{v} \circ \mathbf{v}})$$

gives the same value for the angle $0 \leq \vartheta \leq \pi$ between $\mathbf{v}$ and $\mathbf{w}$ as in the Euclidean plane $\mathbf{E}^2$.

(b) The speed of a curve $\alpha = (\alpha_1, \alpha_2)$ is $\sqrt{\alpha_1'^2 + \alpha_2'^2}/g(\alpha)$.

(c) gU_1, gU_2 is a frame field with dual forms $du/g, dv/g$.

(d) The area forms are $\pm \, du \, dv / g^2$.

Note that $g = 1$ gives the usual Euclidean structure.

2. The *Poincaré half-plane* is the upper half-plane $v > 0$ furnished with the inner product $(\circ)$ obtained by dividing the dot product at each point $\mathbf{p}$ by the square of the distance $v(\mathbf{p}) = p_2$ to the u axis:

$$\mathbf{v} \circ \mathbf{w} = \mathbf{v} \cdot \mathbf{w}/v^2(\mathbf{p}).$$

For the curve $\alpha(t) = (r \cos t, r \sin t), 0 < t < \pi$, find the speed and arc-length function (measure from the top of the semicircle, $t = \pi/2$.)

3. (a) On a geometric surface M, let V and W be vector fields that are linearly independent. Find a frame field for which $E_1 = V/\| V \|$.

(b) Deduce an explicit formula for a frame field on the image of an arbitrary patch $\mathbf{x}$ in M.

4. If dM is an area form on M and ν is an arbitrary 2-form, show that there is a function f such that $\nu = f \, dM$. Deduce that a (connected) orientable geometric surface has exactly two area forms, $\pm \, dM$.

5. Let M be a geometric surface oriented by area form dM. Prove

(a) On each tangent space to M there exists a unique "rotation by $+90°$," that is, a linear operator $J: T_p(M) \to T_p(M)$ such that

$$\| J(\mathbf{v}) \| = \| \mathbf{v} \|, \qquad J(\mathbf{v}) \circ \mathbf{v} = 0,$$

and

$$\overline{dM}(\mathbf{v}, J(\mathbf{v})) > 0 \text{ (if } \mathbf{v} \neq 0).$$

(*Hint:* If E_1, E_2 is a positively oriented frame field,

$$J(E_1) = E_2, J(E_2) = -E_1.)$$

We call these operators—collectively, for all points of M—the *rotation operator*† of M.

(b) J is differentiable $(J(V) \circ J(W)$ is differentiable for any vector fields $V, W)$ and skew-symmetric

$$J(V) \circ W + V \circ J(W) = 0,$$

and $J^2 = -I$ (J applied twice is minus the identity operator).

† Compare the special case in Exercise 14 of Section 4, Chapter II.

(c) If M is oriented instead by $-dM$, then its rotation operator is $-J$.

(d) If M is a surface in $\mathbf{E}^3$ and its orientation is given by the unit normal U, then $J(V) = U \times V$.

The operator J serves as a kind of replacement for the unit normal on surfaces that are not in $\mathbf{E}^3$. In particular, (d) shows how the scheme given in Chapter VI (page 291) for measuring oriented angles applies now to an arbitrary (oriented) geometric surface.

6. If $F\colon M \to N$ is a regular mapping of oriented geometric surfaces, show that the following are equivalent:

(a) F is orientation-preserving and conformal (Definition 4.7 of Chapter VI).

(b) F preserves the rotation operators of M and N; that is,

$$F_*(J(\mathbf{v})) = \bar{J}(F_*(\mathbf{v}))$$

for all tangent vectors $\mathbf{v}$ to M.

(c) F preserves oriented angles; that is, if ϑ is an angle from $\mathbf{v}$ to $\mathbf{w}$, then ϑ is also an angle from $F_*(\mathbf{v})$ to $F_*(\mathbf{w})$.

7. (a) Prove that a regular mapping $F = (f, g)\colon \mathbf{E}^2 \to \mathbf{E}^2$ is orientation-preserving and conformal† if and only if $f_u = g_v, f_v = -g_u$.

If $\mathbf{E}^2$ is considered as the complex plane, with $z = u + iv = (u,v)$, these two equations (the *Cauchy-Riemann equations*) are necessary and sufficient for F to be a complex analytic function $z \to F(z)$.

(b) Given such a complex function F, show that its scale factor $\lambda(z)$ is the magnitude of the (complex) derivative dF/dz.

8. If the origin is deleted from $\mathbf{E}^2$, show that the mapping F in (2) of Example 7.3 of Chapter I is orientation-preserving and conformal. What is the complex function in this case?

9. Let D and E be regions in the plane, furnished with conformal geometric structures given by functions g_1 and g_2, respectively. Let D' and E' be the same regions, with the usual Euclidean structure. If $F\colon D' \to E'$ is a conformal mapping with scale factor λ, prove that $F\colon D \to E$ is conformal with scale factor $\lambda g_1/g_2(F)$.

2 Gaussian Curvature

For arbitrary geometric surfaces, we need a new definition of Gaussian curvature. The definition $K = \det S$ for surfaces in $\mathbf{E}^3$ is meaningless now,

---

† The term *conformal* is often understood to include preservation of orientation.

since it is based on shape operators. But this original definition made K an isometric invariant, so it is reasonable to look to the proof of the *theorema egregium* (specifically to Corollary 2.3 of Chapter VI) to find a satisfactory generalization.

2.1 Theorem   On a geometric surface M there is a unique real-valued function K such that for any frame field on M the *second structural equation*

$$d\omega_{12} = -K\theta_1 \wedge \theta_2$$

holds. K is called the *Gaussian curvature* of M.

Proof. For each frame field E_1, E_2, there is (by the basis formulas of Lemma 2.1 of Chapter VI), a unique function K such that

$$d\omega_{12} = -K\theta_1 \wedge \theta_2.$$

But another frame field $\bar{E}_1$, $\bar{E}_2$ might *a priori* have a different function $\bar{K}$ such that

$$d\bar{\omega}_{12} = -\bar{K}\bar{\theta}_1 \wedge \bar{\theta}_2.$$

What we have to show is consistency: Where the domains of these frame fields overlap, there $K = \bar{K}$. Since such domains cover all of M (Exercise 3 of Section 1), we will then have a single function K on M with the required property. This consistency will follow immediately from Lemma 1.4. First consider the case in which the frame fields have the same orientation, so $\bar{\omega}_{12} = \omega_{12} + d\varphi$. Hence $d\bar{\omega}_{12} = d\omega_{12}$, because $d^2 = 0$. But then

$$K\theta_1 \wedge \theta_2 = \bar{K}\bar{\theta}_1 \wedge \bar{\theta}_2.$$

Since

$$\bar{\theta}_1 \wedge \bar{\theta}_2 = \theta_1 \wedge \theta_2 \neq 0,$$

we conclude that

$$\bar{K} = K.$$

When the orientations are opposite, we get $d\bar{\omega}_{12} = -d\omega_{12}$ but still find that $\bar{K} = K$, since

$$\bar{\theta}_1 \wedge \bar{\theta}_2 = -\theta_1 \wedge \theta_2. \qquad \blacksquare$$

As noted above, Corollary 2.3 of Chapter VI shows that this general definition of Gaussian curvature agrees with the definition $K = \det S$ when M is a surface in $\mathbf{E}^3$. The proof of isometric invariance obtained there is entirely intrinsic in character, and it thus holds for arbitrary geometric surfaces.

Gaussian curvature is the central property of a geometric surface M;

it influences—often decisively—many of the most important properties of M. In Section 6 we shall examine the influence of curvature on geodesics and in Section 8 its influence on the topology of M.

To summarize: Geometrical investigations in terms of a frame field E_1, E_2 are dominated by its structural equations:

$$\begin{cases} d\theta_1 = \omega_{12} \wedge \theta_2, \\ d\theta_2 = \omega_{21} \wedge \theta_1, \end{cases}$$

$$d\omega_{12} = -K\theta_1 \wedge \theta_2.$$

The first structural equations actually define the connection form

$$\omega_{12} = -\omega_{21}$$

of that frame field, while the second structural equation defines the Gaussian curvature K *of the geometric surface* (independent of choice of frame field). It is already clear from Chapter VI, Section 6, how ω_{12} and K may be explicitly computed from these implicit definitions.

2.2 Example   *Gaussian curvature.*

(1) **The Euclidean plane** $\mathbf{E}^2$. If we use the natural frame field U_1, U_2, then the dual 1-forms are merely $\theta_1 = du$, $\theta_2 = dv$. Since $d\theta_1 = d\theta_2 = 0$, the identically zero form $\omega_{12} = 0$ satisfies the first structural equations and hence is the connection form of U_1, U_2. But then $d\omega_{12} = 0$, so $K = 0$. The Euclidean plane is *flat*. This can be no surprise, since $\mathbf{E}^2$ is isometric to a plane in $\mathbf{E}^3$, for which we know that $K = 0$, since its shape operators all vanish.

(2) The plane with conformal inner product

$$\mathbf{v} \circ \mathbf{w} = \frac{\mathbf{v} \cdot \mathbf{w}}{g(\mathbf{p})^2}$$

(See (2) of Example 1.3.)

The natural *Euclidean* frame field U_1, U_2 is no longer a frame field relative to this new inner product. U_1 and U_2 are still orthogonal, but

$$U_1 \circ U_1 = U_2 \circ U_2 = \frac{1}{g^2}.$$

Thus gU_1, gU_2 *is* a frame field. It follows easily that its dual 1-forms are

$$\theta_1 = \frac{du}{g}, \qquad \theta_2 = \frac{dv}{g}.$$

To find the connection form ω_{12}, we first differentiate θ_1 and θ_2.

$$d\theta_1 = d\left(\frac{1}{g}\right) \wedge du$$

$$d\theta_2 = d\left(\frac{1}{g}\right) \wedge dv.$$

Now $d(1/g) = -dg/g^2$ and $dg = g_u\, du + g_v\, dv$. Because $du\, du = dv\, dv = 0$, we find

$$d\theta_1 = \left(\frac{g_v}{g}\, du\right) \wedge \theta_2$$

$$d\theta_2 = -\left(\frac{-g_u}{g}\, dv\right) \wedge \theta_1.$$

(*)

Comparison with the first structural equations then yields

$$\omega_{12} = \frac{1}{g}\left(g_v\, du - g_u\, dv\right)$$

because by (*), this form ω_{12} satisfies the first structural equations; by uniqueness it must be the connection form.

To get the curvature we differentiate once more.

$$d\omega_{12} = d\left(\frac{1}{g}\right) \wedge (g_v\, du - g_u\, dv) + \frac{1}{g}\left(g_{vv}\, dv\, du - g_{uu}\, du\, dv\right).$$

From the above we know that

$$d\left(\frac{1}{g}\right) = \frac{-1}{g^2}\left(g_u\, du + g_v\, dv\right)$$

and that $du = g\theta_1$, $dv = g\theta_2$. A simple computation using these facts then gives

$$d\omega_{12} = (g_u{}^2 + g_v{}^2 - g(g_{uu} + g_{vv}))\, \theta_1 \wedge \theta_2.$$

Thus by the second structural equation we conclude that

$$K = g(g_{uu} + g_{vv}) - (g_u{}^2 + g_v{}^2).$$

The induced inner product discussed on page 308 may be applied in other situations. For example, suppose that $F: M \to N$ is a diffeomorphism of surfaces (Chapter IV, Section 5) and that N is a geometric surface. Then the *induced inner product*

$$\mathbf{v} \circ \mathbf{w} = F_*(\mathbf{v}) \circ F_*(\mathbf{w})$$

on tangent vectors to M, makes M a geometric surface—and F an isometry. M might be called a "new model" of N; however different it may *look*, it is geometrically identical with N.

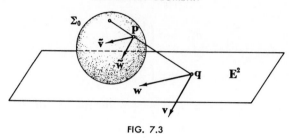

FIG. 7.3

2.3 Example

(1) *The stereographic sphere.* We have proved in Example 5.5 of Chapter IV that stereographic projection P is a diffeomorphism of the punctured sphere Σ_0 onto the Euclidean plane $\mathbf{E}^2$. Now consider Σ_0 merely as a surface, while $\mathbf{E}^2$ is geometric, with its usual dot product. Thus the induced inner product makes Σ_0 a geometric surface which is isometric to $\mathbf{E}^2$ and hence *flat*. If Σ_0 appears round, it is only because we look at it with Euclidean eyes; that is, we erroneously assume it must have the dot product of $\mathbf{E}^3$ as in Chapter V.

(2) *The stereographic plane.* Now let us reverse the process in (1). Consider Σ_0 with its usual geometric structure as a surface in $\mathbf{E}^3$, and let $\mathbf{E}^2$ be merely a surface.

The inverse $P^{-1} \colon \mathbf{E}^2 \to \Sigma_0$ of stereographic projection is also a diffeomorphism. The inner product ($\circ$) induced by P^{-1} on $\mathbf{E}^2$ makes $\mathbf{E}^2$ a geometric surface (the *stereographic plane*) isometric to Σ_0, and thus having a curvature $K = +1$.

We examine this new stereographic plane more closely. Throughout a dot ($\cdot$) will as usual denote the dot product, whether of $\mathbf{E}^2$ or $\mathbf{E}^3$.

If $\mathbf{v}$ and $\mathbf{w}$ are tangent vectors to $\mathbf{E}^2$ at $\mathbf{q} = P(\mathbf{p})$, let $\tilde{\mathbf{v}}$ and $\tilde{\mathbf{w}}$ be the unique tangent vectors to Σ_0 at $\mathbf{p}$ such that $P_*(\tilde{\mathbf{v}}) = \mathbf{v}$, $P_*(\tilde{\mathbf{w}}) = \mathbf{w}$ (Fig. 7.3). Now by Exercise 14 of Section 4, Chapter VI, we know that

$$\mathbf{v} \cdot \mathbf{w} = P_*(\tilde{\mathbf{v}}) \cdot P_*(\tilde{\mathbf{w}}) = \left(1 + \frac{\| \mathbf{q} \|^2}{4} \right)^2 \tilde{\mathbf{v}} \cdot \tilde{\mathbf{w}}.$$

But $(P^{-1})_*$ carries $\mathbf{v}$ and $\mathbf{w}$ back to $\tilde{\mathbf{v}}$ and $\tilde{\mathbf{w}}$, so for the induced inner product of $\mathbf{E}^2$ we find

$$\mathbf{v} \circ \mathbf{w} = (P^{-1})_*(\mathbf{v}) \cdot (P^{-1})_*(\mathbf{w}) = \tilde{\mathbf{v}} \cdot \tilde{\mathbf{w}} = \left(1 + \frac{\| \mathbf{q} \|^2}{4} \right)^{-2} \mathbf{v} \cdot \mathbf{w}.$$

It follows immediately that this inner product is of the conformal type discussed in Example 1.3 with

$$g = 1 + \frac{u^2 + v^2}{4}.$$

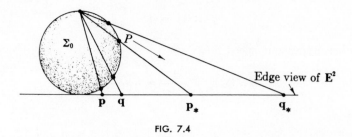

FIG. 7.4

To visualize this unusual "plane," we may imagine that *rulers get longer as they move farther away from the origin*. Since P is now an isometry, the intrinsic distance from **p** to **q** (in Fig. 7.4) is exactly the same as the distance from $\mathbf{p_*}$ to $\mathbf{q_*}$. Also circles $u^2 + v^2 = r^2$, for which r is very *large*, actually have very *small* stereographic arc length since they correspond (under the isometry P) to small circles about the north pole in Σ_0.

2.4 Example   *The hyperbolic plane.*   Let us experiment with a change of sign in the stereographic inner product above, setting

$$g = 1 - \frac{u^2 + v^2}{4}.$$

Since $g > 0$ is necessary, this *hyperbolic inner product* $\mathbf{v} \circ \mathbf{w} = (1/g^2)\mathbf{v} \cdot \mathbf{w}$ is used only on the disc $u^2 + v^2 < 4$ of radius 2 in the plane. The resulting geometric surface is called *the hyperbolic plane H*.

In this case

$$g_u = \frac{-u}{2}, \qquad g_v = \frac{-v}{2}, \quad \text{and} \quad g_{uu} = g_{vv} = \frac{-1}{2},$$

so the general computations in Example 2.2 show that

$$\omega_{12} = \frac{1}{2g} \left(u \, dv - v \, du \right)$$

and that *the hyperbolic plane has constant Gaussian curvature* $K = -1$.

As a point (u, v) approaches the *rim* of H, that is, the circle $u^2 + v^2 = 4$ (not a part of $H!$), $g(u, v)$ approaches zero. Thus in the language used above, rulers must *shrink* as they approach the rim, so that H is a good deal bigger than one's Euclidean intuition may suggest. For example, for ϑ constant let us compute the arc-length function $s(t)$ of the Euclidean line segment

$$\alpha(t) = (t \cos \vartheta, t \sin \vartheta), \qquad 0 \leq t < 2,$$

which runs from the origin almost to the rim. Now $\alpha' = (\cos \vartheta, \sin \vartheta)$, so $\alpha' \circ \alpha' = 1/g(\alpha)^2$. But

$$g(\alpha(t)) = 1 - \frac{t^2}{4},$$

so α has hyperbolic speed

$$\| \alpha'(t) \| = \frac{1}{g(\alpha)} = \frac{1}{(1 - t^2/4)}.$$

Thus

$$s(t) = \int_0^t \frac{dt}{1 - t^2/4} = 2 \tanh^{-1} \frac{t}{2} = \log \frac{2 + t}{2 - t}.$$

Thus as t approaches 2, arc length $s(t)$ from the origin $\alpha(0)$ to $\alpha(t)$ approaches infinity. *This "short" segment α actually has infinite hyperbolic length.* Further properties of the hyperbolic plane will be developed as we go along. We shall see that it—and not the bugle surface (Example 6.6 of Chapter V)—is the true analogue of the sphere for constant negative curvature.

2.5 Example *A flat torus.* Let T be a torus of revolution considered merely as a surface, and let $\mathbf{x} \colon \mathbf{E}^2 \to T$ be its usual parametrization (Example 2.6 of Chapter IV). Now we give T a geometric structure by *defining*

$$\mathbf{x}_u \circ \mathbf{x}_u = 1, \qquad \mathbf{x}_u \circ \mathbf{x}_v = 0, \qquad \mathbf{x}_v \circ \mathbf{x}_v = 1.$$

It is easy to check that this defines—without ambiguity—an inner product on each tangent plane of T.

Because $\mathbf{x}_* (U_1) = \mathbf{x}_u$ and $\mathbf{x}_* (U_2) = \mathbf{x}_v$, it follows immediately that $\mathbf{x}$ is a local isometry of the Euclidean plane $\mathbf{E}^2$ onto the geometric surface T. Since local isometries also preserve Gaussian curvature, T is *flat*. Thus its geometric structure is different from the usual torus in $\mathbf{E}^3$, which has variable curvature.

Because this torus T is compact and flat, Theorem 3.5 of Chapter VI shows that *it can never be found in* $\mathbf{E}^3$. Explicitly, there exists no surface M in $\mathbf{E}^3$ that is isometric to T, for then M would also be compact and flat—but this is forbidden by the theorem. Here we have a proof that the class of geometric surfaces is richer than that of surfaces in $\mathbf{E}^3$. In the course of this chapter we hope to convince the reader that geometric surfaces are the natural objects to study and that surfaces in $\mathbf{E}^3$—however intuitive they may seem at first glance—are no more than an interesting special case.

One should not conclude from the example above that every surface can be given a flat geometric structure. Topological subtleties are involved, as we shall see in Section 8.

2.6 Remark So far we have reserved the dot notation ($\cdot$) for the dot product of Euclidean space, and used a small circle ($\circ$) to emphasize the generality of the inner product of an arbitrary geometric surface. *From now on we shall use a dot for* all *inner products*, reverting to the former convention only when, as in Example 2.4, the two appear in the same context.

EXERCISES

1. Derive the dual forms and connection form $\omega_{12} = du/v$ for the frame field vU_1, vU_2 on the Poincaré half-plane (Ex. 2 of VII.1) and show that this surface has constant negative curvature $K = -1$.

2. For the conformal geometric structure on the entire plane with

$$g = \cosh (uv),$$

   compute the dual forms and connection form of the frame field gU_1, gU_2, and derive the Gaussian curvature K.

3. Find the area A of the disc $u^2 + v^2 \leq r^2$ in the hyperbolic plane. (*Hint:* Find E, F, G for a 2-segment

$$\mathbf{x}(u,v) = (u \cos v, \ u \sin v).)$$

   What is the area of the entire hyperbolic plane?

4. The *hyperbolic plane of pseudo-radius* r is obtained by altering the function g in Example 2.4 to $g = 1 + (u^2 + v^2)/4r^2$. Find its Gaussian curvature.

5. Find the area of the flat torus in Example 2.5. Modify the definition so as to produce a flat torus with arbitrary area $A > 0$.

6. Show that there is a geometric structure on the projective plane such that the natural mapping $P: \Sigma \to \bar{\Sigma}$ is a local isometry. Prove that this geometric surface $\bar{\Sigma}$ cannot be found in $\mathbf{E}^3$. (The same results hold when Σ is a sphere of radius r and $\bar{\Sigma}$ becomes *the projective plane of radius* r.)

7. Show that the plane, furnished with the conformal geometric structure such that $g = \operatorname{sech} u$, is isometric to a helicoid.

8. The identity map $\mathbf{x}(u,v) = (u,v)$ of $\mathbf{E}^2$ is a patch with $\mathbf{x}_u = U_1$, $\mathbf{x}_v = U_2$. Thus if $(\circ)$ is a geometric structure on the plane, this patch has

$$E = U_1 \circ U_1, \qquad F = U_1 \circ U_2, \qquad G = U_2 \circ U_2.$$

   (a) Given any differentiable functions E, F, and G on the plane, such that $E > 0$, $G > 0$, $EG - F^2 > 0$, show that there is a geometric structure on the plane corresponding, as above, to these functions.

   (b) Show that the method used in Example 2.2 is a special case of that of Chapter VI, Section 6, and derive the formula for K in Example 2.2 from Lemma 6.3 of Chapter VI.

(*Hint:* For (a) define

$$\mathbf{v} \circ \mathbf{w} = E v_1 w_1 + F(v_1 w_2 + v_2 w_1) + G v_2 w_2.)$$

3 Covariant Derivative

The covariant derivative ∇ of $\mathbf{E}^3$ (Chapter II, Section 5) is an essential part of Euclidean geometry. We used it, for example, to define the shape operator of a surface in $\mathbf{E}^3$, and in modified form (Chapter II, Section 2) to define the acceleration of a curve in $\mathbf{E}^3$. In this section we will show that *each geometric surface has its own notion of covariant derivative.*

As in Euclidean space, a covariant derivative ∇ on a geometric surface M assigns to each pair of vector fields V, W on M a new vector field $\nabla_V W$, and we must certainly require that it have the usual linear and Leibnizian properties (Corollary 5.4 of Chapter II). Intuitively, the value of $\nabla_V W$ at a point $\mathbf{p}$ will be the rate of change of W in the $V(\mathbf{p})$ direction. Thus if the connection form ω_{12} of a frame field E_1, E_2 is to have its usual geometrical meaning (measuring rates at which E_1 turns toward E_2), we must also require that

$$\omega_{12}(V) = \nabla_V E_1 \cdot E_2. \tag{*}$$

These conditions completely determine $\nabla_V W$ for any vector fields V and W:

3.1 Lemma Assume that ∇ is a covariant derivative on M with the usual linear and Leibnizian properties, and such that (*) holds for a frame field E_1, E_2. Then ∇ obeys the *connection equations*

$$\nabla_V E_1 = \omega_{12}(V)E_2$$
$$\nabla_V E_2 = \omega_{21}(V)E_1.$$

Furthermore if $W = f_1 E_1 + f_2 E_2$ is an arbitrary vector field, then

$$\nabla_V W = \{V[f_1] + f_2 \omega_{21}(V)\}E_1 + \{V[f_2] + f_1 \omega_{12}(V)\}E_2.$$

We call this last expression the *covariant derivative formula*. Note that $V[f_1]$ and $V[f_2]$ only tell how W is changing *relative to* E_1, E_2—the effect of the terms involving connection forms is to compensate for the way E_1, E_2 itself is rotating, so $\nabla_V W$ is an "absolute" rate of change.

Proof. Since $E_1 \cdot E_2 = 0$, by a Leibnizian property of ∇, we get

$$0 = V[E_1 \cdot E_2] = \nabla_V E_1 \cdot E_2 + E_1 \cdot \nabla_V E_2.$$

Hence, by (*),

$$\nabla_V E_2 \cdot E_1 = -\omega_{12}(V) = \omega_{21}(V).$$

But $E_i \cdot E_i = 1$, so the same Leibnizian property gives $2\nabla_V E_i \cdot E_i = 0$ for $i = 1, 2$. Using this information, we derive the connection equations by orthonormal expansion of $\nabla_V E_1$ and $\nabla_V E_2$.

Finally we apply the assumed properties of ∇ to get

$$\nabla_V W = \nabla_V (f_1 E_1 + f_2 E_2) = \nabla_V (f_1 E_1) + \nabla_V (f_2 E_2)$$

$$= V[f_1]E_1 + f_1 \nabla_V E_1 + V[f_2]E_2 + f_2 \nabla_V E_2.$$

Substitution of the connection equations then gives the covariant derivative formula. ∎

This lemma shows how to *define* the covariant derivative of M. Note that the order of events is the reverse of that in Chapter II. There we used the Euclidean covariant derivative to define connection forms; now we shall use the connection form ω_{12} to define the covariant derivative for M.

3.2 Theorem For a geometric surface M, there is one and only one covariant derivative ∇ with the usual linear and Leibnizian properties (Corollary 5.4 of Chapter II) and satisfying equation (*) for every frame field on M.

Proof. The previous lemma shows that there is *at most* one such covariant derivative, for $\nabla_V W$ is given by a formula which does not involve ∇. So what we must prove is that such a covariant derivative ∇ actually exists. The proof split into two parts, and we shall supress some details.

A. *Local definition.* For a fixed frame field E_1, E_2 on a region $\mathcal{O}$, use the formula in Lemma 3.1 as the *definition* of $\nabla_V W$. Routine computations verify that ∇ is linear and Leibnizian, and when W is E_1, we get

$$\nabla_V E_1 = \omega_{12}(V)E_2;$$

hence (*) holds.

B. *Consistency.* For two different frame fields, do the local definitions agree? If $\bar{\nabla}_V W$ derives from $\bar{E}_1$, $\bar{E}_2$ on $\bar{\mathcal{O}}$, we must show that $\nabla_V W = \bar{\nabla}_V W$ holds on the overlap of $\mathcal{O}$ and $\bar{\mathcal{O}}$. For then we have a single covariant derivative on all of M. Because of the linear and Leibnizian properties, it suffices to show that

$$\nabla_V \bar{E}_1 = \bar{\nabla}_V \bar{E}_1, \qquad \nabla_V \bar{E}_2 = \bar{\nabla}_V \bar{E}_2. \tag{1}$$

We use Lemma 1.4, assuming for simplicity that E_1, E_2 and $\bar{E}_1$, $\bar{E}_2$ have the same orientation. Applying ∇_V to the equation

$$\bar{E}_1 = \cos \vartheta \, E_1 + \sin \vartheta \, E_2,$$

the covariant derivative formula gives

$$\nabla_V \bar{E}_1 = \{ V[\cos \vartheta] + \sin \vartheta \, \omega_{21}(V) \} E_1 + \{ V[\sin \vartheta] + \cos \vartheta \, \omega_{12}(V) \} E_2. \quad (2)$$

By Lemma 1.4, $\bar{\omega}_{12} = \omega_{12} + d\vartheta$. Substituting $\omega_{12} = \bar{\omega}_{12} - d\vartheta$ into (2) produces some favorable cancellations, leaving

$$\begin{aligned}
\nabla_V \bar{E}_1 &= \bar{\omega}_{12}(V)\{ - \sin \vartheta \, E_1 + \cos \vartheta \, E_2 \} \\
&= \bar{\omega}_{12}(V) \bar{E}_2 = \bar{\nabla}_V \bar{E}_1.
\end{aligned} \quad (3)$$

In the same way, $\nabla_V \bar{E}_2 = \bar{\nabla}_V \bar{E}_2$ may be derived from

$$\bar{E}_2 = -\sin \vartheta \, E_1 + \cos \vartheta \, E_2. \qquad\blacksquare$$

3.3 Example *The covariant derivative of* $\mathbf{E}^2$. The natural frame field U_1, U_2 has $\omega_{12} = 0$. Thus for a vector field

$$W = f_1 U_1 + f_2 U_2,$$

the covariant derivative formula (Lemma 3.1) reduces to

$$\nabla_V W = V[f_1] U_1 + V[f_2] U_2.$$

This is just Lemma 5.2 of Chapter II (applied to $\mathbf{E}^2$ instead of $\mathbf{E}^3$), so our abstract definition of covariant derivative produces correct results on the Euclidean plane.

The covariant derivative ∇ of a geometric surface M may be modified so as to apply to a *vector field* Y *on a curve* α in M. (As usual, for each t, $Y(t)$ is a tangent vector to M at $\alpha(t)$, as in Fig. 7.6.)

If E_1, E_2 is a vector field on a region of M containing α, we may write

$$Y(t) = y_1(t) E_1(\alpha(t)) + y_2(t) E_2(\alpha(t)),$$

or, briefly,

$$Y = y_1 E_1 + y_2 E_2.$$

Roughly speaking, we want the *covariant derivative* Y' of Y to be $\nabla_{\alpha'} Y$. Thus the covariant derivative formula (Lemma 3.1) shows that we must define

$$Y' = \{ y_1' + y_2 \omega_{21}(\alpha') \} E_1 + \{ y_2' + y_1 \omega_{12}(\alpha') \} E_2.$$

It is a routine matter to check that this notion of covariant derivative is independent of the choice of frame field and has the same linear and Leibnizian properties as in the Euclidean case. Also, as in Chapter II, Section 2, since the velocity α' of a curve in M is a vector field on M, we can take its covariant derivative to obtain the *acceleration* α'' of α.

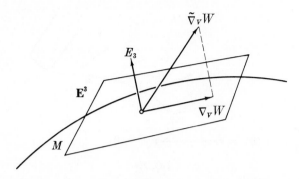

FIG. 7.5

It may be well to look back now at the case of a surface M in $\mathbf{E}^3$. If V and W are tangent vector fields on M, there are two ways to take covariant derivatives: one from the intrinsic geometry of M as a geometric surface, the other the Euclidean covariant derivative of $\mathbf{E}^3$. These two derivatives are generally different, but there is a simple relationship between them.

3.4 Lemma Let V and W be tangent vector fields on a surface M in $\mathbf{E}^3$ (Fig. 7.5). If ∇ is the covariant derivative of M as a geometric surface, and $\tilde{\nabla}$ is the Euclidean covariant derivative, then

$\nabla_V W$ is the component of $\tilde{\nabla}_V W$ tangent to M.

Proof. First suppose that W is one of the vector fields E_1, E_2 of an adapted frame field E_1, E_2 E_3. By the *Euclidean* connection equations (Theorem 7.2 of Chapter II) we have

$$\tilde{\nabla}_V E_1 = \sum_{j=1}^{3} \omega_{ij}(V)E_j = \omega_{12}(V)E_2 + \omega_{13}(V)E_3.$$

But the connection equations (Lemma 3.1) for M give

$$\nabla_V E_1 = \omega_{12}(V)E_2.$$

Thus $\tilde{\nabla}_V E_1$ is $\nabla_V E_1$ plus a vector field normal to M. In other words, $\nabla_V E_1$ is (at each point) the component of $\tilde{\nabla}_V E_1$ tangent to M. The same result, of course, holds for E_2.

In the general case, since W is tangent to M, we may write

$$W = f_1 E_1 + f_2 E_2.$$

Then the required result follows immediately from the special case above, since both covariant derivatives are linear and Leibnizian.   ∎

Thus we have been using the intrinsic covariant derivative of $M \subset \mathbf{E}^3$ all along, although we did not give it formal recognition. It occurs when-

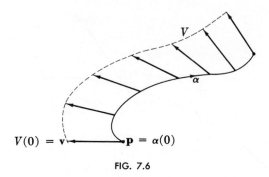

FIG. 7.6

ever we take the tangential component of the Euclidean covariant derivative.

Only the most basic properties of covariant derivatives are shared by all geometric surfaces. In particular, the related notion of *parallelism* (due to Levi Civita) does not always behave as in the Euclidean case. Much of the individual character of Euclidean geometry rests on the fact that a tangent vector $\mathbf{v}_p$ to, say, $\mathbf{E}^2$ may be moved to a *parallel* tangent vector $\mathbf{v}_q$ at any other point $\mathbf{q}$. As we shall see, this phenomenon of "distant parallelism" does not obtain on an arbitrary geometric surface. However it is always possible to define parallelism of a vector field Y *on a curve*. In Euclidean space, this means that Y has constant coefficients relative to the natural frame field, but the infinitesimal characterization $Y' = 0$ makes sense in general.

3.5 Definition A vector field Y on a curve α in geometric surface M is *parallel* provided its covariant derivative vanishes: $Y' = 0$.

Just as in the Euclidean case, a parallel vector field has constant length, for $\| Y \|^2 = Y \cdot Y$, and $(Y \cdot Y)' = 2Y \cdot Y' = 0$.

3.6 Lemma Let α be a curve in a geometric surface M, and let $\mathbf{v}$ be a tangent vector at, say, $\mathbf{p} = \alpha(0)$. Then there is a unique parallel vector field V on α such that $V(0) = \mathbf{v}$ (Fig. 7.6).

Proof. We may suppose that α lies entirely in the domain of a frame field E_1, E_2 on M. (Otherwise we could break α up into segments for which this is the case.) The vector field V must satisfy the conditions

$$V' = 0, \qquad V(0) = \mathbf{v}. \tag{1}$$

Because V has constant length $\| V \| = c$, we may write

$$V = c \cos \varphi \, E_1 + c \sin \varphi \, E_2 \tag{2}$$

where φ is the angle from E_1 to V. Thus the covariant derivative formula gives

$$V' = c \left\{ -\sin \varphi \, \varphi' + \sin \varphi \, \omega_{21}(\alpha') \right\} E_1$$
$$+ \, c \left\{ \cos \varphi \, \varphi' + \cos \varphi \, \omega_{12}(\alpha') \right\} E_2$$

It follows immediately that (1) is equivalent to

$$\varphi' = -\omega_{12}(\alpha'),$$

with $\varphi(0)$ the angle from $E_1(\mathbf{p})$ to $V(0) = \mathbf{v}$. There is only one such function, namely

$$\varphi(t) = \varphi(0) - \int_0^t \omega_{12}(\alpha') \, dt$$

This function φ, substituted in (2), defines the required vector field V. ∎

In the situation stated in Lemma 3.6 we say, for each t, that the vector $V(t)$ at $\alpha(t)$ is obtained from $\mathbf{v}$ at $\mathbf{p} = \alpha(0)$ by *parallel translation along* α.

In $\mathbf{E}^2$, parallel translation of a tangent vector $\mathbf{v}_p$ along a curve segment from $\mathbf{p}$ to $\mathbf{q}$ merely produces the distant-parallelism result $\mathbf{v}_q$, which is thus entirely independent of the choice of the curve. But for an arbitrary geometric surface M, different curves from $\mathbf{p}$ to $\mathbf{q}$ will usually produce different vectors at $\mathbf{q}$. Equivalently: *If a vector $\mathbf{v}$ at $\mathbf{p}$ is parallel-translated around a closed curve α* (starting and ending at $\mathbf{p}$) *the result $\mathbf{v}^*$ is not necessarily the same as $\mathbf{v}$.* This phenomenon is called *holonomy*. If we fix a particular frame field on the curve α, then the proof of Lemma 3.6 shows that parallel translation from $\alpha(a) = \mathbf{p}$ to $\alpha(b) = \mathbf{p}$ along α rotates all vectors through the same angle $\varphi(b) - \varphi(a)$—since φ' is the same for all parallel vector fields. We call this the *holonomy angle* ψ_α of α. (Multiples of 2π may be ignored in ψ_α, since they do not affect the determination of $\mathbf{v}^*$.)

3.7 Example *Holonomy on the sphere Σ of radius r.* Suppose the closed curve α parametrizes a circle on Σ. There is no loss of generality in assuming that α is a circle of latitude, say the u-parameter curve

$$\alpha(u) = \mathbf{x}(u, v_0), \qquad 0 \le u \le 2\pi,$$

where $\mathbf{x}$ is the geographical patch in Σ (Fig. 7.7). According to Example 6.2 of Chapter VI, the associated frame field E_1, E_2 of $\mathbf{x}$ has $\omega_{12} = \sin v \, du$. Now the proof of Lemma 3.6 shows that every parallel vector field on α has angle φ (measured from E_1) satisfying $\varphi' = -\omega_{12}(\alpha')$. It follows that φ' has constant value $-\sin v_0$ on α. Thus the holonomy angle ψ_α of α is

$$\varphi(2\pi) - \varphi(0) = -2\pi \sin v_0$$

Note that only on the equator, $v = 0$, does a vector $\mathbf{v}$ return to itself after parallel translation around α. When v_0 is near $\pi/2$, then α is a small circle around the north pole of Σ. Since φ' is close to -1, parallel vector field V is rotating rapidly *with respect to E_1, E_2*. But the holonomy angle is

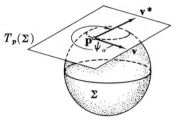

FIG. 7.7

close to -2π, so the actual difference between $\mathbf{v} = V(0)$ and $\mathbf{v}^* = V(2\pi)$ is, as one would expect, quite small.

Gaussian curvature has a strong influence on holonomy, as shown by Exercise 5.

For a patch $\mathbf{x}$ in an arbitrary geometric surface M we shall inevitably use the notation $\mathbf{x}_{uu}$ for the covariant derivative of $\mathbf{x}_u$ along the u-parameter curves of $\mathbf{x}$—with corresponding meanings for $\mathbf{x}_{uv}$, $\mathbf{x}_{vu}$, and $\mathbf{x}_{vv}$. Thus when M is a surface in $\mathbf{E}^3$, we must have a new notation, say $\tilde{\mathbf{x}}_{uu}, \cdots$, for the analogous objects defined in Chapter V, Section 4. There we were using the covariant derivative of $\mathbf{E}^3$, while now we use the covariant derivative of M itself. It is still true in the intrinsic case that $\mathbf{x}_{uv} = \mathbf{x}_{vu}$, but the proof is by no means obvious (Exercise 9).

EXERCISES

1. In the Poincaré half-plane, let α be the curve given in Exercise 2 of Section 1. Express its velocity and acceleration in terms of the frame field

   $$E_1 = vU_1, \ E_2 = vU_2.$$

   (*Hint:* They are collinear.)

2. Let β be the curve

   $$\beta(t) = (ct, st), \ t > 0,$$

   in the Poincaré half-plane, where c and s are constants with $c^2 + s^2 = 1$. (Thus β is a *Euclidean* straight line through the origin.) Express the velocity and acceleration of β in terms of the frame field used in Exercise 1.

3. If V and W are tangent vector fields on a surface in $\mathbf{E}^3$, refine the proof of Lemma 3.4 to show that

$$\tilde{\nabla}_V W = \nabla_V W + S(V) \cdot W \, U,$$

where S is the shape operator derived from $U = \pm E_3$. Hence if α is a curve in M,

$$\tilde{\alpha}'' = \alpha'' + S(\alpha') \cdot \alpha' \, U.$$

4. Show that on the sphere Σ the curve α given in Example 3.7 has (intrinsic) acceleration $\alpha'' = r \cos v_0 \sin v_0 \, E_2$. Compute its Euclidean acceleration, and show that α'' is the component tangent to Σ.

5. Let α be a closed curve in a geometric surface M.
   (a) If α is homotopic to a constant via $\mathbf{x}$ (Ex. 12 of IV.6), show that the holonomy angle of α is $\iint_{\mathbf{x}} K \, dM$. (Assume that $\mathbf{x}(R)$ lies in the domain of a frame field.) When $\mathbf{x}$ is patch-like, this integral is the total curvature of the region $\mathbf{x}(R)$.
   (b) Compute the holonomy angle in Example 3.7 by this method.

6. Let V be a parallel vector field on a curve α in M, and let W be a vector field on α with constant length. Show that W is parallel if and only if the angle between V and W is constant.

7. Show that isometries preserve covariant derivatives in this sense: If Y is a vector field on a curve α in M, and $F \colon M \to \bar{M}$ is an isometry, then

$$F_* (Y') = \bar{Y}', \text{ where } \bar{Y} \text{ is the vector field } F_* (Y) \text{ on } \bar{\alpha} = F(\alpha) \text{ in } \bar{M}.$$

(Simplify matters by assuming that Y can be written $Y = fE_1$, where E_1, E_2 is a frame field on M).

This is the analogue of the Euclidean result (Corollary 4.1 of Chapter III). The general case is given in Exercise 8.

8. Prove that an isometry $F \colon M \to \bar{M}$ preserves the covariant derivatives ∇ and $\bar{\nabla}$. Explicitly, for each vector field X on M, let $\bar{X}$ be the transfered vector field on $\bar{M} \colon \bar{X}(F(\mathbf{p})) = F_*(X(\mathbf{p}))$ for each point of M. Then show that $\bar{\nabla}_{\bar{V}} \bar{W} = \overline{\nabla_V W}$. (*Hint:* If $f_i = W \cdot E_i$ and $\bar{f}_i = \bar{W} \cdot \bar{E}_i$, use Exercise 8 of Section 5, Chapter IV, to show that $V[f_i]$ at $\mathbf{p}$ equals $\bar{V}[\bar{f}_i]$ at $F(\mathbf{p})$.)

9. If $\mathbf{x}$ is an orthogonal patch in a geometric surface M, show that

$$\mathbf{x}_{uv} = \mathbf{x}_{vu}$$

(intrinsic derivatives). (*Hint:* For the associated frame field; compute

$$\omega_{21}(\mathbf{x}_u) = \left(\frac{\mathbf{x}_v}{\sqrt{G}} \right)_u \cdot \frac{\mathbf{x}_u}{\sqrt{E}} = \mathbf{x}_{vu} \cdot \frac{\mathbf{x}_u}{\sqrt{EG}}$$

Then using the formula for ω_{12} given in Chapter VI, Section 6, show that

$$\omega_{21}(\mathbf{x}_u) = \frac{(\sqrt{E})_v}{\sqrt{G}} = \mathbf{x}_u \cdot \frac{\mathbf{x}_{uv}}{\sqrt{EG}}.$$

Find an easier proof in the special case where M is a surface in $\mathbf{E}^3$.

10. (*Continuation*) Show that $\mathbf{y}_{uv} = \mathbf{y}_{vu}$ for an arbitrary patch $\mathbf{y}$. (*Hint:* There exists an orthogonal patch $\mathbf{x}$ such that $\mathbf{x} = \mathbf{y}(\bar{u},\bar{v})$.)

11. If there exists a nonvanishing vector field W on M such that $\nabla_V W = 0$ for all V, show that M is flat. Find such a vector field on a cylinder in $\mathbf{E}^3$.

4 Geodesics

Geodesics in an arbitrary geometric surface generalize *straight lines* in Euclidean geometry. We have seen that a straight line $\alpha(t) = \mathbf{p} + t\mathbf{q}$ is characterized infinitesimally by vanishing of acceleration; thus

4.1 Definition A curve α in a geometric surface M is a *geodesic* of M provided its acceleration is zero; $\alpha'' = 0$.

In other words, the velocity α' of a geodesic is parallel: geodesics never turn. Recall that α' parallel implies $\| \alpha' \|$ constant; so geodesics have constant speed.

Because acceleration is preserved by isometries (Exercise 7 of Section 3), it follows that geodesics are isometric invariants. (A direct proof appears in Exercise 1 of Section 5, Chapter VI). In fact, if $F: M \to N$ is merely a local isometry, then F carries each geodesic α of M to a geodesic $F(\alpha)$ of N—for F is locally an isometry, as discussed in Chapter VI, Section 4.

The general definition of geodesics given above agrees with that of 5.7 in Chapter V when M is a surface in $\mathbf{E}^3$, for we can deduce from Lemma 3.4 that the *intrinsic* acceleration of a curve α in $M \subset \mathbf{E}^3$ is the component tangent to M of its *Euclidean* acceleration. Thus the former is zero if and only if the latter is normal to M.

Suppose that $\alpha: I \to M$ is a curve in an arbitrary geometric surface M and E_1, E_2 is a frame field on M. Throughout this section we use the notation

$$\alpha' = v_1 E_1 + v_2 E_2 \qquad \text{and} \qquad \alpha'' = A_1 E_1 + A_2 E_2$$

for the velocity and acceleration of α. From Section 3 we know that these components of acceleration are

$$A_1 = v_1' + v_2 \omega_{21}(\alpha')$$
$$A_2 = v_2' + v_1 \omega_{12}(\alpha')$$

and are real-valued function on the interval I. *Our main criterion for α to be geodesic is thus $A_1 = A_2 = 0$.* Using orthogonal coordinates we now rewrite these equations in a more informative way.

4.2 Theorem   Let $\mathbf{x}$ be an orthogonal coordinate patch in a geometric surface M. A curve $\alpha(t) = \mathbf{x}(a_1(t), a_2(t))$ is a geodesic of M if and only if

$$a_1'' + \frac{1}{2E}\{E_u a_1'^2 + 2E_v a_1' a_2' - G_u a_2'^2\} = 0$$

$$a_2'' + \frac{1}{2G}\{-E_v a_1'^2 + 2G_u a_1' a_2' + G_v a_2'^2\} = 0.$$

We shall subsequently refer to these equations as $A_1 = 0$ and $A_2 = 0$. Note that they are symmetric, in the sense that the reversals $1 \leftrightarrow 2$, $u \leftrightarrow v$, $E \leftrightarrow G$ turn each one into the other. *In this context we shall always understand that the functions E, G and their partial derivatives E_u, E_v, $\cdots$ are evaluated on (a_1, a_2), and hence become functions on the domain I of α.*

Proof. The velocity of α is $\alpha' = a_1' \mathbf{x}_u + a_2' \mathbf{x}_v$, so in terms of the associated frame field of $\mathbf{x}$ (Chapter VI, Section 6), we have

$$\alpha' = (a_1' \sqrt{E})E_1 + (a_2' \sqrt{G})E_2.$$

Thus the acceleration components A_1, A_2 defined above become

$$\begin{aligned} A_1 &= (a_1' \sqrt{E})' + (a_2' \sqrt{G})\omega_{21}(\alpha') \\ A_2 &= (a_2' \sqrt{G})' + (a_1' \sqrt{E})\omega_{12}(\alpha'). \end{aligned} \tag{1}$$

Using the formula for ω_{12} from Chapter VI, Section 6, we find

$$\omega_{12}(\alpha') = \omega_{12}(a_1' \mathbf{x}_u + a_2' \mathbf{x}_v) = -\frac{(\sqrt{E})_v}{\sqrt{G}} a_1' + \frac{(\sqrt{G})_u}{\sqrt{E}} a_2' \tag{2}$$

When this is substituted in (1), the geodesic equations $A_1 = A_2 = 0$ become

$$(a_1' \sqrt{E})' + (\sqrt{E})_v a_1' a_2' - \frac{\sqrt{G}(\sqrt{G})_u}{\sqrt{E}} a_2'^2 = 0$$

$$(a_2' \sqrt{G})' - \frac{\sqrt{E}(\sqrt{E})_v}{\sqrt{G}} a_1'^2 + (\sqrt{G})_u a_1' a_2' = 0. \tag{3}$$

Standard calculus computations will transform (3) to the form stated in the theorem. We merely remind the reader that in a Leibnizian expansion such as

$$(a_1' \sqrt{E})' = a_1'' \sqrt{E} + a_1' \frac{E'}{2\sqrt{E}},$$

the notation E is short for $E(a_1, a_2)$, so

$$E' = E_u a_1' + E_v a_2'.$$ ∎

4.3 Theorem   Given a tangent vector $\mathbf{v}$ to M at a point $\mathbf{p}$, there is a unique geodesic α of M such that

$$\alpha(0) = \mathbf{p}, \qquad \alpha'(0) = \mathbf{v}.$$

Thus there are lots of geodesics in any geometric surface, and each is completely determined by its initial position and velocity. In $\mathbf{E}^2$ for example, the geodesic determined by $\mathbf{v}$ at $\mathbf{p}$ is the straight line $\alpha(t) = \mathbf{p} + t\mathbf{v}$.

Proof.   Let $\mathbf{x}$ be an orthogonal patch in M with $\mathbf{p} = \mathbf{x}(u_0, v_0)$, and write $\mathbf{v} = c_0 \mathbf{x}_u + d_0 \mathbf{x}_v$. The geodesic equations in Theorem 4.2 have the form

$$
\begin{aligned}
a_1'' &= f_1(a_1, a_2, a_1', a_2') \\
a_2'' &= f_2(a_1, a_2, a_1', a_2')
\end{aligned}
\tag{1}
$$

Furthermore, α will satisfy the given initial conditions if and only if

$$
\begin{aligned}
a_1(0) &= u_0 & a_1'(0) &= c_0 \\
a_2(0) &= v_0 & a_2'(0) &= d_0.
\end{aligned}
\tag{2}
$$

Now the fundamental existence and uniqueness theorem of differential equations asserts that there is an interval I about 0 on which are defined unique functions a_1, a_2 which satisfy (1) and (2). Thus $\alpha = \mathbf{x}(a_1, a_2)$ is the only geodesic defined on I such that $\alpha(0) = \mathbf{p}$, $\alpha'(0) = \mathbf{v}$. ∎

This proof is not entirely satisfactory, because the interval I may be unnecessarily small. We describe briefly a way to make it as large as possible. Suppose that $\alpha_1 \colon I_1 \to M$ and $\alpha_2 \colon I_2 \to M$ are geodesics satisfying the same initial conditions at $t = 0$. Using the uniqueness property above, we can deduce that $\alpha_1 = \alpha_2$ on the common part of I_1 and I_2. Applying this consistency result to all such geodesics, we obtain a single *maximal* geodesic $\alpha \colon I \to M$ satisfying the initial conditions. (The interval I is the largest possible.) Intuitively this means we simply let the geodesic run as far as it can.

4.4 Definition   A geometric surface M is *geodesically complete* provided every maximal geodesic is defined on the whole real line $\mathbf{R}$.

Briefly: geodesics run forever. A constant curve is trivially a geodesic, but excluding this case, every geodesic has constant nonzero speed. Thus geodesic completeness means that all nontrivial (maximal) geodesics are infinitely long—in both directions. For example, $\mathbf{E}^2$ is certainly complete, and the explicit computations in Example 5.8 of Chapter V show that

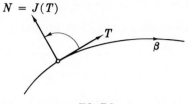

$$N = J(T)$$

FIG. 7.8

spheres and cylinders in $\mathbf{E}^3$ are. More generally, all compact geometric surfaces are complete, as are all surfaces in $\mathbf{E}^3$ of the form $M: g = c$ (consequences of Theorem 15, Chapter 10, of Hicks [5]). Removal of even a single point from a complete surface will destroy the property since geodesics formerly passing through the point will be obliged to stop.

There is a Frenet theory of curves in a geometric surface M which generalizes that for curves in the plane (Exercise 8 of Section 3 in Chapter II). Because M has only two dimensions, torsion cannot be defined. If M is oriented, however, curvature can be given a geometrically meaningful sign, as follows. If $\beta: I \to M$ is a unit-speed curve in an oriented geometric surface, then $T = \beta'$ is the *unit tangent vector field* of β. To get the *principal normal vector field* N, we "rotate T through $+90°$," defining

$$N = J(T),$$

where J is the rotation operator from Exercise 5, Section 1 (Fig. 7.8). Then the *geodesic curvature* κ_g of β is the real-valued function on I for which the *Frenet formula* $T' = \kappa_g N$ holds. Thus κ_g is not restricted to nonnegative values as in the case of curves in $\mathbf{E}^3$: $\kappa_g > 0$ means that T—hence β—is turning in the *positive* direction as given by the orientation of M, and $\kappa_g < 0$ means negative turning.

4.5 Lemma Let β be a unit-speed curve in a region oriented by a frame field E_1, E_2. If φ is an angle function from E_1 to β' along β, then

$$\kappa_g = \frac{d\varphi}{ds} + \omega_{12}(\beta').$$

Proof. By definition of angle function (Exercise 15 in Chapter VI, Section 7), we have

$$T = \beta' = \cos \varphi \, E_1 + \sin \varphi \, E_2.$$

Since the orientation derives from this frame field,

$$J(E_1) = E_2 \qquad \text{and} \qquad J(E_2) = -E_1,$$

so

$$N = J(T) = -\sin\varphi\, E_1 + \cos\varphi\, E_2.$$

Using the derivative formula on page 320, we find

$$T' = \beta'' = \{-\sin\varphi\,\varphi' + \sin\varphi\,\omega_{21}(\beta')\}E_1$$
$$+ \{\cos\varphi\,\varphi' + \cos\varphi\,\omega_{12}(\beta')\}E_2.$$

But $\kappa_g = T' \cdot N = T' \cdot J(T)$, so using the formulas for T' and $J(T)$, we get

$$\kappa_g = (\cos^2\varphi + \sin^2\varphi)(\varphi' + \omega_{12}(\beta')) = \varphi' + \omega_{12}(\beta').\qquad\blacksquare$$

For example, in $\mathbf{E}^2$ the natural frame field has $\omega_{12} = 0$, and φ becomes the usual slope angle of the curve β. Thus the result reduces to $\kappa_g = d\varphi/ds$, which in elementary calculus is often taken as the definition of curvature.

For an arbitrary-speed regular curve α in M, the Frenet apparatus T, N, κ_g is defined—just as in Chapter II, Section 4—by reparametrization. Furthermore the same proof as for Lemma 4.2 of Chapter II shows

$$\alpha' = vT \qquad \alpha'' = \frac{dv}{dt}T + \kappa_g v^2 N \qquad\qquad (^*)$$

where $v = \|\alpha'\|$ is the speed function of α.

4.6 Lemma   A regular curve α in M is a geodesic if and only if α has constant speed and geodesic curvature $\kappa_g = 0$.

Proof. Since $v > 0$, we have $\alpha'' = 0$ if and only if $(dv/dt) = \kappa_g = 0$. $\blacksquare$

The equations $(^*)$ also show that α *has geodesic curvature zero if and only if α' and α'' are always collinear.* Such curves are sometimes called geodesics: to get a geodesic in the strict sense of Definition 4.1 it suffices to reparametrize α to a constant speed curve. (Proof: κ_g is unaffected by reparametrization.) In contexts where parametrization is of some importance we shall call a curve with $\kappa_g = 0$ a *pre-geodesic*.

Computation of explicit formulas for the geodesics of a given geometric surface is rarely a simple task. Our purpose, however, is not to collect formulas, but to study the general behavior of geodesics. Before continuing this study we are going to examine an important special case in which a considerable amount of concrete information about geodesics can often be obtained with only a minimum of computation.

4.7 Definition   A *Clairut parametrization* $\mathbf{x}: D \to M$ is an orthogonal parametrization for which E and G depend only on u, that is, $F = 0$ and $E_v = G_v = 0$.

For example, the usual parametrization of a surface of revolution is a Clairut parametrization.

4.8 Lemma   If $\mathbf{x}$ is a Clairut parametrization, then

1. All the u-parameter curves of $\mathbf{x}$ are pre-geodesics, and
2. A v-parameter curve, $u = u_0$, is a geodesic if and only if $G_u(u_0) = 0$.

Proof. For (1) it suffices by a remark above to show that $\mathbf{x}_u$ and $\mathbf{x}_{uu}$ are collinear. Since $\mathbf{x}_u$ and $\mathbf{x}_v$ are orthogonal, this is equivalent to

$$\mathbf{x}_v \cdot \mathbf{x}_{uu} = 0.$$

But the following equations imply this result:

$$0 = E_v = (\mathbf{x}_u \cdot \mathbf{x}_u)_v = 2\mathbf{x}_u \cdot \mathbf{x}_{uv}$$

$$0 = F_u = (\mathbf{x}_u \cdot \mathbf{x}_v)_u = \mathbf{x}_{uu} \cdot \mathbf{x}_v + \mathbf{x}_u \cdot \mathbf{x}_{vu}.$$

Similarly, for (2), the v-parameter curve, $u = u_0$, is pre-geodesic if and only if $\mathbf{x}_{vv}(u_0, v) \cdot \mathbf{x}_u(u_0, v) = 0$. The following equations show that this is the case if and only if $G_u(u_0) = 0$.

$$0 = F_v = \mathbf{x}_{uv} \cdot \mathbf{x}_v + \mathbf{x}_u \cdot \mathbf{x}_{vv}$$

$$G_u(u_0) = G_u(u_0, v) = 2\mathbf{x}_{vu}(u_0, v) \cdot \mathbf{x}_v(u_0, v) \qquad \text{for all } v.$$

(Recall that $\mathbf{x}_{uv} = \mathbf{x}_{vu}$.) *We have not used the condition* $G_v = 0$; its only effect is to show that v-parameter pre-geodesics are in fact geodesics, since it means that v-parameter curves have constant speed. ∎

In the case of a surface of revolution this lemma provides an intrinsic proof that meridians are geodesics and that a parallel, $u = u_0$, is geodesic if and only if $h'(u_0) = 0$. (See Exercise 3 of Chapter V, Section 5).

Because of the preceding lemma we shall think of a Clairut parametrization as a "flow" whose streamlines are its u-parameter geodesics, and we shall measure the behavior of arbitrary geodesics relative to this flow.

4.9 Lemma If $\alpha = \mathbf{x}(a_1, a_2)$ is a unit speed geodesic, and $\mathbf{x}$ is a Clairut parametrization, then the function

$$c = G(a_1)a_2' = \sqrt{G(a_1)} \sin \varphi$$

is *constant*, where φ is the angle from $\mathbf{x}_u$ to α'. Hence α cannot leave the region for which $G \geq c^2$.

We call the constant c thus associated with each geodesic α the *slant* of α, since—in combination with G—it determines the angle φ at which α is cutting across the u-parameter streamlines of $\mathbf{x}$ (see Fig. 7.9).

Proof. Because $E_v = G_v = 0$ for a Clairut parametrization, the equation $A_2 = 0$ of Theorem 4.2 reduces to

$$a_2'' + \frac{G_u}{G} a_1' a_2' = 0$$

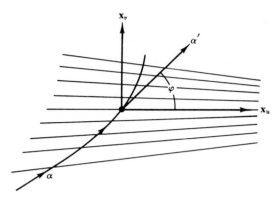

FIG. 7.9

But this is equivalent to the constancy of $c = Ga_2{}'$, since

$$(Ga_2{}')' = G'\, a_2{}' + G\, a_2{}'' = G_u a_1{}'\, a_2{}' + G\, a_2{}''$$

To show that $c = \sqrt{G}\,\sin\varphi$, compare the two equations

$$\alpha' \cdot \mathbf{x}_v = (a_1{}'\, \mathbf{x}_u + a_2{}'\, \mathbf{x}_v)\cdot \mathbf{x}_v = Ga_2{}' = c$$

$$\alpha' \cdot \mathbf{x}_v = \|\,\alpha'\,\|\,\|\, \mathbf{x}_v\,\|\,\cos\!\left(\frac{\pi}{2} - \varphi\right) = \sqrt{G}\,\sin\varphi$$

It follows immediately from $|\sin\varphi| \leq 1$ that $G \geq c^2$.   ∎

If α is moving in a direction in which G is increasing, then the constancy of $c = \sqrt{G}\,\sin\varphi$ shows that φ is decreasing: α is forced to turn more in the direction of the flow. On the other hand, if G is decreasing along α, then α cuts across the u-parameter geodesics at ever-increasing angles. Some interesting consequences are presented in Exercises 11 and 12. This interpretation is particularly simple on a surface of revolution (Exercise 13).

We can add to the lemma above the equation

$$a_1{}' = \pm\frac{\sqrt{G - c^2}}{\sqrt{EG}} \qquad (1)$$

In fact, since α has unit speed we have $1 = \alpha' \cdot \alpha' = E a_1{}'^2 + G a_2{}'^2$. In this equation substitute

$$a_2{}' = \frac{c}{G} \qquad (2)$$

(from Lemma 4.9), and solve for $a_1{}'$ to obtain equation (1).

Conversely, a straightforward computation shows that if $a_1{}'$ is nonzero, the equations (1) and (2) imply that $\alpha = \mathbf{x}(a_1, a_2)$ is a unit speed geodesic. Furthermore, $a_1{}'$ nonzero is a necessary and sufficient condition for an arbitrary curve α to have a reparametrization of the form

$$\beta(u) = \mathbf{x}(u, v(u)).$$

The point of all this is that we can now give a relatively simple criterion for a curve of this form to be a pre-geodesic—and thereby determine *the routes of the geodesics in a region with a Clairut parametrization.* (The special parametrization of β misses essentially only the geodesics given by (2) of Lemma 4.8; see Exercise 12.)

4.10 Theorem A curve $\beta(u) = \mathbf{x}(u, v(u))$, where $\mathbf{x}$ is a Clairut parametrization, is a pre-geodesic if and only if

$$\frac{dv}{du} = \frac{\pm c\,\sqrt{E}}{\sqrt{G}\,\sqrt{G - c^2}}$$

The constant c is then the slant of β.

Proof. The argument is simply an exercise in change of parametrization. Let α be a unit speed reparametrization of β derived just as in Chapter II from an arc length function s for β. Thus β is pre-geodesic if and only if α is geodesic. Let a_1 be the inverse function of s (so $a_1{}'$ does not vanish). Then

$$\alpha = \beta(a_1) = \mathbf{x}(a_1, v(a_1)),$$

and we set $a_2 = v(a_1)$. The remarks above show that α is a geodesic (of slant c) if and only if

$$a_1{}' = \frac{\pm\sqrt{G - c^2}}{\sqrt{EG}}, \qquad a_2{}' = \frac{c}{G} \qquad (E,G \text{ evaluated on } a_1) \qquad (1)$$

If these equations hold, then by elementary calculus

$$\frac{dv}{du} = \frac{a_2{}'(s)}{a_1{}'(s)} = \frac{\pm c\,\sqrt{E}}{\sqrt{G}\,\sqrt{G - c^2}} \qquad (2)$$

where the substitution of s (inverse function of a_1) makes E and G merely functions of u. Conversely, if (2) holds, we deduce (1) using the equation $Ea_1{}'^2 + Ga_2{}'^2 = 1$ which expresses the fact that α has unit speed. ∎

Since the formula above for dv/du depends only on u, by the fundamental theorem of calculus it can be written in integral form as

$$v(u) = v(u_0) \pm \int_{u_0}^{u} \frac{c\,\sqrt{E}\,du}{\sqrt{G}\,\sqrt{G - c^2}}$$

Thus for a Clairut parametrization—in particular, for a surface of revolution—the computation of pre-geodesics is reduced to a single integration. This is, of course, a vastly simpler criterion than the second-order differential equations of Theorem 4.2. Unfortunately, however, the integration can rarely be carried out in terms of elementary functions.

4.11 Example Routes of Geodesics

(1) *The Euclidean plane* $\mathbf{E}^2$. We begin with a surface whose geodesics we already know, but to illustrate the preceding result we shall find their routes in terms of the polar parametrization

$$\mathbf{x}(u, v) \ = \ (u \cos v, \, u \sin v).$$

Since $E = 1$, $F = 0$, and $G = u^2$, this is a Clairut parametrization. The u-parameter geodesics are just the radial lines through the origin. All others may be parametrized as $\beta(u) = x(u, v(u))$, where by Theorem 4.10

$$\frac{dv}{du} = \frac{\pm c}{u \sqrt{u^2 - c^2}} = \pm \frac{d}{du}\left(\cos^{-1} \frac{c}{u} \right)$$

Hence $v - v_0 = \pm \cos^{-1}(c/u)$, or $u \cos(v - v_0) = c$, which is the polar equation of a straight line. The slant c has geometrical significance as the distance from the line to the origin.

(2) *The hyperbolic plane* H. Polar coordinates are a more natural choice in this case since the function g giving the geometric structure of H depends only on distance to the origin.† Thus if

$$\mathbf{x}(u, v) \ = \ (u \cos v, \, u \sin v), \, 0 < u < 2,$$

then $g(\mathbf{x}) = 1 - u^2/4$. (We write simply g henceforth.) Now

$$E = \mathbf{x}_u \circ \mathbf{x}_u = \frac{1}{g^2}, \qquad F = 0, \qquad G = \mathbf{x}_v \circ \mathbf{x}_v = \frac{u^2}{g^2};$$

thus $\mathbf{x}$ is a Clairut parametrization. By Lemma 4.8 the u-parameter curves —Euclidean lines through the origin—are routes of geodesics of H. By Theorem 4.10, $\beta(u) = \mathbf{x}(u, v(u))$ is pre-geodesic provided

$$\frac{dv}{du} = \frac{\pm (cg/u^2)}{\sqrt{1 - (cg/u)^2}}. \tag{1}$$

To carry out the required integration, set

$$w = \frac{a}{u}\left(1 + \frac{u^2}{4} \right), \qquad \text{where } a = \frac{c}{\sqrt{1 + c^2}}.$$

Then a straightforward computation yields

† An even better choice, the *hyperbolic* polar coordinates discussed in Example 5.5, gives rise to the substitution used later in this example.

$$\frac{dv}{du} = \frac{\mp \, dw/du}{\sqrt{1 - w^2}}. \tag{2}$$

Hence

$$v - v_0 = \pm\cos^{-1} w, \quad \text{or} \quad \cos (v - v_0) = w = \frac{a}{u}\left(1 + \frac{u^2}{4}\right).$$

Thus

$$u^2 + 4 - \frac{4u}{a} \cos (v - v_0) = 0 \tag{3}$$

Using the law of cosines in a diagram similar to that in Fig. 7.10, one finds the polar equation of a circle of radius r, centered at $\mathbf{x}(u_0, v_0)$.

$$u^2 + u_0^2 - 2u_0u \cos (v - v_0) = r^2. \tag{4}$$

Comparison with equation (3) shows that the route C of β is a Euclidean circle with $u_0^2 - r^2 = 4$. Since $u_0 > 2$, the center of C lies outside the hyperbolic plane $H\colon x^2 + y^2 < 4$. One can see from Fig. 7.10 that the circle C is orthogonal to the rim $x^2 + y^2 = 4$ of H. Of course, β lies in the open arc of C inside H, and we deduce from Theorem 4.2 that β fills this arc.

Conclusion. The routes of the geodesics of the hyperbolic plane H are the portions in H of: *all Euclidean straight lines through the origin, and all Euclidean circles orthogonal to the rim of H.*

The argument in Example 2.4 suggests that the geodesics of H have infinite length (a formal proof is given in Exercise 1 of Section 5); thus H is geodesically complete.

The geodesics of the hyperbolic plane bear comparison with those of the Euclidean plane. Around 300 B.C. Euclid established a celebrated set of axioms for the straight lines of his plane. The goal was to derive its

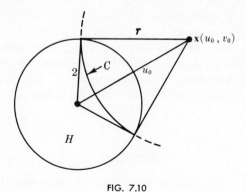

FIG. 7.10

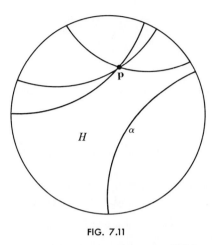

FIG. 7.11

geometry from axioms so overwhelmingly reasonable as to be "self-evident." The most famous of these is equivalent to the *parallel postulate:* If **p** is a point not on a line α, then there is a unique line β through **p** which does not meet α. Over the centuries this postulate began to seem somewhat less self-evident than the others. For example, the axiom that two points determine a unique straight line might be checked by laying down a (perhaps long, but still finite) straight edge touching both points. But for the parallel postulate one would have to travel the whole infinite length of β to be sure it never touches α. Thus tremendous efforts were expended in trying to deduce the parallel postulate from the other axioms. The hyperbolic plane H offers the most convincing proof that this cannot be done. For if we replace "line" by "route of a geodesic," then every Euclidean axiom holds in H *except* the parallel postulate. For example, given any two points it is easy to see that one and only one geodesic route runs through them. But it is clear from Fig. 7.11 that in H there are always an infinite number of geodesic routes through **p** that do not meet α. When the implications of this discovery were worked out, what was destroyed was not merely the modest hope of deducing the parallel postulate, but the whole idea that $\mathbf{E}^2$ is, in some philosophical sense, an Absolute, whose properties are "self-evident." It had become but one geometric surface among the infinitely many others discovered by Riemann.

EXERCISES

1. Show that a reparametrization $\alpha(h)$ of a nonconstant geodesic α is again a geodesic if and only if h has the form $h(t) = at + b$.

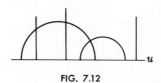

FIG. 7.12

2. Denote by γ_v the unique geodesic in M with initial velocity $\mathbf{v}$. For any number a, show that $\gamma_{av}(t) = \gamma_v(at)$ for all t.

3. Let V be a vector field on a geodesic α. Show that V is parallel if and only if $\| V \|$ is constant and V makes a constant angle with α'.

4. In the sphere Σ, let $\mathbf{n}$ be the north pole, $\mathbf{p}_1$ and $\mathbf{p}_2$ points on the equator. Consider the broken curve β following a meridian from $\mathbf{n}$ to $\mathbf{p}_1$, the equator from $\mathbf{p}_1$ to $\mathbf{p}_2$, then a meridian from $\mathbf{p}_2$ back to $\mathbf{n}$. Prove that the holonomy angle of β is the angle at $\mathbf{n}$ between the two meridians.

5. Find the routes of the geodesics in the stereographic sphere, (1) of Example 2.3.

6. In the Poincaré half-plane, show that the routes of the geodesics are: all semicircles with centers on the u-axis, and all vertical lines. (*Hint:* $\mathbf{x}(u, v) = (u, v)$ is a Clairut patch "relative to v," so in the text equations, reverse u and v, and E and G.) See Fig. 7.12.

7. Let α be a unit speed curve such that α' is never collinear with E_1, where E_1, E_2 is a frame field. If $\alpha'' = A_1E_1 + A_2E_2$, show that the single equation $A_1 = 0$ implies that α is a geodesic.

8. In the projective plane of radius r (Exercise 6 of Section 2), prove:
   (a) The geodesics are simple closed curves of length πr.
   (b) There is a unique geodesic route through any two distinct points.
   (c) Two distinct geodesic routes meet in exactly one point.
   (*Hint:* every geodesic in $\bar{\Sigma}$ is the image under the projection P: $\Sigma \rightarrow \bar{\Sigma}$ of a geodesic in the sphere Σ.)

9. If α is a curve in M with speed $v > 0$, prove:
   (a) The geodesic curvature κ_g of α is $\alpha'' \cdot J(\alpha')/v^3$. Hence if M is a surface in $\mathbf{E}^3$, $\kappa_g = U \cdot \alpha' \times \alpha''/v^3$.
   (b) If α has unit speed, then in Ex. 7 of Chapter V, Section 5, the vector field V is the unit normal N of α, and the function g is geodesic curvature κ_g.

10. Let M be the plane with the origin deleted, and furnish M with the conformal geometric structure for which $g = r = \sqrt{u^2 + v^2}$. Find the Gaussian curvature of M and the routes of its geodesics. Show that M is isometric to a surface in $\mathbf{E}^3$.

11. Let $\mathbf{x}$ be a Clairut parametrization, and let $\alpha = \mathbf{x}(a_1, a_2)$ be a unit

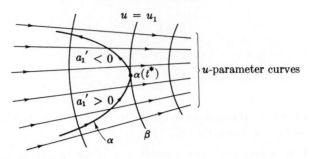

FIG. 7.13

speed geodesic with slant c. Suppose that α starts at the point

$$\alpha(0) \,=\, \mathbf{x}(a_1(0),\, a_2(0)) \,=\, \mathbf{x}(u_0,\, v_0)$$

and that $a_1'(0) > 0$. If there is a number $u > u_0$ such that $G(u) = c^2$, let u_1 be the smallest such number. Then the v-parameter curve

$$\beta(v) \,=\, \mathbf{x}(u_1,\, v)$$

is called a *barrier curve* for α. Prove that:
(a) α comes arbitrarily close to β
(b) If β is a geodesic, α does not meet β (thus α asymptotically approaches β).

12. (*Continuation*) If the barrier curve β is not a geodesic, it can be shown that α does meet β.† If $\alpha(t^*)$ is the meeting point, show that $a_1'(t^*) = 0$ and that a_1' changes sign at t^*. Thus α bounces off β as shown in Fig. 7.13. (*Hint:* prove that $a_1''(t^*) < 0$.)

13. Let α be a geodesic on a surface of revolution.
    (a) Show that the slant of α is $c = h \sin \varphi$, where $h(t)$ is the distance from $\alpha(t)$ to the axis of revolution, and φ gives the angles at which α cuts the meridians of M.
    (b) Deduce that α cannot cross a parallel of radius $|c|$.

14. Let α be a geodesic of slant c on the paraboloid of revolution

$$M: z \,=\, x^2 + y^2.$$

Find the minimum value of $z(\alpha)$, that is, the lowest height to which α descends. (*Hint:* Use a Monge patch.)

15. Prove that no geodesic on the bugle surface (V.6.6) can be defined on the whole real line.

16. On a torus of revolution, let α be a geodesic that at some point is

† We assume, of course, that α remains in the region parametrized by $\mathbf{x}$.

tangent to the top circle ($u = \pi/2$). Show that α remains always on the outer half of the torus ($-\pi/2 \leqq u \leqq \pi/2$) and travels around the torus oscillating between the top circle and bottom circle.

17. Let C be a catenoid (Example 6.1 of Chapter V) with $c = 1$, and let α be the geodesic such that

$$\alpha(0) = \mathbf{x}(u_0, v_0), \qquad u_0 \neq 0,$$

and $\alpha'(0)$ makes angle φ_0 with the meridians. (Note that φ_0 and $\pi - \varphi_0$ determine different parametrizations of the same geodesic.) For which values of φ_0 does α cross the minimal circle, $u = 0$, of C?

18. A *Liouville parametrization* $\mathbf{x} \colon D \to M$ is an orthogonal parametrization for which $E = G = U + V$, where U is a function of u only and V a function of v only. If $\alpha = \mathbf{x}(a_1, a_2)$ is a unit speed geodesic expressed in terms of such a parametrization, show that

$$U(a_1) \sin^2 \varphi - V(a_2) \cos^2 \varphi$$

is a constant, where φ is the angle from $\mathbf{x}_u$ to α'.

19. Let E_1, E_2 be a frame field on a geometric surface M. For $i = 1, 2$, let $\kappa_i(\mathbf{p})$ be the geodesic curvature at $\mathbf{p}$ of the integral curve of E_i through $\mathbf{p}$.
   (a) Prove that $K = E_1[\kappa_2] - E_2[\kappa_1] - \kappa_1^2 - \kappa_2^2$.
   (b) Test this formula on an arbitrary surface of revolution, using the frame field in Example 6.4 of Chapter VI.
   (*Hint:* for (a), prove $\omega_{12}(E_i) = \kappa_i$)

5 Length-Minimizing Properties of Geodesics

The previous section considered geodesics as *straightest* curves; now we investigate their character as *shortest* curves. The basic problem is, roughly speaking, to find the shortest route from one point to another in a geometric surface. For $\mathbf{E}^2$ the solution is simple: Given any two points $\mathbf{p}$ and $\mathbf{q}$, there is a unique straight-line segment from $\mathbf{p}$ to $\mathbf{q}$, and this is shorter than any other curve from $\mathbf{p}$ to $\mathbf{q}$ (Exercise 11 of Section 2 of Chapter II). For an arbitrary geometric surface M, the situation is more interesting. In the first place, there may be no shortest curve from $\mathbf{p}$ to $\mathbf{q}$ (Exercise 3 of Section 4 of Chapter VI). And even if there is one, *it may not be unique*. For example, we shall soon prove the expected result that on a sphere, all semicircles from the north pole to the south pole have the same shortest length. To make the terminology precise, we use the notion of intrinsic distance (Chapter VI, Section 4).

5.1 Definition Let α be a curve segment from $\mathbf{p}$ to $\mathbf{q}$ in M. Then

(1) α is *a shortest curve segment* from $\mathbf{p}$ to $\mathbf{q}$ provided $L(\alpha) = \rho(\mathbf{p}, \mathbf{q})$

(2) α is *the shortest curve segment* from $\mathbf{p}$ to $\mathbf{q}$ provided that

$$L(\alpha) = \rho(\mathbf{p}, \mathbf{q})$$

and that any other shortest segment from $\mathbf{p}$ to $\mathbf{q}$ is merely a reparametrization of α.

In the first case we shall also say that α *minimizes arc length from* $\mathbf{p}$ *to* $\mathbf{q}$; the definition means that if β is any other curve segment from $\mathbf{p}$ to $\mathbf{q}$, then $L(\beta) \geq L(\alpha)$.

In the second case, we say that α *uniquely minimizes arc length*. "Uniqueness" must be interpreted liberally enough to allow for reparametrization, since monotone reparametrization (Exercise 10 of Section 2 of Chapter II) does not change arc length.

All such shortest curves will turn out to be geodesics (Lemma 5.8). Our first main result (Theorem 5.6) will show that *short enough* geodesic segments behave as well in an arbitrary geometric surface as they do in $\mathbf{E}^2$. Some preparatory work is needed first.

In the Euclidean plane, if one is interested in the distance to the origin, it is natural to use polar coordinates, for then the distance from $\mathbf{0}$ to

$$\mathbf{x}(u, v) = (u \cos v, u \sin v)$$

is simply u. We shall now generalize this parametrization to the case of an arbitrary geometric surface M. As for $\mathbf{E}^2$, the u-parameter curves will be geodesics radiating out from some fixed point $\mathbf{p}$ of M. Such geodesics may conveniently be described as follows: If $\mathbf{v}$ is a unit tangent vector at $\mathbf{p}$, let $\gamma_\mathbf{v}$ be the unique geodesic which starts at $\mathbf{p}$ with initial velocity $\mathbf{v}$. Now we assemble all these geodesics into a single mapping as follows:

5.2 Definition Let $\mathbf{e}_1, \mathbf{e}_2$ be a frame at the point $\mathbf{p}$ of M. Then

$$\mathbf{x}(u, v) = \gamma_{\cos v\mathbf{e}_1 + \sin v\mathbf{e}_2}(u)$$

is the *geodesic polar mapping* of M with *pole* $\mathbf{p}$.

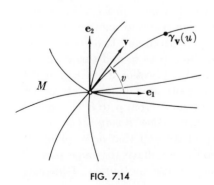

FIG. 7.14

Here the domain of $\mathbf{x}$ is the largest region of $\mathbf{E}^2$ on which the formula makes sense. A choice of v fixes a unit tangent vector

$$\mathbf{v} = \cos v\mathbf{e}_1 + \sin v\mathbf{e}_2$$

at $\mathbf{p}$ (Fig. 7.14). Then the u-parameter curve

$$u \rightarrow \mathbf{x}(u, v) = \gamma_\mathbf{v}(u)$$

is the radial geodesic with initial velocity $\mathbf{v}$. Since $\| \mathbf{v} \| = 1$, this geodesic has unit speed, so that the length of $\gamma_\mathbf{v}$ from $\mathbf{p} = \gamma_\mathbf{v}(0)$ to $\gamma_\mathbf{v}(u)$ is just u.

In the special case where $\mathbf{e}_1$, $\mathbf{e}_2$ is the natural frame at the origin $\mathbf{0}$ of $\mathbf{E}^2$, the geodesic polar mapping becomes

$$\mathbf{x}(u, v) = \gamma_{\cos v \mathbf{e}_1 + \sin v \mathbf{e}_2}(u)$$

$$= \mathbf{0} + u(\cos v \mathbf{e}_1 + \sin v \mathbf{e}_2)$$

$$= (u \cos v, u \sin v).$$

Thus $\mathbf{x}$ *is* a generalization of polar coordinates in the plane.

The pole $\mathbf{p}$ is a trouble spot for a geodesic polar mapping. To clarify the situation near $\mathbf{p}$, we define (in the situation described in Definition 5.2) a new mapping

$$\mathbf{y}(u, v) = \gamma_{u\mathbf{e}_1 + v\mathbf{e}_2}(1).$$

Differential equations theory shows that $\mathbf{y}$ is differentiable, and it is easy to check that $\mathbf{y}$ is regular at the origin. Thus by the inverse function theorem, $\mathbf{y}$ is a diffeomorphism of some disc D_ϵ: $u^2 + v^2 < \epsilon^2$ onto a neighborhood $\mathfrak{N}_\epsilon$ of $\mathbf{p}$. We call $\mathfrak{N}_\epsilon$ a *normal neighborhood* of $\mathbf{p}$. In the special case $M = \mathbf{E}^2$, $\mathbf{y}$ is just the identity map $\mathbf{y}(u, v) = (u, v)$, so for arbitrary M, $\mathbf{y}$ is a generalization of the natural (rectangular) coordinates of $\mathbf{E}^2$.

5.3 Lemma For a sufficiently small number $\epsilon > 0$, let S_ϵ be the strip $0 < u < \epsilon$ in $\mathbf{E}^2$. Then a geodesic polar mapping $\mathbf{x}: S_\epsilon \rightarrow M$ with pole $\mathbf{p}$ parametrizes a normal neighborhood $\mathfrak{N}_\epsilon$ of $\mathbf{p}$—omitting $\mathbf{p}$ itself, (see Fig. 7.15).

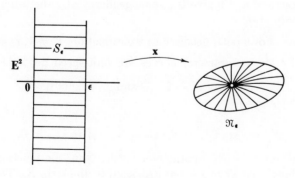

FIG. 7.15

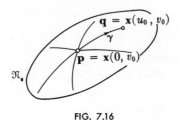

FIG. 7.16

Proof. Note that $\mathbf{x}$ bears to $\mathbf{y}$ the usual relationship of polar to rectangular coordinates; that is,

$$\mathbf{x}(u, v) = \gamma_{\cos v\mathbf{e}_1 + \sin v\mathbf{e}_2}(u) = \gamma_{u\cos v\mathbf{e}_1 + u\sin v\mathbf{e}_2} \quad (1)$$

$$= \mathbf{y}(u\cos v, u\sin v)$$

where we use the identity $\gamma_{\mathbf{v}}(u) = \gamma_{u\mathbf{v}}(1)$ from Exercise 2 of Section 4. Now this formula expresses $\mathbf{x}$ as the composition of two regular mappings:

(1) The Euclidean polar mapping $(u, v) \rightarrow (u\cos v, u\sin v)$ which wraps the strip S_ϵ around the disc D_ϵ, and

(2) The one-to-one mapping $\mathbf{y}$ of D_ϵ onto $\mathfrak{N}_\epsilon$.

Thus $\mathbf{x}$ is regular and carries S_ϵ in usual polar-coordinate fashion onto the neighborhood $\mathfrak{N}_\epsilon$—omitting only the pole. ∎

We draw a fundamental consequence: If $\mathbf{q} = \mathbf{x}(u_0, v_0)$ is any point in a normal neighborhood $\mathfrak{N}$ of $\mathbf{p}$, *then there is only one unit speed geodesic from $\mathbf{p}$ to $\mathbf{q}$ which lies entirely in $\mathfrak{N}$,* namely, the radial geodesic

$$\gamma(u) = \mathbf{x}(u, v_0) \qquad 0 \leq u \leq u_0.$$

(*Proof:* Any unit-speed geodesic starting at $\mathbf{p}$ is, by the uniqueness of geodesics, a u-parameter curve of the polar parametrization. As suggested by Fig. 7.16, all except $v = v_0 + 2\pi n$ lead out of $\mathfrak{N}$ without hitting $\mathbf{q}$, and different choices of n still give the same geodesic γ by the usual ambiguity of polar coordinates).

5.4 Lemma   For a polar geodesic parametrization, $E = 1, F = 0, G > 0$.

Proof. Since the u-parameter curves are unit-speed geodesics, we have

$$E = \mathbf{x}_u \cdot \mathbf{x}_u = 1, \qquad \text{and} \qquad \mathbf{x}_{uu} = 0.$$

Thus

$$F_u = (\mathbf{x}_u \cdot \mathbf{x}_v)_u = \mathbf{x}_u \cdot \mathbf{x}_{vu} = \mathbf{x}_u \cdot \mathbf{x}_{uv} = \tfrac{1}{2}E_v = 0,$$

so F is constant on each u-parameter curve. Now the functions E, F, G are well-defined even when $\mathbf{x}$ is not restricted to the strip S_ϵ. The v-parameter curve $v \rightarrow \mathbf{x}(0, v)$ is simply the constant curve at the pole $\mathbf{p}$, so

$$\mathbf{x}_v(0, v) = 0.$$

But then $F(0, v) = 0$ for all v, and since $F_u = 0$, we conclude that F is identically zero. Because $\mathbf{x}$ (now restricted once more to the strip S_ϵ) is a parametrization, that is a regular map, we know that $EG - F^2 = EG$ is never zero. Hence $G > 0$.   ∎

5.5 Example We explicitly work out geodesic polar parametrizations in two classic cases.

(1) *The unit sphere* Σ *in* $\mathbf{E}^3$. For simplicity let $\mathbf{p}$ be the north pole $(0, 0, 1)$. To get the geodesics radiating out from $\mathbf{p}$ as in Fig. 7.17, we change the geographical parametrization to

$$\mathbf{x}(u, v) = (\sin u \cos v, \sin u \sin v, \cos u).$$

Each u-parameter curve is indeed a unit-speed parametrization of a great circle, hence is geodesic. For $u = 0$ we find

$$\mathbf{x}_u(0, v) = (\cos v, \sin v, 0)$$

$$= \cos v\mathbf{e}_1 + \sin v\mathbf{e}_2$$

where

$$\mathbf{e}_1 = U_1(\mathbf{p}), \qquad \mathbf{e}_2 = U_2(\mathbf{p}).$$

Thus by the uniqueness of geodesics,

$$\mathbf{x}(u, v) = \gamma_{\cos v\mathbf{e}_1 + \sin v\,\mathbf{e}_2}(u)$$

which shows that $\mathbf{x}$ as defined above is the polar geodesic mapping (Definition 5.2). It is easy to see that the largest possible normal neighborhood $\mathfrak{N}_\epsilon$ of $\mathbf{p}$ occurs when $\epsilon = \pi$, for on the strip S_π, $\mathbf{x}$ is a polar parametrization of all the sphere except the north and south poles.

(2) *The hyperbolic plane* H (Example 2.4). We choose $\mathbf{p} = (0, 0)$ and

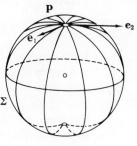

FIG. 7.17

$\mathbf{e}_1 = U_1(\mathbf{p})$, $\mathbf{e}_2 = U_2(\mathbf{p})$. (Since the function g is 1 at the origin, this *is* a frame.) We know from Example 4.11 that the geodesics of H through the origin follow Euclidean straight lines. Thus for any number v, the curve

$$\alpha(t) = (t \cos v, t \sin v)$$

is at least a pre-geodesic of the type we are looking for. In Example 2.4 we found the arc length function $s(t) = 2 \tanh^{-1}(t/2)$ for such a curve; thus

$$s \to \alpha\left(2 \tanh \frac{s}{2}\right) = \left(2 \tanh \frac{s}{2} \cos v, 2 \tanh \frac{s}{2} \sin v\right)$$

is the unit-speed reparametrization of α. Shifting notation from s to u, we obtain

$$\mathbf{x}(u, v) = \left(2 \tanh \frac{u}{2} \cos v, 2 \tanh \frac{u}{2} \sin v\right).$$

Since the u-parameter curves of $\mathbf{x}$ are unit-speed geodesics and

$$\mathbf{x}_u(0, v) = \cos v\mathbf{e}_1 + \sin v\mathbf{e}_2,$$

we conclude as in (1) that $\mathbf{x}$ is a geodesic polar mapping. The normal neighborhood in this case is the entire surface H.

5.6 Theorem For each point $\mathbf{q}$ of a normal neighborhood $\mathfrak{N}_\epsilon$ of $\mathbf{p}$ the radial geodesic segment in $\mathfrak{N}_\epsilon$ from $\mathbf{p}$ to $\mathbf{q}$ uniquely minimizes arc length.

Proof. Let $\mathbf{x}$ be the polar parametrization of the normal neighborhood $\mathfrak{N}_\epsilon$. If

$$\mathbf{q} = \mathbf{x}(u_0, v_0),$$

the radial geodesic segment is

$$\gamma(u) = \mathbf{x}(u, v_0), \qquad 0 \le u \le u_0.$$

Now let α be an arbitrary curve segment from $\mathbf{p}$ to $\mathbf{q}$ in M; we may arrange for α to be defined on the same interval as γ.

We begin by proving

$$L(\gamma) \le L(\alpha). \tag{1}$$

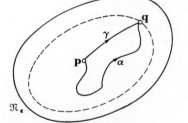

FIG. 7.18

First consider the case (Fig. 7.18) where α stays in the neighborhood $\mathfrak{N}_\epsilon$. We may assume that once having left $\mathbf{p}$, α never returns to $\mathbf{p}$—if it did, throwing away this loop would only shorten α. Thus it is possible to write

$$\alpha(t) = \mathbf{x}(a_1(t), a_2(t)).$$

Since $\alpha(0) = \mathbf{p}$ and $\alpha(u_0) = \mathbf{q}$, we have

$$a_1(0) = 0 \qquad a_1(u_0) = u_0$$
$$a_2(0) = v_0 \qquad a_2(u_0) = v_0 + 2\pi n. \tag{2}$$

(The term $2\pi n$ results again from the nonuniqueness of angles in polar coordinates.)

Since for $\mathbf{x}$ we have $E = 1$ and $F = 0$, the speed of α is

$$\| \alpha' \| = \sqrt{a_1'^2 + G a_2'^2}.$$

Now

$$\sqrt{a_1'^2 + G a_2'^2} \geqq \sqrt{a_1'^2} = |a_1'| \geqq a_1' \tag{3}$$

Hence

$$L(\alpha) = \int_0^{u_0} \sqrt{a_1'^2 + G a_2'^2}\, dt \geqq \int_0^{u_0} a_1'\, dt \tag{4}$$
$$= a_1(u_0) - a_1(0) = u_0$$

where the last step uses (2). But the radial geodesic γ has unit speed, so

$$L(\gamma) = \int_0^{u_0} dt = u_0,$$

and we conclude that

$$L(\gamma) \leqq L(\alpha).$$

If α does not stay in $\mathfrak{N}_\epsilon$ we have strict inequality, $L(\gamma) < L(\alpha)$. For α must cross the *polar circle*, $u = u_0$—indicated by a dashed line in Fig. 7.18 —to escape from $\mathfrak{N}_\epsilon$.† But by the proof above it must already have arc length at least $u_0 = L(\gamma)$ when it reaches the circle.

Now we prove the uniqueness assertion:

$$\text{If } L(\alpha) = L(\gamma), \text{ then } \alpha \text{ is a reparametrization of } \gamma \tag{5}$$

The argument above shows that if $L(\alpha) = L(\gamma)$, then α stays inside $\mathfrak{N}_\epsilon$ and the inequality in (4) becomes equality. The latter implies that $\sqrt{a_1'^2 + G a_2'^2} = a_1'$. Since $G > 0$ we conclude from (3) that

$$a_1' \geqq 0, \qquad a_2' = 0. \tag{6}$$

Thus a_2 has constant value v_0 (so $n = 0$ in (2)) and

$$\alpha(t) = \mathbf{x}(a_1(t), v_0) = \gamma(a_1(t)),$$

showing that α is in fact a monotone reparametrization of γ. ∎

† A careful proof will involve the Hausdorff axiom (Ex. 5 of IV.8) which we assume throughout this chapter.

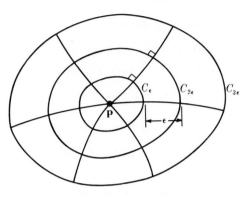

FIG. 7.19

This fundamental result shows, as we remarked earlier, that if points **p** and **q** are close enough together, then—as in Euclidean space for *arbitrary* points—there is a unique geodesic segment from **p** to **q** which is shorter than any curve from **p** to **q**. (Unlike the Euclidean case however, there may be many other *non*shortest geodesics from **p** to **q**.) If **x** is a geodesic polar parametrization at **p** we shall also call the route C_ϵ of the v-parameter curve, $u = \epsilon$, the *polar circle* of radius ϵ at **p**. (Fig. 7.19). Theorem 5.6 shows that C_ϵ does in fact consist of all points at distance ϵ from **p**.

In special cases where large normal neighborhoods are available, this local information may be decisive.

5.7 Example   *Length-minimizing properties of geodesics on the sphere* Σ *of radius r.* By a mere change of scale we can conclude from Example 5.5 that each point **p** of Σ has normal neighborhood $\mathfrak{N}_{\pi r}$: all of Σ except the point, $-\mathbf{p}$, antipodal to the pole **p**. Hence Theorem 5.6 implies:

(a) If two points **p** and **q** of Σ are not antipodal (that is, $\mathbf{q} \neq -\mathbf{p}$), then there is a unique shortest curve γ from **p** to **q**. But we know all the geodesics of Σ: γ can only be the one that folllows the shorter arc of the great circle through **p** and **q**.

(b) Intrinsic distance ρ on Σ is given by the formula

$$\rho(\mathbf{p}, \mathbf{q}) = r\vartheta$$

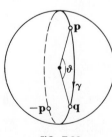

FIG. 7.20

where ϑ $(0 \leq \vartheta \leq \pi)$ is the angle from **p** to **q** in E^3 (Fig. 7.20). If **p** and **q** are not antipodal, this follows from (a), since

$$\rho(\mathbf{p}, \mathbf{q}) = L(\gamma) = r\vartheta.$$

As **q** moves toward the antipodal point $-\mathbf{p}$ of **p** we deduce by continuity that $\rho(\mathbf{p}, -\mathbf{p}) = r\pi$. Hence

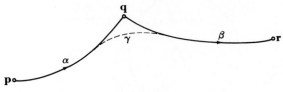

FIG. 7.21

(c) There are infinitely many minimizing geodesics from a point **p** on Σ to the antipodal point $-\mathbf{p}$, namely (constant-speed parametrizations of) semicircles from **p** to $-\mathbf{p}$. (*Proof:* These all have length $r\pi = \rho(\mathbf{p}, -\mathbf{p})$.)

(d) No geodesic segment γ of length $L(\gamma) > \pi r$ can minimize arc length between its end points. This follows immediately from the fact that intrinsic distance ρ never exceeds πr. It is geometrically clear, since if γ starts at **p**, its length exceeds πr as soon as γ passes the antipodal point $-\mathbf{p}$. But then the other arc γ^* of the same great circle is shorter than γ.

Suppose that α is a curve segment in M from **p** to **q**, and β is a curve segment from **q** to **r**. Now α and β cannot generally be united to form a single (differentiable) curve from **p** to **r**, since there may be a "corner" at **q** as in Fig. 7.21. Using the techniques of advanced calculus, one can "round off" this corner, obtaining a curve segment γ from **p** to **r** which (to state the weakest theorem) is only slightly longer than α and β. Explicitly, for each $\epsilon > 0$ there is a γ such that $L(\gamma) \leq L(\alpha) + L(\beta) + \epsilon$.

It follows that intrinsic distance satisfies the *triangle inequality*. In fact, given points **p**, **q**, and **r** the definition of intrinsic distance shows that for any $\epsilon > 0$ there exist curves α and β as above such that

$$L(\alpha) \leq \rho(\mathbf{p}, \mathbf{q}) + \epsilon, \qquad L(\beta) \leq \rho(\mathbf{q}, \mathbf{r}) + \epsilon.$$

Rounding off the corner at **q** costs at most another ϵ: We get a curve segment γ from **p** to **r** such that

$$\rho(\mathbf{p}, \mathbf{r}) \leq L(\gamma) \leq \rho(\mathbf{p}, \mathbf{q}) + \rho(\mathbf{q}, \mathbf{r}) + 3\epsilon.$$

But since ϵ is arbitrary, we conclude that

$$\rho(\mathbf{p}, \mathbf{r}) \leq \rho(\mathbf{p}, \mathbf{q}) + \rho(\mathbf{q}, \mathbf{r}).$$

5.8 Lemma   If α is a shortest curve segment in M from **p** to **q**, then α is geodesic.

Proof. We shall prove that if $\alpha \colon [a, b] \to M$ is a curve segment from **p** to **q** which is *not* a geodesic, then $L(\alpha) > \rho(\mathbf{p}, \mathbf{q})$. But if α is not geodesic, then at some time t_0 the acceleration $\alpha''(t_0)$ is not zero. By continuity α'' is nonzero near t_0, so we may assume $t_0 < b$. For $\epsilon > 0$ sufficiently small, $\alpha(t_0 + \epsilon)$ lies in a normal neighborhood of $\alpha(t_0)$, and the segment of α from

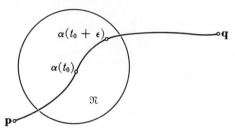

FIG. 7.22

t_0 to $t_0 + \epsilon$ is not a geodesic, since $\alpha''(t_0) \neq 0$ (Fig. 7.22). But then by Theorem 5.6 its length, $L_{t_0, t_0+\epsilon}$, is strictly greater than the intrinsic distance from $\alpha(t_0)$ to $\alpha(t_0 + \epsilon)$. Thus by the triangle inequality,

$$L(\alpha) = L_{a, t_0} + L_{t_0, t_0 + \epsilon} + L_{t_0 + \epsilon, b}$$
$$> \rho(\mathbf{p}, \alpha(t_0)) + \rho(\alpha(t_0), \alpha(t_0 + \epsilon)) + \rho(\alpha(t_0 + \epsilon), \mathbf{q})$$
$$\geq \rho(\mathbf{p}, \mathbf{q}). \qquad\blacksquare$$

This result is not too surprising: A shortest road can never turn. Nor will it have any corners, for a somewhat more sophisticated argument shows that a (possibly broken) shortest curve must in fact be an (unbroken) geodesic.

We have now reached the main result of this section.

5.9 Theorem   Given any two points $\mathbf{p}$ and $\mathbf{q}$ in a geodesically complete geometric surface M, there is a shortest geodesic segment from $\mathbf{p}$ to $\mathbf{q}$.

Proof.   The scheme is an ingenious one, worked out successively by several mathematicians. (See Theorem 10.9, p. 62, of Milnor [7].) We begin by selecting a candidate for shortest curve from $\mathbf{p}$ to $\mathbf{q}$. Let

$$\beta(v) = \mathbf{x}(a, v), \qquad 0 \leq v \leq 2\pi,$$

parametrize the polar circle C of radius a in a normal neighborhood of $\mathbf{p}$. By Exercise 6 it follows that the function $v \to \rho(\beta(v), \mathbf{q})$ is continuous on the closed interval $[0, 2\pi]$; hence the function takes on its minimum value at say v_0. Let γ be the parameter curve, $v = v_0$. Since M is geodesically complete, $\gamma(u)$ is defined for all $u \geq 0$. We will show that γ hits $\mathbf{q}$—in fact, that

$$\gamma(r) = \mathbf{q} \qquad \text{where } r = \rho(\mathbf{p}, \mathbf{q}). \tag{1}$$

(This situation is illustrated in Fig. 7.23.) Since γ has unit speed, it will follow that

$$L(\gamma) = r = \rho(\mathbf{p}, \mathbf{q}),$$

thus proving the theorem.

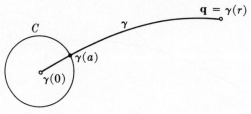

FIG. 7.23

To establish (1), we use a variant of the standard induction argument in which integers are replaced by real numbers. For each number $u \geqq 0$ consider the assertion

$$\alpha(u): \qquad \rho(\gamma(u), \mathbf{q}) = r - u \tag{2}$$

where, as above, $r = \rho(\mathbf{p}, \mathbf{q})$. This says that γ (unit speed) is efficient: After traveling distance u, the distance to $\mathbf{q}$ has been reduced by precisely u. *The proof will be finished if we can prove that $\alpha(r)$ is true*, for then

$$\rho(\gamma(r), \mathbf{q}) = 0,$$

so by Exercise 5, $\gamma(r) = \mathbf{q}$. We make a start on this by showing that $\alpha(a)$ is true for a as above; that is,

$$\rho(\gamma(a), \mathbf{q}) = r - a. \tag{3}$$

According to Theorem 5.6, $\rho(\mathbf{p}, \gamma(a)) = a$; hence by the triangle inequality

$$r = \rho(\mathbf{p}, \mathbf{q}) \leqq a + \rho(\gamma(a), \mathbf{q}).$$

To get (3) we must reverse this inequality. By definition of intrinsic distance, for any $\epsilon > 0$, there is a curve segment α from $\mathbf{p}$ to $\mathbf{q}$ such that

$$L(\alpha) \leqq \rho(\mathbf{p}, \mathbf{q}) + \epsilon.$$

Now α must hit the polar circle C, say at $\alpha(t_0)$, and we observe that the portion of α from $\mathbf{p}$ to $\alpha(t_0)$ has length $L_1 \geqq a$, and the remainder of α has length

$$L_2 \geqq \rho(\alpha(t_0), \mathbf{q}) \geqq \rho(\gamma(a), \mathbf{q}).$$

(The latter since $\gamma(a)$ was a nearest point to $\mathbf{q}$ on C.) Thus

$$a + \rho(\gamma(a), \mathbf{q}) \leqq L_1 + L_2 = L(\alpha) \leqq \rho(\mathbf{p}, \mathbf{q}) + \epsilon.$$

Since ϵ was arbitrary, we obtain the inequality

$$a + \rho(\gamma(a), \mathbf{q}) \leqq \rho(\mathbf{p}, \mathbf{q})$$

required to prove (3).

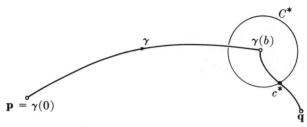

FIG. 7.24

Now we turn to the inductive step of the proof. Since ρ cannot be negative, $\mathcal{Q}(u)$ is nonsense for $u > r$. Thus the set of numbers a for which $\mathcal{Q}(a)$ is true has a least upper bound b, with $b \leq r$. Since the functions involved in the assertion $\mathcal{Q}(a)$ are continuous, it follows from the definition of least upper bound that $\mathcal{Q}(b)$ is true.

Here is the plan of the rest of the proof: *Assume $b < r$ and deduce a contradiction.* Then (since $b \leq r$) we must have $b = r$; hence $\mathcal{Q}(r)$ is true, as required.

Let C^* be a polar circle of radius $a^* < r - b$ in a normal neighborhood of $\gamma(b)$. By reproducing the argument for the circle C, we obtain a point $\mathbf{c}^*$ such that

$$\rho(\mathbf{c}^*, \mathbf{q}) = \rho(\gamma(b), \mathbf{q}) - a^*. \tag{3'}$$

(See Fig. 7.24.) But $\mathcal{Q}(b)$ reads $\rho(\gamma(b), \mathbf{q}) = r - b$, so

$$\rho(\mathbf{c}^*, \mathbf{q}) = r - b - a^*. \tag{4}$$

The main step that remains is the proof of

$$\mathbf{c}^* = \gamma(b + a^*). \tag{5}$$

This is not too difficult. By the triangle inequality,

$$\rho(\mathbf{p}, \mathbf{c}^*) + \rho(\mathbf{c}^*, \mathbf{q}) \geq \rho(\mathbf{p}, \mathbf{q}) = r.$$

Using (4) we get

$$\rho(\mathbf{p}, \mathbf{c}^*) \geq b + a^*.$$

But there is a broken curve from $\mathbf{p}$ to $\mathbf{c}^*$ whose length is precisely $b + a^*$. In fact, referring to Fig. 7.24, we can travel on γ from $\mathbf{p}$ to $\gamma(b)$ with arc length b, then from $\gamma(b)$ to $\mathbf{c}^*$ on a radial geodesic with arc length a^*. Thus by the remark preceding this theorem, this curve is *not* actually broken. Hence it is γ all the way, so $\gamma(b + a^*)$ is precisely $\mathbf{c}^*$.

Finally, we substitute (5) in (4), obtaining

$$\rho(\gamma(b + a^*), \mathbf{q}) = r - (b + a^*).$$

This says that $\mathcal{C}(b + a^*)$ is true, and since $b + a^*$ is strictly larger than the upper bound b, we have the required contradiction. ∎

EXERCISES

1. In the hyperbolic plane H, find the intrinsic distance from the origin **0** to an arbitrary point **p**. Deduce that all geodesics of H have infinite length, hence H is complete. (*Hint:* Use the triangle inequality.)

2. For the Poincaré half-plane (Exercise 6 of Section 4):
   (a) Find an equation $F(x, y, c) = 0$ for the routes of the semicircular geodesics through the point $(0, 1)$.
   (b) Find an equation $G(x, y, a) = 0$ for the polar circles centered at $(0, 1)$. (*Hint:* they are the orthogonal trajectories of the curves in (a).)
   (c) Make a sketch showing several curves from each family.

3. At the point **p** $= (r, 0, 0)$ of the cylinder $M: x^2 + y^2 = r^2$, let

$$\mathbf{e}_1 = (0, 1, 0) \qquad \text{and} \qquad \mathbf{e}_2 = (0, 0, 1).$$

   Find an explicit formula for the mapping **y** (p. 341) in this case. What is the largest normal neighborhood of the point **p**?

4. Test the scheme used to prove Theorem 5.9 in the special case $M = \mathbf{E}^2$. Explicitly: Starting from the geodesic polar mapping

$$\mathbf{x}(u, v) = (p_1 + u \cos v, p_2 + u \sin v) \qquad \text{at } \mathbf{p}.$$

   follow the first paragraph of the proof of Theorem 5.9 to determine the geodesic γ.

5. *Intrinsic distance is a metric on M.* Show that
   (a) A normal neighborhood $\mathfrak{N}_\epsilon$ of **p** consists of all points **q** of M such that $\rho(\mathbf{p}, \mathbf{q}) < \epsilon$.
   (b) ρ satisfies the three metric properties: (i) $\rho \geqq 0$; and $\rho(\mathbf{p}, \mathbf{q}) = 0$ if and only if $\mathbf{p} = \mathbf{q}$, (ii) $\rho(\mathbf{p}, \mathbf{q}) = \rho(\mathbf{q}, \mathbf{p})$, and (iii) the triangle inequality.
   (*Hint:* It is necessary to use the Hausdorff axiom for the same purpose as in the footnote on page 345.)

6. *Intrinsic distance is continuous.* In a geometric surface M, define $\mathbf{p}_i \to \mathbf{p}$ to mean that the sequence of real numbers $\rho(\mathbf{p}, \mathbf{p}_i)$ converges to 0. Prove that if $\mathbf{p}_i \to \mathbf{p}$ and $\mathbf{q}_i \to \mathbf{q}$, then $\rho(\mathbf{p}_i, \mathbf{q}_i)$ converges to $\rho(\mathbf{p}, \mathbf{q})$.

7. Let α and β be two different unit-speed geodesics which start at the same point $\alpha(0) = \beta(0)$. If α and β meet again after having traveled

the same distance $r > 0$, that is, $\alpha(r) = \beta(r)$, prove that neither α or β minimizes arc length past r. (Use the fact that broken geodesics cannot minimize arc length.)

8. (*Continuation*). On the cylinder $M: x^2 + y^2 = r^2$, prove:
   (a) A geodesic starting at (a, b, c) cannot minimize arc length after it passes through the antipodal line $t \to (-a, -b, t)$.
   (b) If **q** is not on the antipodal line of **p**, show that there is a unique shortest geodesic from **p** to **q**.
   Derive a formula for intrinsic distance on the cylinder.

9. Show that the converse of Theorem 5.9 is false: Give an example of a geometric surface M such that any two points can be joined by a minimizing geodesic, but M is not geodesically complete.

10. Let $\gamma: [a, b] \to M$ parametrize a portion of a meridian of a surface of revolution M. Prove that γ uniquely minimizes arc length. (*Hint:* Express a competitive curve α as $\mathbf{x}(a_1, a_2)$ where $\mathbf{x}$ is a canonical parametrization, and follow the scheme of Theorem 5.6.)

11. Let M be an augmented surface of revolution (Ex. 12 of IV.1).
    (a) If M has only one intercept **p** (on the axis of revolution) show that every geodesic γ of M starting at **p** uniquely minimizes arc length.
    (b) If M has a second intercept **q** show that the assertion in (a) holds if and only if γ does not reach **q**.
    (*Hint:* No computations are needed.)

6 Curvature and Conjugate Points

We briefly examine the influence of Gaussian curvature K of a geometric surface M on the geodesics of M.

6.1 Definition A geodesic segment γ from **p** to **q** *locally minimizes* arc length from **p** to **q** provided that for any curve segment α from **p** to **q** which is sufficiently near γ we have $L(\alpha) \geqq L(\gamma)$.

To clarify the phrase "sufficiently near," we first define α to be ε-*close* to γ provided there is a reparametrization $\tilde{\alpha}$ of α, on the same interval I as γ, such that $\rho(\tilde{\alpha}(t), \gamma(t)) < \varepsilon$ for all t in I (Fig. 7.25). Then we change the ending of Definition 6.1 to "provided there exists an $\varepsilon > 0$ such that for any α which is ε-close to γ, we have $L(\alpha) \geqq L(\gamma)$." This local minimization is *strict* (or *unique*) provided we get strictly inequality

$$L(\alpha) > L(\gamma)$$

unless α is a reparametrization of γ.

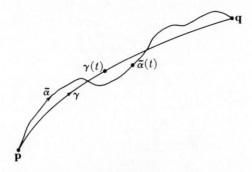

FIG. 7.25

To get an intuitive picture of this definition we shall imagine that γ is an elastic string—or rubber band—which (1) is constrained to lie in M; (2) is under tension; and (3) has its end points pinned down at $\mathbf{p}$ and $\mathbf{q}$.

Because γ is a geodesic, it is in equilibrium: If it were not geodesic, its tension would pull it into a new shorter position. But is the equilibrium stable? If γ is pulled aside slightly to a new curve α and released, will it return to its original position γ? Evidently γ is (strictly) stable if and only if γ is a (strict) local minimum in the sense above, for if α is *longer* than γ, its tension will pull it back to γ.

Investigation of local minimization depends on the notion of *conjugate points*. If γ is a unit-speed geodesic starting at $\mathbf{p}$, then γ is a u-parameter curve, $v = v_0$, of a geodesic polar mapping $\mathbf{x}$ with pole $\mathbf{p}$. We know that along γ the function $G = \mathbf{x}_v \cdot \mathbf{x}_v$ is zero at $u = 0$, but is nonzero immediately thereafter (Lemma 5.4). A point $\gamma(s) = \mathbf{x}(s, v_0)$ with $s > 0$ is a *conjugate point of* $\gamma(0) = \mathbf{p}$ *on* γ provided $G(s, v_0) = 0$. (Such points may or may not exist.)

The geometric meaning of conjugacy rests on the interpretation of $\sqrt{G} = \| \mathbf{x}_v \|$ as the rate at which the (radial geodesic) u-parameter curves are spreading apart. Roughly speaking, for fixed $\epsilon > 0$, if $\sqrt{G} = \| \mathbf{x}_v \|$ is large, then the distance from $\mathbf{x}(u, v)$ to $\mathbf{x}(u, v + \epsilon)$ is large: The radial geodesics are spreading rapidly. When $\sqrt{G}$ is small, this distance is small, and the radial geodesics are pulling back together again. Thus when G *vanishes* at a conjugate point

$$\gamma(s_1) = \mathbf{x}(s_1, v_0),$$

it suggests that for v near v_0, the u-parameter curves have all reached this same point after traveling (at unit speed) the same distance s_1 (Fig. 7.26). Unfortunately *this meeting may not actually occur.* (G controls only the first derivative terms, and higher-order terms may still be nonzero even though G vanishes.)

The Euclidean plane $\mathbf{E}^2$ should give the "standard" rate at which radial

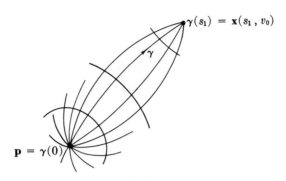

$$\gamma(s_1) = \mathbf{x}(s_1, v_0)$$

$$\gamma$$

$$\mathbf{p} = \gamma(0)$$

FIG. 7.26

geodesics spread apart, and for $\mathbf{x}(u, v) = (u \cos v, u \sin v)$, we have

$$\sqrt{G} = u.$$

In particular, there are no conjugate points. Let us compare the cases discussed in Example 5.5, the unit sphere Σ and hyperbolic plane H. For Σ, we find

$$\sqrt{G} = \sin u.$$

Thus radial geodesics starting, say, at the north pole $\mathbf{p}$ of Σ, spread less rapidly than in $\mathbf{E}^2$, since $\sin u < u$ for $u > 0$. Indeed one can see in Fig. 7.17 that after passing the equator they begin to crowd closer together. All have their first conjugate point after traveling distance π, since $\sqrt{G}(\pi, v) = \sin \pi = 0$. In this case, of course, the meeting of geodesics actually takes place—at the south pole of Σ.

For the hyperbolic plane, we know that the geodesics radiating out from the origin are just Euclidean straight lines, but they are spreading more rapidly than in $\mathbf{E}^2$, as one may surmise from the fact that in H "rulers shrink as they approach the rim." To prove it we use the data in (2) of Example 5.5 to compute

$$\sqrt{G} = \sinh u.$$

Thus $\sqrt{G} > u$ for $u > 0$, and again there are no conjugate points.

6.2 Theorem   If γ is a geodesic segment from $\mathbf{p}$ to $\mathbf{q}$ such that there are no conjugate points of $\mathbf{p} = \gamma(0)$ on γ, then γ locally minimizes arc length (strictly) from $\mathbf{p}$ to $\mathbf{q}$.

Proof.   Let $\mathbf{x}$ be a geodesic polar mapping at $\mathbf{p}$, and restrict its domain to the region in $\mathbf{E}^2$ on which $G > 0$. Because there are no conjugate points of $\mathbf{p}$ on γ, we may write $\gamma(u) = \mathbf{x}(u, v_0)$ for $0 \leq u \leq u_0$. (Thus we allow $u = 0$ in this equation, as usual, even though $G = 0$ there.) Then let α

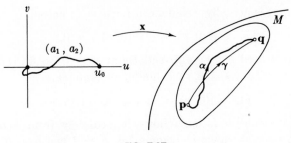

FIG. 7.27

be another curve segment from **p** to **q** with α also defined on the interval $[0, u_0]$. Now our proof rests on the fact that if α *is sufficiently close to* γ (as defined earlier), then α has an expression

$$\alpha(t) = \mathbf{x}(a_1(t), a_2(t))$$

which is so close to that of γ that

$$a_1(0) = 0, \quad a_1(u_0) = u_0.$$

(Fig. 7.27).

A complete proof of this rather plausible assertion is not trivial. There is no trouble at $u = 0$, since we can replace a short initial segment of α by a radial geodesic—with no loss of generality, since this will not lengthen α. Then a_1 and a_2 are constructed in step-by-step fashion using the fact that **x** is a regular mapping, and hence is locally a diffeomorphism.

Then exactly as in the proof of Theorem 5.6 we have

$$L(\alpha) = \int_0^{u_0} \sqrt{a_1'^2 + G a_2'^2} \, dt \geqq \int_0^{u_0} a_1' \, dt$$

$$= a_1(u_0) - a_1(0) = u_0 = L(\gamma)$$

and if $L(\alpha) = L(\gamma)$, α is merely a reparametrization of γ.   ∎

Our task now is to free the notion of conjugate point from dependence on geodesic polar mappings. To do so we examine the "spreading coefficient" $\sqrt{G}$ more closely.

6.3 Theorem Let **x** be a geodesic polar mapping defined on a region where $G > 0$. Then $\sqrt{G} = \| \mathbf{x}_v \|$ satisfies the *Jacobi differential equation*

$$(\sqrt{G})_{uu} + K \sqrt{G} = 0$$

subject to the initial conditions

$$\sqrt{G}(0, v) = 0 \qquad (\sqrt{G})_u(0, v) = 1 \qquad \text{for all } v.$$

The restriction $G > 0$ is needed to ensure that $\sqrt{G}$ is differentiable.

Now $\sqrt{G}(u, v)$ is actually well-defined for $u = 0$; indeed,

$$\sqrt{G}(0, v) = \| \mathbf{x}_v(0, v) \| = 0.$$

However, $\sqrt{G}$ need not be *differentiable* at $u = 0$, so we shall interpret $(\sqrt{G})_u(0, v)$ and $(\sqrt{G})_{uu}(0, v)$ as limits, for example,

$$(\sqrt{G})_u(0, v) = \lim_{u \to 0} (\sqrt{G})_u(u, v).$$

Proof. The Jacobi equation follows immediately from Lemma 6.3 of Chapter VI, since as shown in Lemma 5.4, $E = 1$ and $F = 0$ for $\mathbf{x}$. Thus by the remarks above, it suffices to prove

$$\lim_{u \to 0} (\sqrt{G})_u(u, v) = 1 \qquad (u > 0).$$

We need only consider a single radial geodesic $\gamma(u) = \mathbf{x}(u, v_0)$, setting

$$g(u) = \sqrt{G}(u, v_0) \qquad \text{for } u > 0.$$

Again, since $E = 1$ and $F = 0$, we obtain a frame field

$$E_1 = \gamma' = \mathbf{x}_u, \qquad E_2 = \mathbf{x}_v/g \qquad \text{on } \gamma \text{ for } u > 0.$$

Since γ is a geodesic, E_1 is parallel and by Exercise 3 of Section 4, so is E_2. By parallelism, E_2 is thus well-defined at $u = 0$ (Fig. 7.28). Now

$$E_1(0) = \mathbf{x}_u(0, v_0) = \cos v_0 \, \mathbf{e}_1 + \sin v_0 \, \mathbf{e}_2$$

hence

$$E_2(0) = -\sin v_0 \, \mathbf{e}_1 + \cos v_0 \, \mathbf{e}_2.$$

Furthermore, since E_2 is parallel and $\mathbf{x}_v = gE_2$ on γ, we get

$$\mathbf{x}_{uv} = \mathbf{x}_{vu} = g'E_2 \qquad \text{on } \gamma \text{ for } u > 0.$$

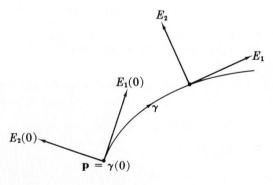

FIG. 7.28

Taking limits as $u \to 0$ yields

$$\mathbf{x}_{uv}(0, v_0) = (\lim_{u \to 0} g'(u)) E_2(0).$$

But

$$\mathbf{x}_u(0, v) = \cos v \, \mathbf{e}_1 + \sin v \, \mathbf{e}_2 \qquad \text{for all } v;$$

hence

$$\mathbf{x}_{uv}(0, v_0) = -\sin v_0 \, \mathbf{e}_1 + \cos v_0 \, \mathbf{e}_2 = E_2(0).$$

Thus the last equation implies $\lim_{u \to 0} g'(0) = 1$; that is,

$$\lim_{u \to 0} (\sqrt{G})_u(u, v_0) = 1$$

for arbitrary v_0.   ∎

In terms of the spreading apart of radial geodesics, the initial conditions above show that as they first leave the pole $\mathbf{p}$ in *any* geometric surface, they are spreading at the same rate as in the Euclidean plane $\mathbf{E}^2$. For there we found $\sqrt{G} = u$; hence

$$\sqrt{G}(0, v) = 0, \qquad (\sqrt{G})_u(0, v) = 1.$$

However, the Jacobi equation, written $(\sqrt{G})_{uu} = -K \sqrt{G}$, *shows that thereafter the rate of spreading depends on Gaussian curvature.* For $K < 0$, radial geodesics spread apart *faster* than in $\mathbf{E}^2$. (We observed this earlier in the hyperbolic plane.) For $K > 0$ the rate of spreading is less than in $\mathbf{E}^2$ (as on the sphere).

In particular, to locate conjugate points, it is no longer necessary to explicitly construct geodesic polar mappings, as we have done heretofore. We can find $\sqrt{G}$ on a geodesic γ by simply solving the Jacobi equation on γ, subject to the given initial conditions. Explicitly, Theorem 6.3 implies the following result.

6.4 Corollary Let γ be a unit-speed geodesic starting at the point $\mathbf{p}$ in M. Let g be the unique solution of the *Jacobi equation on γ*,

$$g'' + K(\gamma) g = 0$$

such that $g(0) = 0$, $g'(0) = 1$. Then the first conjugate point of $\gamma(0) = \mathbf{p}$ on γ (if it exists) is $\gamma(s_1)$, where s_1 is the smallest positive number such that $g(s_1) = 0$.

6.5 Example *Conjugate points*

(1) Let γ be a unit-speed geodesic starting at any point $\mathbf{p}$ of the sphere Σ of radius r. The Jacobi equation for γ is thus $g'' + g/r^2 = 0$, which has the general solution

$$g(s) = A \sin \frac{s}{r} + B \cos \frac{s}{r}.$$

The initial conditions $g(0) = 0$, $g'(0) = 1$, then give $g(s) = r \sin (s/r)$. The first zero of this function with $s_1 > 0$ occurs at $s_1 = \pi r$. Thus *the first conjugate point of $\gamma(0) = $ **p** on γ is at the antipodal point of* **p**. (This agrees with our earlier computation for the unit sphere by means of geodesic polar mappings.)

(2) Let γ be a unit-speed parametrization of the outer equator of a torus of revolution T with radii $R > r > 0$. Now γ is a geodesic and on γ we know that K has the constant value $1/r(R + r)$. Thus by Corollary 6.4, the first conjugate point $\gamma(s_1)$ of $\gamma(0) = $ **p** on γ will occur at *exactly the same distance s_1 along γ as if γ were on a sphere with this curvature K*. It follows that $s_1 = \pi \sqrt{r(R + r)}$.

6.6 Corollary There are no conjugate points on any geodesic in a surface with curvature $K \leq 0$. Hence every geodesic segment on such a surface is locally minimizing.

Proof. Apply Corollary 6.4 to a geodesic γ in M. Since $g(0) = 0$ and $g'(0) = 1$, we have $g(s) \geq 0$ for $s \geq 0$, at least up to the first conjugate point (if it exists). But $K \leq 0$ implies $g'' = -Kg \geq 0$, so g' is an increasing function; in fact $g' \geq 1$. Hence $g(s) \geq s$ up to the first conjugate point— which can thus never occur. The final assertion then follows from Theorem 6.2. ∎

For example, on a circular cylinder C $(K = 0)$ the helical geodesic γ from **p** to **q** indicated in Fig. 7.29 is indeed *stable*, as one can verify by direct experiment. Although locally minimizing, it is certainly not minimizing. Evidently the straight-line segment σ provides a considerably shorter way to get from **p** to **q**.

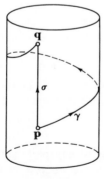

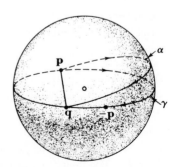

FIG. 7.29   FIG. 7.30

To carry the study of conjugate points much further, it is necessary to use the calculus of variations (see Milnor [7]). We shall quote just one result, which supplements Theorem 6.2. *As soon as a geodesic γ starting at* **p** *passes the first conjugate point of* **p** *on* γ, *it no longer locally minimizes arc length.* This is fairly easy to see on a sphere Σ. In Fig. 7.30 the geodesic γ from **p** to **q** is only slightly longer than the first conjugate distance πr. If the plane of the great circle of γ is rotated slightly about an axis through the end points **p** and **q**, it will slice from Σ a curve segment α which, as one can verify analytically, is strictly shorter than γ. (Note that the only shorter *geodesic* from **p** to **q** is *not* near γ.)

Theorem 6.3 can also be used to give a rather intuitive description of Gaussian curvature in an arbitrary geometric surface.

6.7 Lemma   If **x** is a geodesic polar mapping with pole **p**, then

$$\sqrt{\overline{G}}\,(u, v) = u - K(\mathbf{p})\frac{u^3}{6} + o(u^3) \qquad (u \geqq 0).$$

Throughout, $o(u^n)$ denotes a function of u and v $(u > 0)$ such that $\lim_{u \to 0} o(u^n)/u^n = 0$. In the formula then, if u is small enough, $o(u^3)$ is negligible compared to the first two terms.

Proof. As before, consider $g(u) = \sqrt{\overline{G}}\,(u, v)$ on a radial geodesic $\gamma(u) = \mathbf{x}(u, v)$. As a solution of the Jacobi equation on γ, g is differentiable at $u = 0$. Thus it has a Taylor expansion

$$g(u) = g(0) + g'(0)u + g''(0)\frac{u^2}{2} + g'''(0)\frac{u^3}{6} + o(u^3).$$

The initial conditions in Corollary 6.4 are $g(0) = 0$, $g'(0) = 1$; hence from the Jacobi equation we get $g''(0) = 0$. Differentiating the Jacobi equation gives

$$g''' + K(\gamma)'g + K(\gamma)g' = 0.$$

Hence

$$g'''(0) = -K(\gamma(0)) = -K(\mathbf{p}).$$

Substitution in the Taylor expansion then gives the required result.   ∎

Suppose that the inhabitants of a geometric surface M want to determine the Gaussian curvature of M at a point **p**. By measuring a short distance ε in all directions from **p**, they obtain the polar circle C_ε of radius ε.

Now if $M = \mathbf{E}^2$, the circumference of C_ε is just $L(C_\varepsilon) = 2\pi\varepsilon$. But for $K > 0$ the radial geodesics from **p** are not spreading as rapidly, so C_ε should be shorter than $2\pi\varepsilon$; and for $K < 0$ they are spreading *more* rapidly, so C_ε should be longer than $2\pi\varepsilon$.

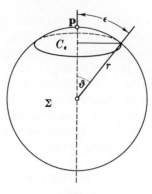

FIG. 7.31

The relation between $L(C_\varepsilon)$ and K can be measured with some precision. For $\varepsilon > 0$ small enough, C_ε is parametrized by $v \to \mathbf{x}(\varepsilon, v)$, where $\mathbf{x}$ is a geodesic polar parametrization at $\mathbf{p}$. Thus

$$L(C_\varepsilon) = \int_0^{2\pi} \sqrt{G}\,(\varepsilon, v)\, dv.$$

Hence by the preceding lemma,

$$L(C_\varepsilon) = 2\pi \left(\varepsilon - K(\mathbf{p}) \frac{\varepsilon^3}{6} + o(\varepsilon^3) \right). \qquad (*)$$

Thus if surveyors in M measure $L(C_\varepsilon)$ very carefully for ε small, they can determine approximately what the Gaussian curvature of M is at $\mathbf{p}$. Taking limits yields

6.8 Corollary $K(\mathbf{p}) = \lim_{\varepsilon \to 0} (3/\pi\varepsilon^3)\,(2\pi\varepsilon - L(C_\varepsilon))$.

We can easily test formula $(*)$ for a sphere Σ of radius r in $\mathbf{E}^3$. As in Fig. 7.31, the polar circle C_ε with center $\mathbf{p}$ is actually a Euclidean circle of Euclidean radius $r \sin \vartheta$, where $\vartheta = \varepsilon/r$. Thus by the Taylor series of the sine function,

$$L(C_\varepsilon) = 2\pi \left(r \sin \frac{\varepsilon}{r} \right) = 2\pi \left(\varepsilon - \frac{\varepsilon^3}{6r^2} + o(\varepsilon^3) \right)$$

which gives yet another *proof* that Σ has Gaussian curvature $K = 1/r^2$.

EXERCISES

1. Let $\mathbf{x}$ be the polar parametrization of the hyperbolic plane given in Example 5.5. Derive $\sqrt{G}\,(u, v) = \sinh u$ in two different ways: by computing $\mathbf{x}_v \circ \mathbf{x}_v$, and by solving the Jacobi equation.

2. If C_ε is a polar circle around a point $\mathbf{p}$ of M, call the region enclosed by C_ε the *polar disc* D_ε of radius ε.

   (a) Show that the area of the polar disc is

$$A(D_\varepsilon) = \pi \left[\varepsilon^2 - K(\mathbf{p}) \frac{\varepsilon^4}{12} + o(\varepsilon^4) \right]$$

hence

$$K(\mathbf{p}) = \frac{12}{\pi} \lim_{\varepsilon \to 0} \frac{\pi \varepsilon^2 - A(D_\varepsilon)}{\varepsilon^4}.$$

   (b) Use this formula to find the Gaussian curvature of a sphere of radius r.

3. At the origin $\mathbf{0}$ in the hyperbolic plane, find the length of the polar circle C_ε and the area of the polar disc D_ε $(0 < \varepsilon < 2)$. Deduce from each result that $K(\mathbf{0}) = -1$.

4. Let M be an augmented surface of revolution (Ex. 12 of IV.1).

   (a) If M crosses A at only one point $\mathbf{p}$ (as on a paraboloid of revolution), show that $\mathbf{p}$ has no conjugate points on any geodesic.

   (b) If M crosses A at two points, $\mathbf{p}$ and $\mathbf{q}$, (as on an ellipsoid of revolution), show that $\mathbf{p}$ and $\mathbf{q}$ are conjugate on every geodesic joining them. (*Hint:* Theorem 6.3 of Chapter V provides a solution of the Jacobi equation.)

The following exercises deal with a useful variant of the polar geodesic parametrization in which the pole $\mathbf{p}$ is replaced by an arbitrary regular curve.

5. Let $\beta: I \to M$ be a regular curve in M, and let X be a (nonvanishing) vector field on β such that β' and X are linearly independent at each point. Define

$$\mathbf{x}(u, v) = \gamma_{X(v)}(u);$$

   thus the u-parameter curves of $\mathbf{x}$ are geodesics cutting across β with initial velocities given by X (Fig. 7.32). Prove:

   (a) $\mathbf{x}$ is a regular mapping on some region D containing the interval $(0, v)$, v in I.

   (b) By a suitable choice of β and X, this parametrization $\mathbf{x}$ becomes (i) the identity map of $\mathbf{E}^2$ (natural coordinates), (ii) the canonical parametrization of a surface of revolution, and (iii) a ruled parametrization of a ruled surface (Chapter V, Section 5).

6. (*Continuation*). If β is a unit-speed curve, and X is the unit normal N of β (Section 4), show that for $\mathbf{x}$: $E = 1$, $F = 0$, and $\sqrt{G}$ is the solution

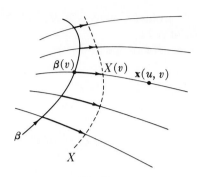

FIG. 7.32

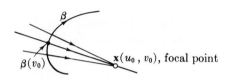

FIG. 7.33

of the Jacobi equation $(\sqrt{G})_{uu} + K\sqrt{G} = 0$ such that

$$\sqrt{G}\,(0, v) = 1 \qquad \text{and} \qquad (\sqrt{G})_u(0, v) = -\kappa_g(v).$$

The natural choice of X in the preceding example means that G for this parametrization is geometrically significant. If $G(u_0, v_0) = 0$, then (by analogy with conjugate points) we say that $\mathbf{x}(u_0, v_0)$ is a *focal point* of β along the normal geodesic $v = v_0$. Here light rays emerging orthogonally from β tend to meet (Fig 7.33).

7. (a) If β is a circle of latitude on a sphere Σ, show that the north and south poles of Σ are the only focal points of β.

   (b) If β is a curve in the Euclidean plane, show that its focal points are exactly its centers of curvature; that is, the points on its evolute. (See Ex. 15 of II.4.)

7 Mappings that Preserve Inner Products

We have already seen that a local isometry $F: M \to N$ carries geodesics of M to geodesics of N. Using the notation γ_v for the geodesic with initial velocity $\mathbf{v}$, we can be more explicit.

7.1 Lemma If $F: M \to N$ is a local isometry, and $\mathbf{v}$ is a tangent vector to M at $\mathbf{p}$, then

$$F(\gamma_v) = \gamma_{F_*(v)}$$

Proof. By the remark above, $\bar{\gamma} = F(\gamma_v)$ is a geodesic of N. Its initial velocity is the tangent vector

$$\bar{\gamma}'(0) = F_*(\gamma_v'(0)) = F_*(\mathbf{v})$$

to N at $F(\mathbf{p})$. Thus by the uniqueness of geodesics (Theorem 4.3), $\bar{\gamma}$ is precisely $\gamma_{F_*(v)}$. ∎

It follows that a local isometry is completely determined by its effect on just one frame.

7.2 Theorem   Let F and G be local isometries from M to N. If for some one frame $\mathbf{e}_1$, $\mathbf{e}_2$ at a point $\mathbf{p}$ of M we have

$$F_*(\mathbf{e}_1) = G_*(\mathbf{e}_1), \qquad F_*(\mathbf{e}_2) = G_*(\mathbf{e}_2),$$

then $F = G$.

Proof. If M is geodesically complete, the proof is particularly easy. If $\mathbf{q}$ is an arbitrary point of M, then by Theorem 5.9 there is a vector $\mathbf{v}$ at the special point $\mathbf{p}$ such that $\gamma_v(r) = \mathbf{q}$. From the hypothesis on F_* and G_*, we deduce by linearity that F_* and G_* agree on $\mathbf{v} = c_1\mathbf{e}_1 + c_2\mathbf{e}_2$. Thus the preceding lemma shows that

$$F(\gamma_v) = \gamma_{F_*(v)} = \gamma_{G_*(v)} = G(\gamma_v).$$

Hence, in particular,

$$F(\mathbf{q}) = F(\gamma_v(r)) = G(\gamma_v(r)) = G(\mathbf{q})$$

or all points $\mathbf{q}$ of M.

The proof for M arbitrary is a refinement. Using Lemma 5.3 it is possible to get a broken geodesic β from $\mathbf{p}$ to $\mathbf{q}$ and deduce $F(\beta) = G(\beta)$ by applying the argument above to each unbroken segment of β. ∎

We shall now use the fact that local isometries preserve geodesics to *construct* some local isometries. The goal is to demonstrate a family resemblance among geometric surfaces of the same constant curvature. Given any number K, there is a particularly simple geometric surface $M(K)$ whose Gaussian curvature has constant value K.

(1) If $K > 0$, let $M(K)$ be the sphere Σ of curvature K (hence radius $1/\sqrt{K}$).

(2) If $K = 0$, let $M(K)$ be the Euclidean plane $\mathbf{E}^2$.

(3) If $K < 0$, let $M(K)$ be the hyperbolic plane H of curvature K (hence pseudo-radius $1/\sqrt{-K}$: see Exercise 4 of Section 2).

We shall call $M(K)$ *the standard geometric surface* of constant curvature K. Of course, there are many other constant curvature surfaces; these are distinguished by the fact that they are geodesically complete and simply connected (p. 176).

7.3 Theorem Let N be a geodesically complete geometric surface with *constant* Gaussian curvature K. Then there is a local isometry F of the standard surface $M(K)$ onto N.

The first mapping in Example 4.6 of Chapter VI is an instance of this theorem, as is the local isometry (Exercise 6 of Section 2) of a sphere onto a projective plane.

Proof. The Case $K < 0$. We use the language of Example 2.4, where $K = -1$. A mere change of scale (Exercise 4 of Section 2) takes care of arbitrary $K < 0$. As in (2) of Example 5.5, let $\mathbf{p}$ be the origin of $H = M(-1)$, with $\mathbf{e}_1 = U_1(\mathbf{p})$ and $\mathbf{e}_2 = U_2(\mathbf{p})$. Let $\bar{\mathbf{e}}_1, \bar{\mathbf{e}}_2$ be a frame at an arbitrary point of N. Then let $\mathbf{x}$ and $\bar{\mathbf{x}}$ be the resulting geodesic polar mappings of H and N.

For the surface N, we assert that

(1) $\bar{\mathbf{x}}$ is defined on the entire right half-plane S: $u > 0$ (a consequence of geodesic completeness).

(2) Its image $\bar{\mathbf{x}}(S)$ covers all of N except possibly the pole $\bar{\mathbf{p}}$ (a consequence of Theorem 5.9 and the definition of geodesic polar mappings).

(3) $\bar{\mathbf{x}}\colon S \to N$ is a regular mapping. (By Lemma 5.4, $\bar{E} = 1$, and $\bar{F} = 0$, but we have observed earlier that the Jacobi equation for $K = -1$ gives $\sqrt{\bar{G}} = \sinh u$, so $\bar{E}\bar{G} - \bar{F}^2 = \sinh^2 u > 0$ on S.)

This general result is thus valid for $\mathbf{x}\colon S \to H$ as well, but here we know more. By Example 5.5 the whole surface H is a normal neighborhood of the pole $\mathbf{p}$; thus $\mathbf{x}$ has only the usual ambiguities of polar coordinates; the equation $\mathbf{x}(u, v) = \mathbf{q}$ determines u uniquely, and v uniquely but for addition of some multiple of 2π ($\mathbf{q} \neq \mathbf{p}$). From this extra information we conclude that the formula

$$F(\mathbf{x}(u, v)) = \bar{\mathbf{x}}(u, v)$$

is *consistent* and thus defines a mapping F of H onto all of N.

(To prove differentiability of F at the pole $\mathbf{p}$ we must resort, as in the proof of Lemma 5.3, to the mappings $\mathbf{y}$ and $\bar{\mathbf{y}}$ corresponding to $\mathbf{x}$ and $\bar{\mathbf{x}}$.) It is easy to show that F is a local isometry by using the criterion of Lemma 4.5 of Chapter VI. Indeed, from (3) above, we have

$$E = 1 = \bar{E}, \qquad F = 0 = \bar{F}, \qquad G = \sinh^2 u = \bar{G}, \qquad \text{for } u > 0,$$

and at the pole $\mathbf{p}$ the preservation of inner products is an honest consequence of continuity.

The Case $K = 0$. This is a word-for-word copy of the preceding argument except that

$$M(K) = \mathbf{E}^2 \qquad \text{and} \qquad G = \bar{G} = u^2.$$

The Case $K > 0$. Here a new idea is required, since the largest normal neighborhood $\mathfrak{N}$ of a point $\mathbf{p}$ in the sphere $\Sigma = M(K)$ is not all of Σ: The antipodal point $-\mathbf{p}$ is omitted.

Arguing as in the case $K < 0$, we get a local isometry $F_1 \colon \mathfrak{N} \to N$. Now repeat this argument once more at a point $\mathbf{p}^*$ in Σ different from both $\mathbf{p}$ and $-\mathbf{p}$. We obtain another local isometry $F_2 \colon \mathfrak{N}^* \to N$ where $\mathfrak{N}^*$ is all of Σ except $-\mathbf{p}^*$. The frames determining F_2 are chosen so that the derivative maps of F_1 and F_2 agree at $\mathbf{p}^*$. Thus by Theorem 7.2, F_1 and F_2 are identical on the overlap of $\mathfrak{N}$ and $\mathfrak{N}^*$. But $\mathfrak{N}$ and $\mathfrak{N}^*$ cover the whole sphere Σ, so taken together F_1 and F_2 constitute a single local isometry $F \colon \Sigma \to N$. Because Σ is compact and N is connected, Exercise 6 of Section 7 in Chapter IV shows that F carries Σ onto M. ∎

An isometry $F \colon M \to M$ of a geometric surface onto itself may be viewed as a *symmetry* of M. Every feature of the geometry of M is the same at each point $\mathbf{p}$ as at $F(\mathbf{p})$, since this geometry consists of isometric invariants. The results of Exercise 9 of Section 4, Chapter VI, show at once that the set $\mathscr{I}(M)$ of all isometries $F \colon M \to M$ forms a group, just as do the set of all isometries of Euclidean space (Exercise 7 of Section 1 of Chapter III). We call $\mathscr{I}(M)$ the *isometry group* of M.

This group $\mathscr{I}(M)$ is, of course, intrinsic to M, and when M is a surface in $\mathbf{E}^3$, should not be confused with the group $\mathbf{S}(M)$ of Euclidean symmetries of M (Exercise 7 of Section 8, Chapter VI). A Euclidean symmetry $\mathbf{F}$ of $M \subset \mathbf{E}^3$ is an isometry of $\mathbf{E}^3$ such that $\mathbf{F}(M) = M$; these exist when the shape of M in $\mathbf{E}^3$ is symmetric in the ordinary sense of the word. Each Euclidean symmetry $\mathbf{F}$ of M gives rise to an isometry $\mathbf{F} \mid M \colon M \to M$, but in general this process does not give *all* isometries of $M \subset \mathbf{E}^3$ (Exercise 9).

For an arbitrary geometric surface M, the isometry group $\mathscr{I}(M)$ gives a novel algebraic description of M. Roughly speaking, the more symmetrical M is the larger $\mathscr{I}(M)$ is. For example, although we shall not carry out the proof, the ellipsoid

$$M \colon \frac{x^2}{a^2} + \frac{y^2}{b^2} + \frac{z^2}{c^2} = 1 \qquad (a > b > c)$$

has exactly eight elements in its isometry group, these all arising from its Euclidean symmetries as described above: three reflections (one in each coordinate plane), three 180° rotations (one around each coordinate axis), the isometry $\mathbf{p} \to -\mathbf{p}$, and, of course, the identity map of M.

The smallest possible isometry group $\mathscr{I}(M)$ occurs when the identity map of M is the only isometry of M. We can produce such a geometric surface by putting a bump on the ellipsoid in such a way as to destroy all seven of its nontrivial isometries.

By contrast, a geometric surface M has the maximum possible sym-

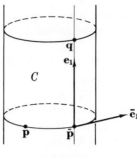

FIG. 7.34

metry when every possible isometry permitted by Theorem 7.2 actually exists. That is, given frames $\mathbf{e}_1$, $\mathbf{e}_2$ and $\bar{\mathbf{e}}_1$, $\bar{\mathbf{e}}_2$ at any two points of M, there exists an isometry $F: M \to M$ such that

$$F_*(\mathbf{e}_1) = \bar{\mathbf{e}}_1, \qquad F_*(\mathbf{e}_2) = \bar{\mathbf{e}}_2.$$

In this case, we shall say that M is *frame-homogeneous;* any two frames on M are symmetrically positioned.

Thus what we proved in Theorem 2.3 of Chapter III is that $\mathbf{E}^3$ is frame-homogeneous, and the same proof is valid for arbitrary $\mathbf{E}^n$, in particular for $\mathbf{E}^2$. In the exercises for this section, we shall see that *every standard surface $M(K)$ of constant curvature is frame-homogeneous.*

7.4 Definition A geometric surface M is *point-homogeneous* (or merely *homogeneous*) provided that given any two points $\mathbf{p}$ and $\mathbf{q}$ of M there is an isometry $F: M \to M$ such that $F(\mathbf{p}) = \mathbf{q}$.

A frame-homogeneous surface, is of course, homogeneous—but not conversely. A circular cylinder C in $\mathbf{E}^3$ furnishes an example. In fact, if F is a rotation of $\mathbf{E}^3$ about the axis of C, or a translation of $\mathbf{E}^3$ along this axis, then F carries C onto C, producing an isometry of C. Hence given any points $\mathbf{p}$ and $\mathbf{q}$ of C, we can first rotate to bring $\mathbf{p}$ to $\bar{\mathbf{p}}$ on the same ruling as $\mathbf{q}$, then translate $\bar{\mathbf{p}}$ to $\mathbf{q}$. The composition of these two isometries is an isometry carrying $\mathbf{p}$ to $\mathbf{q}$. On the other hand, C is not frame-homogeneous: all its *points* are geometrically equivalent, but not all its *frames*. (*Proof:* For the unit vectors shown in Fig. 7.34, no isometry could carry $\mathbf{e}_1$ to $\bar{\mathbf{e}}_1$, for by Lemma 7.1, F would have to send the one-to-one geodesic $\gamma_{\mathbf{e}_1}$ to the periodic geodesic $\gamma_{\bar{\mathbf{e}}_1}$—an impossibility, since F is one-to-one.)

Homogeneity is a very strong restriction.

7.5 Theorem If a geometric surface M is homogeneous, then M is geodesically complete and has constant Gaussian curvature.

Proof. Constancy of curvature follows immediately from the definition of homogeneity and the fact that isometries preserve curvature. The proof

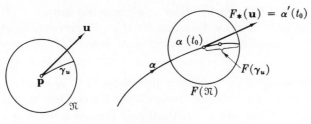

FIG. 7.35

of completeness is more interesting. If M is not geodesically complete, there is a *maximal* unit-speed geodesic α defined only on an interval, say, I: $t < a$, which is not the whole real line. Let us show that this is impossible. By Lemma 5.3, all geodesics emanating from some arbitrary point $\mathbf{p}$ of M run at least for some fixed distance $\varepsilon > 0$. Choose t_0 in I such that $a - t_0 < \varepsilon/2$. Because M is homogeneous, there is an isometry $F: M \to M$ such that $F(\mathbf{p}) = \alpha(t_0)$. Now for some unit vector $\mathbf{u}$ at $\mathbf{p}$, $F_*(\mathbf{u}) = \alpha'(t_0)$. Thus the geodesic segment $F(\gamma_u)$ has initial velocity

$$F_*(\gamma_u{}'(0)) = F_*(\mathbf{u}) = \alpha'(t_0)$$

and runs for distance ε at unit speed (Fig. 7.35). But then a shift of para-metrization enables us to apply Theorem 4.3 and thereby define α on the interval $I^*: t < t_0 + \varepsilon$. But $t_0 + \varepsilon > a$, so this contradicts the maximality of the interval I, and thus proves M is geodesically complete. ∎

As the title of this section suggests, (local) isometries are not the only inner-product-preserving mappings of importance in geometry. We shall take a brief look at the other main types.

7.6 Definition   Let $F: M \to \mathbf{E}^3$ be a mapping of a geometric surface into $\mathbf{E}^3$. If the derivative map F_* preserves inner products of tangent vectors, then F is an *isometric immersion*. If F is also one-to-one, then F is an *isometric imbedding*. An isometric imbedding F such that the inverse function $F^{-1}: F(M) \to M$ is continuous is said to be *proper*.

This definition is unduly restrictive. Evidently we could replace $\mathbf{E}^3$—or even M—by any Riemannian manifold (p. 308).

7.7 Lemma   If $F: M \to \mathbf{E}^3$ is a proper isometric imbedding of a geometric surface M into $\mathbf{E}^3$, then the image $F(M)$ is a surface in $\mathbf{E}^3$ and the function $F: M \to F(M)$ is an isometry.

Proof. If $\mathbf{x}: D \to M$ is a proper patch in M, then the composite mapping $F(\mathbf{x}): D \to \mathbf{E}^3$ is a patch that lies in $F(M)$. Furthermore, $F(\mathbf{x})$ is a proper patch. In fact, its inverse function $F(\mathbf{x}(D)) \to D$ is just $\mathbf{x}^{-1}F^{-1}$, which is continuous since $\mathbf{x}^{-1}$ and F^{-1} are continuous. Thus we can easily check

Definition 1.2 of Chapter IV. Now as a geometric surface, $F(M)$ uses the dot product of $\mathbf{E}^3$, and by definition $F\colon M \to \mathbf{E}^3$ preserves inner products. Hence when considered as a mapping of M onto $F(M)$, F preserves inner products.   ∎

Thus the study of the geometry of surfaces in $\mathbf{E}^3$ is exactly the same as the study of *proper isometric imbeddings* of geometric surfaces into $\mathbf{E}^3$. This rather technical fact is important only because it suggests a considerable generalization of our work in Chapters V and VI. We could just as well have studied the far larger class of *isometric immersions* into $\mathbf{E}^3$, dropping the one-to-one and properness restrictions. There is, in fact, no real difficulty involved, except for complications of notation.

As in the special case discussed on page 308, the image $F(M)$ of an isometric immersion $F\colon M \to \mathbf{E}^3$ may cut across itself; nevertheless, we shall think of it as a kind of defective surface in $\mathbf{E}^3$. If we define the shape operator of such an *immersed surface*, this should at least suggest how to generalize the rest of Chapters V and VI.

Because inner products are preserved, an isometric immersion F is regular. Thus $F_*(T_p(M))$ is a two-dimensional subspace of $T_{F(p)}(\mathbf{E}^3)$; it plays the role of a tangent plane for $F(M)$ at $F(\mathbf{p})$. A *unit normal function* U assigns to each point $\mathbf{p}$ (in some region of M) a unit vector orthogonal to $F_*(T_p(M))$. If α is a curve in M, then U_α is a vector field on $F(\alpha)$ in $\mathbf{E}^3$. Then if $\mathbf{v}$ is the initial velocity of α, we define $S(\mathbf{v})$ to be the unique vector in $T_p(M)$ such that

$$F_*(S(\mathbf{v})) = -U_\alpha{}'(0)$$

This *shape operator* S is again a symmetric linear operator on $T_p(M)$.

Most of our earlier results hold up rather well under generalization. For example, if the Gaussian curvature K of M is defined intrinsically as in Section 2—then by reorganizing the logic in Chapter VI, Section 2, we can show that $K = \det S$.

A theorem such as Theorem 3.7 of Chapter VI becomes more informative: If M is a compact surface with constant curvature $K\,(>0)$, and $F\colon M \to \mathbf{E}^3$ is an isometric immersion, then F is an isometry of M onto a Euclidean sphere Σ of radius $1/\sqrt{K}$ in $\mathbf{E}^3$.

In other words, even if we give $F(M)$ permission to cut across itself, this cannot happen: $F(M)$ can only be an ordinary round sphere in $\mathbf{E}^3$.

We have seen that there are geometric surfaces M which cannot be isometrically imbedded in $\mathbf{E}^3$—for example, the flat torus (Example 2.5) or the projective plane, Exercise 6 of Section 2. In this case it is natural to try to imbed M in a higher dimensional Euclidean space $\mathbf{E}^n$. The larger n is, the less difficult the task becomes. (Roughly speaking, with more dimensions available for M to curve through, there is a better chance that a shape can be found for M which is compatible with its intrinsic geometry. See Chapter VI, Section 9.)

Thus, although there are no flat toruses in $\mathbf{E}^3$, they *can* be found in $\mathbf{E}^4$.

7.8 Example An isometric imbedding of a flat torus in $\mathbf{E}^4$. Start with the mapping $\bar{\mathbf{x}}\colon \mathbf{E}^2 \to \mathbf{E}^4$ such that

$$\bar{\mathbf{x}}(u, v) = (\cos u,\, \sin u,\, \cos v,\, \sin v).$$

If $\mathbf{x}$ is the parametrization of the flat torus T given in Example 2.5, then the formula

$$F(\mathbf{x}(u, v)) = \bar{\mathbf{x}}(u, v)$$

is consistent; in fact it defines a *one-to-one* mapping $F\colon T \to \mathbf{E}^4$. The proof consists in observing that

$$\mathbf{x}(u, v) = \mathbf{x}(u_1, v_1) \Leftrightarrow u_1 = u + 2\pi m, v_1 = v + 2\pi n \Leftrightarrow \bar{\mathbf{x}}(u, v) = \bar{\mathbf{x}}(u_1, v_1).$$

Reading the implication arrows from left to right we get the required consistency; the reverse direction shows that F is one-to-one.

Then F is an isometric imbedding provided F_* preserves inner products. In the usual fashion we compute

$$\bar{\mathbf{x}}_u = (-\sin u,\, \cos u,\, 0,\, 0)$$

$$\bar{\mathbf{x}}_v = (0,\, 0,\, -\sin v,\, \cos v)$$

Hence

$$\bar{E} = 1, \qquad \bar{F} = 0, \qquad \bar{G} = 1.$$

These functions agree with E, F, and G for $\mathbf{x}$, so exactly the same argument used to prove Lemma 4.5 of Chapter VI shows that F_* preserves inner products.

The general situation here is not well understood. Although every compact geometric surface can be isometrically imbedded in $\mathbf{E}^{17}$, it remains a possibility that 17 can be replaced by as low a dimension as 4.

EXERCISES

1. Let $F\colon M \to N$ be a local isometry, and suppose that M is geodesically complete. Show that F is onto if and only if N is geodesically complete.†

2. Prove that a geodesically complete geometric surface with constant pos-

---

† Though the proof is not elementary, it is known that both of these properties are consequences of the geodesic completeness of M.

itive curvature is compact. (The result still holds if merely $K \geq c > 0$. See Myers' theorem in Hicks [5].)

3. Suppose that in M any two points can be joined by *at least* one geodesic, and that in N any two points can be joined by *at most* one geodesic. Prove that every local isometry $F: M \to N$ of such surfaces is one-to-one.

4. Let $F: M \to M$ be an isometry that is not the identity mapping. If a unit speed curve is *fixed* under F, that is,

$$F(\alpha(s)) = \alpha(s) \qquad \text{for all } s,$$

show that α is a geodesic of M.

5. Let $\mathbf{x}$ and $\bar{\mathbf{x}}$ be geodesic polar parametrizations of normal neighborhoods $\mathfrak{N}_\varepsilon$ and $\overline{\mathfrak{N}_\varepsilon}$ (same ε) in two geometric surfaces. If $K(\mathbf{x}) = \bar{K}(\bar{\mathbf{x}})$ on the common domain S_ε of $\mathbf{x}$ and $\bar{\mathbf{x}}$, prove that $\mathfrak{N}_\varepsilon$ and $\overline{\mathfrak{N}_\varepsilon}$ are isometric.

6. Prove that the sphere Σ and hyperbolic plane H are frame homogeneous. (*Hint:* for Σ derive the required isometries from orthogonal transformations of $\mathbf{E}^3$; for H use Theorem 7.3 and a preceding exercise.)

7. Show that the flat torus (Example 2.5) is homogeneous, but not frame homogeneous, and that an ordinary torus of revolution in $\mathbf{E}^3$ is not homogeneous.

8. Prove:
   (a) For the right circular cylinder $C: x^2 + y^2 = r^2$ in $\mathbf{E}^3$, every isometry $F: C \to C$ has the form

   $$F(\mathbf{p}) = (p_1 \cos \vartheta \pm p_2 \sin \vartheta, \, p_1 \sin \vartheta \pm p_2 \cos \vartheta, \, \varepsilon p_3 + a)$$

   where $\varepsilon = \pm 1$.
   (b) Every isometry of a sphere or right circular cylinder in $\mathbf{E}^3$ is the restriction of an isometry of $\mathbf{E}^3$.

9. Let M be the cylinder in $\mathbf{E}^3$ whose cross-sectional curve is the ellipse $4x^2 + y^2 = 4$. (Any other closed noncircular curve could be used.) Show that there is an isometry of M which is *not* the restriction of an isometry of $\mathbf{E}^3$. (*Hint:* parametrize M by $\mathbf{x}(u, v) = \alpha(u) + vU_3$, where α is a periodic unit-speed parametrization of the ellipse.)

10. In the sphere Σ of radius r, let T be a triangle whose sides are geodesic segments of lengths a, b, and c (all less than πr). Let ϑ be the angle of T at the vertex $\mathbf{p}$ opposite side a.
    (a) Prove the law of cosines:

    $$\cos \frac{a}{r} = \cos \frac{b}{r} \cos \frac{c}{r} + \sin \frac{b}{r} \sin \frac{c}{r} \cos \vartheta.$$

(b) Show that this formula approximates the usual Euclidean law of cosines when r is large compared to a, b, c.

(*Hint:* to determine $\cos \vartheta$ find unit vectors $\mathbf{u}_b$, $\mathbf{u}_c$ at $\mathbf{p}$ tangent to the sides b and c.)

11. Prove that the projective plane (Exercise 6 of Section 2) is frame homogeneous. (*Hint:* if $F: \Sigma \to \Sigma$ is an isometry of the sphere $\Sigma \subset \mathbf{E}^3$, then $F(-\mathbf{p}) = -F(\mathbf{p})$, hence there is a mapping $\bar{F}: \bar{\Sigma} \to \bar{\Sigma}$ such that $PF = \bar{F}P$.)

12. Show that the isometry groups of isometric surfaces are isomorphic.

13. If M is a surface in $\mathbf{E}^3$ that does not lie in a plane, show that the function $\mathbf{F} \to \mathbf{F} \mid M$ is an isomorphism of the Euclidean symmetry group $\mathcal{S}(M)$ onto a subgroup of the isometry group $\mathcal{I}(M)$.

Isometries of the hyperbolic plane may be constructed explicitly by recognizing a point of the plane as a complex number

$$z = u + iv = (u, v),$$

and using exercises of Section 1. Thus if $\mid z \mid$ denotes the magnitude of z,

$$\mid z \mid^2 = z\bar{z} = u^2 + v^2,$$

the hyperbolic plane may be described as the disc $\mid z \mid < 2$ with conformal geometric structure given as in Example 1.3 by $g(z) = 1 - \mid z \mid^2/4$.

14. (*A translation of the hyperbolic plane.*) For a fixed real number $c = (c, 0)$ in H, let T be the mapping $T(z) = 4\,[(z + c)/(cz + 4)]$ defined on H.
    (a) Show that $T(H) \subset H$, and that $T: H \to H$ is one-to-one and onto.

    If H' denotes the same disc, $\mid z \mid < 2$, but with the usual Euclidean structure, then Exercise 7 of Section 1 shows that $T: H' \to H'$ is a conformal mapping with scale factor $\lambda(z) = \mid dT/dz \mid$.
    (b) Show that this scale factor is

$$\lambda(z) = 4\,\frac{4 - c^2}{\mid cz + 4 \mid^2}.$$

    (c) Deduce that $T: H \to H$ is an isometry of the hyperbolic plane. (*Hint:* Use Ex. 9 of VII.1.)

    These methods can be used to show that H is frame-homogeneous, and—carried somewhat further—to give an elegant derivation of the geodesics of H.

15. (*The Poincaré half-plane P is isometric to the hyperbolic plane H.*) In terms of complex numbers, P is the half-plane $\operatorname{Im} z > 0$ with conformal geometric structure $g(z) = \operatorname{Im} z$. ($\operatorname{Im} z$ is the imaginary part v of $z = u + iv$.) Let F be the mapping

$$F(z) = \frac{z + 2i}{iz + 2},$$

defined on H. Show that

(a) $\operatorname{Im} F(z) = (4 - |z|^2)/|iz + 2|^2$.

(b) F is a one-to-one mapping of H onto P. (Compute F^{-1} explicitly.)

(c) Relative to Euclidean structures, F is conformal, with scale factor
$\lambda(z) = 4/|iz + 2|^2$.

(d) $F: H \to P$ is an isometry.

Make a sketch of H and P indicating the images in P of each of the four quadrants of H.

8   The Gauss-Bonnet Theorem

We have seen that the Gaussian curvature K of a geometric surface M has a strong influence on other *geometric* features of M such as parallel translation, geodesics, isometries, and, of course, the shape of M if it happens to be in $\mathbf{E}^3$. Now we will show that the influence of Gaussian curvature penetrates to the ultimate topological conformation of M—to properties completely independent of the particular geometric structure on M.

The main step in this proof is a theorem which relates the total curvature of a 2-segment to the total amount that its boundary curve turns.

For an arbitrary curve α in M, the geodesic curvature tells its rate of turning *relative to arc length*. Thus to find the total amount α turns, we integrate with respect to arc length:

8.1 Definition   Let $\alpha: [a, b] \to M$ be a regular curve segment in an oriented geometric surface M. The total geodesic curvature $\int_\alpha \kappa_g \, ds$ of α is

$$\int_{s(a)}^{s(b)} \kappa_g(s) \, ds$$

where $\kappa_g(s)$ is the geodesic curvature of a unit-speed reparametrization of α.

The total geodesic curvature of α in M is thus an analogue of the total Gaussian curvature of a surface in $\mathbf{E}^3$. For example, let C be a circle of radius r in $\mathbf{E}^2$, where $\mathbf{E}^2$ has its natural orientation. If α is a curve making one counterclockwise trip around C, then α has constant geodesic curvature $\kappa_g = 1/r$. Thus

$$\int_\alpha \kappa_g \, ds = \frac{1}{r} 2\pi r = 2\pi,$$

regardless of the size of the circle. A *clockwise* trip around C will have total curvature -2π, for in general: If the orientation of M is kept fixed, then the total geodesic curvature of a curve segment α is unaffected by orientation-*preserving* reparametrization, but has its sign changed by an orientation-*reversing* reparametrization. (The former is clear from the definition; the latter can be deduced, for example, from the following lemma.)

8.2 Lemma Let $\alpha\colon [a, b] \to M$ be a regular curve segment in a region of M oriented by a frame field E_1, E_2. Then

$$\int_\alpha \kappa_g \, ds = \varphi(b) - \varphi(a) + \int_\alpha \omega_{12}$$

where φ is an angle function from E_1 to α' on α, and ω_{12} is the connection form of E_1, E_2.

Proof. None of these terms are affected by orientation-preserving reparametrization; thus we may assume that α is a unit-speed curve. But then the result follows immediately by integrating the formula in Lemma 4.5 ∎

For the integration theory in Chapter VI, Section 7, we used 2-segments $\mathbf{x}\colon R \to M$ which were one-to-one and regular on the interior R° of R. Now we shall impose the more stringent requirement that $\mathbf{x}$ be one-to-one and regular on the boundary of R as well. (This is equivalent to saying that $\mathbf{x}\colon R \to M$ is the restriction to R of a patch defined on some open set containing R.)

When $\mathbf{x}$ is a one-to-one regular 2-segment, its edge curves α, β, γ, δ (Definition 6.4, Chapter IV) are one-to-one regular, and we shall think of the boundary $\partial\mathbf{x} = \alpha + \beta - \gamma - \delta$ as a single broken curve enclosing the rectangular region $\mathbf{x}(R)$. We now want to define the total geodesic curvature of $\partial\mathbf{x}$. The definition of geodesic curvature shows that the total geodesic curvature of a curve is simply the total angle that its unit tangent T turns (relative to arc length). But to go all the way around

$$\partial\mathbf{x} = \alpha + \beta - \gamma - \delta$$

we must get not merely the total turning on the edge curves; that is,

$$\int_{\partial x} \kappa_g \, ds = \int_\alpha \kappa_g \, ds + \int_\beta \kappa_g \, ds + \int_{-\gamma} \kappa_g \, ds + \int_{-\delta} \kappa_g \, ds$$

$$= \int_\alpha \kappa_g \, ds + \int_\beta \kappa_g \, ds - \int_\gamma \kappa_g \, ds - \int_\delta \kappa_g \, ds$$

but also *the angles through which a unit tangent T on $\partial\mathbf{x}$ would have to turn at the four corners* of the rectangular region $\mathbf{x}(R)$ (Fig. 7.37). For

$$R\colon a \leq u \leq b, \quad c \leq v \leq d$$

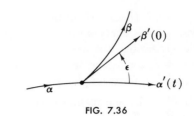

FIG. 7.36

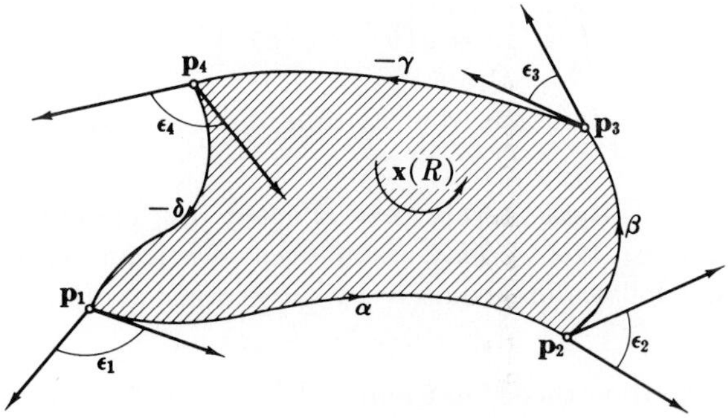

FIG. 7.37

these "corners"

$$\mathbf{p}_1 = \mathbf{x}(a, c), \qquad \mathbf{p}_2 = \mathbf{x}(b, c), \qquad \mathbf{p}_3 = \mathbf{x}(b, d), \qquad \mathbf{p}_4 = \mathbf{x}(a, d)$$

are called the *vertices* of $\mathbf{x}(R)$.

In general, if a regular curve segment α in an oriented region ends at the starting point of another segment β, say $\alpha(1) = \beta(0)$, then the *turning angle* ε from α to β is the oriented angle from $\alpha'(1)$ to $\beta'(0)$ which is smallest in absolute value (Fig. 7.36). For a 2-segment we use the orientation *determined by* $\mathbf{x}$, that is, the area form dM such that $dM(\mathbf{x}_u, \mathbf{x}_v) > 0$, to establish some terminology which is familiar in the case of a polygon in the plane.

8.3 Definition Let $\mathbf{x}: R \to M$ be a one-to-one regular 2-segment, with vertices $\mathbf{p}_1$, $\mathbf{p}_2$, $\mathbf{p}_3$, $\mathbf{p}_4$. The *exterior angle* ε_j of $\mathbf{x}$ at $\mathbf{p}_j$ ($1 \leqq j \leqq 4$) is the turning angle at $\mathbf{p}_j$ derived from the edge curves α, β, $-\gamma$, $-\delta$, α, $\cdots$ in order of occurrence in $\partial\mathbf{x}$. The *interior angle* ι_j of $\mathbf{x}$ at $\mathbf{p}_j$ is $\pi - \varepsilon_j$ (Fig. 7.37).

This definition is given with more general applications in mind; in the case at hand, exterior angles can easily be expressed in terms of the usual coordinate angle ϑ from $\mathbf{x}_u$ to $\mathbf{x}_v$ ($0 < \vartheta < \pi$) by

$$\varepsilon_1 = \pi - \vartheta_1 \qquad \varepsilon_2 = \vartheta_2 \qquad \varepsilon_3 = \pi - \vartheta_3 \qquad \varepsilon_4 = \vartheta_4$$

where ϑ_j is the coordinate angle at the vertex $\mathbf{p}_j$. For example, let us con-

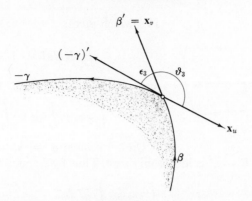

FIG. 7.38

sider the situation at $\mathbf{p}_3$, as shown in Fig. 7.38. By the definition of the edge curves, β' is $\mathbf{x}_v$, but $(-\gamma)'$ is $-\mathbf{x}_u$, since $-\gamma$ is an orientation-reversing reparametrization of γ. Thus $\varepsilon_3 + \vartheta_3 = \pi$. (Analytical proofs may be based on the definition of oriented angle in Section 7, Chapter VI.)

We are now ready to prove the fundamental result of this section.

8.4 Theorem   Let $\mathbf{x}: R \to M$ be a one-to-one regular 2-segment in a geometric surface M. If dM is the area form on $\mathbf{x}(R)$ determined by $\mathbf{x}$, then

$$\underbrace{\iint_{\mathbf{x}} K\, dM}_{\substack{\text{total Gaussian} \\ \text{curvature of } \mathbf{x}}} + \underbrace{\int_{\partial \mathbf{x}} \kappa_g\, ds + (\varepsilon_1 + \varepsilon_2 + \varepsilon_3 + \varepsilon_4)}_{\substack{\text{total geodesic} \\ \text{curvature of } \partial \mathbf{x}}} = 2\pi$$

(The geodesic curvature and exterior angles use the orientation of $\mathbf{x}(R)$ given by dM, where $dM(\mathbf{x}_u, \mathbf{x}_v) > 0$. Note that M itself need not be oriented—or even orientable.)

We call this result the *Gauss-Bonnet formula with exterior angles*. Since $\varepsilon_j = \pi - \iota_j$ for $1 \le j \le 4$, the formula may be rewritten

$$\iint_{\mathbf{x}} K\, dM + \int_{\partial \mathbf{x}} \kappa_g\, ds = (\iota_1 + \iota_2 + \iota_3 + \iota_4) - 2\pi$$

in terms of the *interior angles* of $\mathbf{x}(R)$.

Proof.   Let $E_1 = \mathbf{x}_u/\sqrt{E}$ on the region $\mathbf{x}(R)$. Then let E_2 be the unique vector field such that E_1, E_2 is a frame field with $dM(E_1, E_2) = +1$. In this case (compare with p. 292), the second structural equation becomes

$$d\omega_{12} = -K\theta_1 \wedge \theta_2 = -K\, dM$$

The power for this proof is supplied by Stokes' theorem (6.5 of Chapter

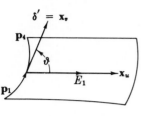

FIG. 7.39

IV) which gives

$$\iint_{\mathbf{x}} K \, dM + \int_{\partial \mathbf{x}} \omega_{12} = 0. \qquad (1)$$

Let us now use Lemma 8.2 to evaluate

$$\int_{\partial \mathbf{x}} \omega_{12} = \int_{\alpha} \omega_{12} + \int_{\beta} \omega_{12} - \int_{\gamma} \omega_{12} - \int_{\delta} \omega_{12}. \qquad (2)$$

On α we have $\alpha' = \mathbf{x}_u = \sqrt{E} \, E_1$, so the angle φ from E_1 to α' is identically zero. Thus by Lemma 8.2,

$$\int_{\alpha} \omega_{12} = \int_{\alpha} \kappa_g \, ds. \qquad (3)$$

Next we try a harder case, say $\int_{\delta} \omega_{12}$. Here the angle φ from

$$E_1 = \frac{\mathbf{x}_u}{\sqrt{E}} \text{ to } \delta' = \mathbf{x}_v$$

is precisely the coordinate angle ϑ from $\mathbf{x}_u$ to $\mathbf{x}_v$ (See Fig. 7.39.) Hence by Lemma 8.2 we get

$$\int_{\delta} \kappa_g \, ds = \vartheta_4 - \vartheta_1 + \int_{\delta} \omega_{12}$$

where, as above, $0 < \vartheta_j < \pi$ is the coordinate angle at the vertex $\mathbf{p}_j$ of $\mathbf{x}$ $(1 \le j \le 4)$. But since

$$\vartheta_1 = \pi - \varepsilon_1 \quad \text{and} \quad \vartheta_4 = \varepsilon_4,$$

This becomes

$$\int_{\delta} \omega_{12} = \pi - \varepsilon_1 - \varepsilon_4 + \int_{\delta} \kappa_g \, ds. \qquad (4)$$

In an entirely similar way we find

$$\int_{\beta} \omega_{12} = -\pi + \varepsilon_2 + \varepsilon_3 + \int_{\beta} \kappa_g \, ds \qquad (5)$$

and

$$\int_{\gamma} \omega_{12} = \int_{\gamma} \kappa_g \, ds. \qquad (6)$$

Thus (2) becomes

$$\int_{\partial \mathbf{x}} \omega_{12} = \int_{\alpha} \kappa_g \, ds + \int_{\beta} \kappa_g \, ds - \int_{\gamma} \kappa_g \, ds - \int_{\delta} \kappa_g \, ds - 2\pi + (\varepsilon_1 + \varepsilon_2 + \varepsilon_3 + \varepsilon_4)$$

$$= \int_{\partial \mathbf{x}} \kappa_g \, ds - 2\pi + (\varepsilon_1 + \varepsilon_2 + \varepsilon_3 + \varepsilon_4).$$

Substitution in (1) then yields the required formula.   ∎

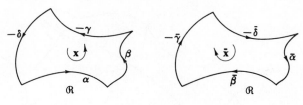

FIG. 7.40

The Gauss-Bonnet formula actually depends not on the particular mapping $\mathbf{x}: R \to M$, but only on its image $\mathcal{R} = \mathbf{x}(R)$. Explicitly, if $\bar{\mathbf{x}}$ is another one-to-one regular 2-segment $\mathbf{x}$ with the same image $\mathcal{R}$, then each of the six terms in the Gauss-Bonnet formula for $\bar{\mathbf{x}}$ has exactly the same numerical value as the corresponding term for $\mathbf{x}$. This should be no surprise if $\mathbf{x}$ and $\bar{\mathbf{x}}$ have the same orientation, that is, determine the same area form on $\mathcal{R}$. But suppose they have opposite orientation (as in Fig. 7.40) so that $dM_x = -dM_{\bar{x}}$. To take the trickiest case, consider corresponding edge curves such as α and $\bar{\beta}$ in the Fig. 7.40. Now α and $\bar{\beta}$ run in opposite directions: $\bar{\beta}$ is an orientation-reversing reparametrization of α. But the geodesic curvatures of α and $\bar{\beta}$ are computed in terms of the opposite area forms dM_x and $dM_{\bar{x}}$. Thus there are *two* sign changes, so

$$\int_\alpha \kappa_g \, ds = \int_{\bar{\beta}} \kappa_g \, ds.$$

The remainder of this section will be devoted to applications of the Gauss-Bonnet formula. The basic idea is to extend it to more general regions—in particular to entire geometric surfaces. To do so we must look at some fundamental properties of surfaces which do not involve geometry.

A *rectangular decomposition* $\mathcal{D}$ of a surface M is a finite collection of one-to-one regular 2-segments $\mathbf{x}_1, \cdots, \mathbf{x}_f$ whose images cover M in such a way that if any two overlap they do so in either a single common vertex or a single common edge.

Evidently a rectangular decomposition is a special kind of paving (Definition 7.3 of Chapter VI), but the regions $\mathbf{x}_i(R_i)$ are now really "rectangular" (since $\mathbf{x}_i$ is one-to-one regular on all of R_i) and they are required to fit together very neatly, as in Fig. 7.41 (compare the paving in Fig. 6.17).

8.5 Theorem   Every compact surface M has a rectangular decomposition.

(Hence in particular M has a paving.) This result is certainly plausible, for if M is made of paper, we could just take a pair of scissors and cut out rectangular pieces until all of M is gone. A general proof is given in Lefschetz [8] (use Exercise 10).

We shall understand that a rectangular decomposition $\mathcal{D}$ carries with it not only its rectangular regions $\mathbf{x}_i(R_i)$—called *faces*—but also the *vertices* and *edges* of these regions.

FIG. 7.41

8.6 Theorem If $\mathfrak{D}$ is a rectangular decomposition of a compact surface M, let v, e, and f be the number of vertices, edges, and faces in $\mathfrak{D}$. Then the integer $v - e + f$ is the same for all rectangular decompositions of M. This integer $\chi(M)$ is called the *Euler-Poincaré characteristic* of M.

The natural proof of this famous theorem is purely topological, however it is an easy consequence of Theorem 8.8.

These results may easily be generalized. First, instead of an entire surface, we could deal with a *polygonal region*, one which can be decomposed into rectangular regions $\mathbf{x}_i(R_i)$ fitting together neatly (as above). Second, we could everywhere replace rectangles R by polygons. (A polygon P is the bounded region in $\mathbf{E}^2$ enclosed by a simple polygonal curve—P including this curve.) Combining both generalizations we get the concept of *polygonal decomposition* $\mathfrak{D}$ of a (polygonal) region $\mathfrak{R}$ in M. The Euler-Poincaré characteristic $\chi(\mathfrak{R})$ of $\mathfrak{R}$ is still independent of the choice of polygonal decomposition $\mathfrak{D}$.

8.7 Example *Euler-Poincaré characteristic.*

(1) A sphere Σ has $\chi(\Sigma) = 2$. By "inflating" a cube as in Fig. 7.42 we get a rectangular decomposition $\mathfrak{D}_1$ of Σ. $\mathfrak{D}_1$ has $v = 8$, $e = 12$, $f = 6$—and thus $\chi = 2$. Inflating a prism gives a polygonal decomposition $\mathfrak{D}_2$ with $v = 6$, $e = 9$, $f = 5$—again $\chi = 2$ (Fig. 7.42).

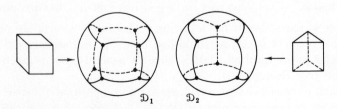

$\mathfrak{D}_1$ $\mathfrak{D}_2$

FIG. 7.42

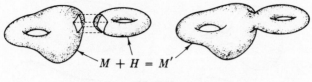

$$M + H = M'$$

FIG. 7.43

(2) A torus T has $\chi(T) = 0$. Picture T as a torus of revolution, and cut it along any three meridians and three parallels. This gives a rectangular decomposition $\mathfrak{D}$ for which $v = 9$, $e = 18$, $f = 9$, hence $\chi = 0$.

(3) Adding a handle to a compact surface reduces its Euler-Poincaré characteristic by 2.

Roughly speaking, a "handle" is a torus with the interior of one face removed. (We suppose that M and the torus are given in some rectangular decomposition.) To add a handle to M, remove the interior of a face of M also, and to the resulting rim smoothly attach the rim of the handle, so that the vertices and edges of the two rims coincide (Fig. 7.43).

This operation produces a new surface M' already equipped with a rectangular decomposition. The Euler-Poincaré characteristic of M' is

$$\chi(M) - 2,$$

since its decomposition has exactly two faces less than M and the torus combined. (Coalescing the two rims eliminates four vertices and four edges as well, but this has no effect on χ.)

It is easy to see that *diffeomorphic surfaces have the same Euler-Poincaré characteristic*, for if $\mathbf{x}_1, \cdots, \mathbf{x}_f$ is a decomposition of M and $F: M \to \bar{M}$ is a diffeomorphism, then $F(\mathbf{x}_1), \cdots, F(\mathbf{x}_f)$ is a decomposition of $\bar{M}$ with exactly the same v, e, and f.

For example, no matter how wildly we distort the sphere

$$\Sigma: x^2 + y^2 + z^2 = 1,$$

the resulting surface M will still have Euler-Poincaré characteristic 2. So long as no geometric structures are involved, the word "sphere" might be used to mean "a surface diffeomorphic to Σ." To avoid any possible confusion we shall stay with the longer terminology.

Suppose we start with the sphere Σ and successively add h handles ($h = 0, 1, 2, \cdots$) to obtain a new surface $\Sigma(h)$. What is remarkable about the operation of adding handles is that *every compact orientable surface M is diffeomorphic to some $\Sigma(h)$*. In this case we shall say that M itself *has h handles*. By (3) of Example 8.7, it follows that

$$\chi(M) = \chi(\Sigma(h)) = \chi(\Sigma) - 2h = 2 - 2h.$$

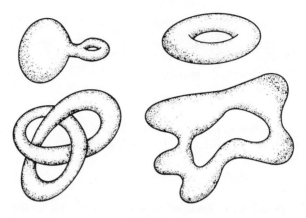

FIG. 7.44

In Fig. 7.44, for example, all four surfaces have just one handle, and thus all have $\chi = 0$.

Although we have used the concepts of calculus in this brief discussion of the Euler-Poincaré characteristic, our remarks remain valid if differentiability is everywhere replaced by continuity. The Euler-Poincaré characteristic is, in fact, a *topological invariant*.†

Returning now to the subject of *geometric* surfaces, we prove a spectacular consequence of Theorem 8.4.

8.8 Theorem   (Gauss-Bonnet)   If M is a compact orientable geometric surface, then the total Gaussian curvature of M is $2\pi\chi(M)$, where $\chi(M)$ is the Euler-Poincaré characteristic of M.

Proof.   Fix an orientation of M with area form dM, and let $\mathfrak{D}$ be a rectangular decomposition of M whose 2-segments $\mathbf{x}_1, \cdots, \mathbf{x}_f$ are all positively oriented. Thus $\mathfrak{D}$ is in particular an oriented paving of M as defined in Chapter VI, Section 7. By definition the total curvature of M is

$$\iint_M K \, dM = \sum_{i=1}^{f} \iint_{\mathbf{x}_i} K \, dM \tag{1}$$

We shall apply the Gauss-Bonnet formula to each summand. (This is permissible, since on each region $\mathbf{x}_i(R_i)$ the area form dM is the one determined by $\mathbf{x}_i$.) In terms of interior angles, this formula reads

† A topological invariant is a property preserved by every *homeomorphism* (that is, continuous function with a continuous inverse). A diffeomorphism is a homeomorphism, but not conversely. However it is a peculiarity of low dimensions that two surfaces are diffeomorphic if (and only if) they are homeomorphic.

$$\iint_{\mathbf{x}_i} K \, dM = -\int_{\partial \mathbf{x}_i} \kappa_g \, ds - 2\pi + (\iota_1 + \iota_2 + \iota_3 + \iota_4) \tag{2}$$

Now consider what happens when (2) is substituted in (1).

Because M is a surface—locally like $\mathbf{E}^2$—each edge of the decomposition $\mathfrak{D}$ will occur in exactly *two* faces, say $\mathbf{x}_i(R_i)$ and $\mathbf{x}_j(R_j)$. Let α_i and α_j be the parametrizations of this edge occurring in $\partial \mathbf{x}_i$ and $\partial \mathbf{x}_j$, respectively.

Because these regions have the same orientation as M itself, α_i and α_j are orientation-reversing reparametrizations of each other, as in Fig. 7.45. Thus

$$\int_{\alpha_i} \kappa_g \, ds + \int_{\alpha_j} \kappa_g \, ds = 0.$$

It follows that

$$\sum_{i=1}^{f} \int_{\partial \mathbf{x}_i} \kappa_g \, ds = 0 \tag{3}$$

for we have just seen that the integrals over edge curves will cancel in pairs. (As usual, we write v, e, and f for the number of vertices, edges, and faces in the decomposition.)

Thus the substitution of (2) in (1) yields

$$\iint_{M} K \, dM = -2\pi f + \mathcal{I} \tag{4}$$

where $\mathcal{I}$ is the sum of all interior angles of all the faces in the decomposition. But the sum of the interior angles at each vertex is just 2π (Fig. 7.46), so $\mathcal{I} = 2\pi v$. Thus

$$\iint_{M} K \, dM = -2\pi f + 2\pi v \tag{5}$$

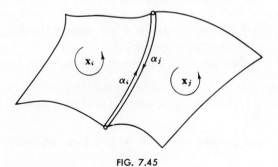

FIG. 7.45

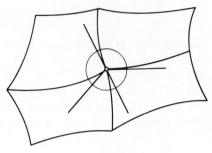

FIG. 7.46

A simple combinatorial observation will complete the proof. The faces of the decomposition $\mathcal{D}$ are rectangular: Each face has four edges. But each edge belongs to exactly two faces. Thus $4f$ *counts e twice*; that is, $4f = 2e$. Equivalently, $-f = f - e$, so (5) becomes

$$\iint_M K \, dM = 2\pi(v - e + f) = 2\pi\chi(M). \qquad \blacksquare$$

Because the Euler-Poincaré characteristic is a topological invariant, this theorem shows that *total curvature is a topological invariant*.

Explicitly we let M and $\bar{M}$ be geometric surfaces that are merely *diffeomorphic*.[†] Then the Gaussian curvatures K and $\bar{K}$ of M and $\bar{M}$ can be quite different—but *their total curvatures are identical*, for (being diffeomorphic) M and $\bar{M}$ have the same Euler-Poincaré characteristic; hence

$$\iint_M K \, dM = 2\pi\chi(M) = 2\pi\chi(\bar{M}) = \iint_{\bar{M}} \bar{K} \, d\bar{M}.$$

We have already seen instances of this theorem. For example, the torus in Example 2.5 has $K = 0$, hence total curvature zero. Alternatively, this same surface acquires from $\mathbf{E}^3$ its usual geometric structure as a torus of revolution, for which the curvature is variable—but we found in Chapter VI, Section 7, that its *total* curvature is also zero. (The diffeomorphism in this case is just the identity mapping.)

In general it suffices to count handles to find total curvature.

8.9 Corollary If M is a compact orientable surface with h handles $(h = 0, 1, 2, \cdots)$, then for any geometric structure on M, the total curvature is $4\pi(1 - h)$.

Proof. We have already seen that M has Euler-Poincaré characteristic $2 - 2h$. $\qquad \blacksquare$

† See footnote, page 380.

The Gauss-Bonnet theorem (Theorem 8.8 or 8.9) provides a way to attack some seemingly formidable problems. For example, (1) of Example 2.3 shows that if one single point is removed from a sphere Σ, there exists a geometric structure on the punctured sphere with $K = 0$. But *there can be no geometric structure on an* entire *sphere* Σ *for which* $K \leqq 0$, since then

$$\iint_{\Sigma} K \, d\Sigma \leqq 0,$$

contradicting $2\pi\chi(\Sigma) = 4\pi$. Reversing this argument, we find that *a compact orientable geometric surface with* $K > 0$ *must be diffeomorphic to a sphere.* Its total curvature is positive—but in Corollary 8.9, h is a nonnegative integer, so it must be zero. Thus the surface has no handles; it is diffeomorphic to the sphere $\Sigma = \Sigma(0)$. Further results of this type are given in the exercises.

The Gauss-Bonnet theorem is proved by cutting M into rectangular regions, and applying the Gauss-Bonnet formula to each. The scheme works because all these regions are consistently oriented by an orientation of M itself, so that the integrals $\int \kappa_g \, ds$ on the boundaries of these regions cancel in pairs. Here in essence is the fundamental idea of *algebraic topology;* indeed considerations of this kind led Poincaré to its invention (see Lefschetz [8]). By applying this scheme to suitable regions in M we can get a more general form of the Gauss-Bonnet theorem (Exercise 8). A corollary (Exercise 11) shows how to extend Theorem 8.4 from rectangles to arbitrary polygons. To see how the notion of *boundary* generalizes in such situations we shall give a direct (and logically unnecessary) proof of Exercise 11 in the special case of a *triangle*, that is, the one-to-one regular image Δ of an ordinary triangle T in $\mathbf{E}^2$ (Fig. 7.47).

Corollary 8.10　If Δ is a triangle in a geometric surface M, then

$$\iint_{\Delta} K \, dM + \int_{\partial\Delta} \kappa_g \, ds = 2\pi - (\varepsilon_1 + \varepsilon_2 + \varepsilon_3) = (\iota_1 + \iota_2 + \iota_3) - \pi.$$

(We explain this notation in the course of the proof.)

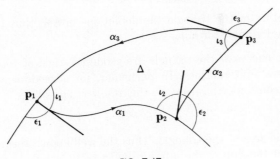

FIG. 7.47

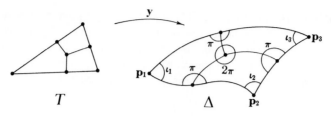

FIG. 7.48

Proof. Let dM be an arbitrary area form on the region Δ.

We get a rectangular decomposition of $\Delta = \mathbf{y}(T)$ as follows. Cut T into three quadrilaterals, as indicated in Fig. 7.48; then changes of variables in $\mathbf{y}$ will exhibit their images as rectangular regions $\mathbf{x}_1(R_1)$, $\mathbf{x}_2(R_2)$, $\mathbf{x}_3(R_3)$ constituting a rectangular decomposition of Δ. As usual we arrange for each $\mathbf{x}_i$ to be positively oriented. Thus, by the Gauss-Bonnet formula with interior angles, the total curvature of Δ is

$$\iint_\Delta K \, dM = \sum_{i=1}^{3} \iint_{\mathbf{x}_i} K \, dM = -\sum_{i=1}^{3} \int_{\partial \mathbf{x}_i} \kappa_g \, ds - 6\pi + \mathscr{I}$$

where $\mathscr{I}$ is the sum of all interior angles.

Of the twelve edges in $\partial \mathbf{x}_1$, $\partial \mathbf{x}_2$, and $\partial \mathbf{x}_3$, the six interior ones cancel in pairs (at least $\int \kappa_g \, ds$ does, on them). The remaining six *combine* in pairs to give the curves α_1, α_2, α_3 (Fig. 7.47) constituting the *boundary* $\partial \Delta$ of the oriented triangle Δ. Hence

$$\sum_{i=1}^{3} \int_{\partial \mathbf{x}_i} \kappa_g \, ds = \int_{\partial \Delta} \kappa_g \, ds = \int_{\alpha_1} \kappa_g \, ds + \int_{\alpha_2} \kappa_g \, ds + \int_{\alpha_3} \kappa_g \, ds$$

In the sum $\mathscr{I}$, the interior angles ι_1, ι_2, ι_3 at $\mathbf{p}_1$, $\mathbf{p}_2$, $\mathbf{p}_3$ are those of the triangle Δ itself. The others, which occur at the artificially introduced vertices, evidently add up to 5π; thus we find

$$\iint_\Delta K \, dM + \int_{\partial \Delta} \kappa_g \, ds = (\iota_1 + \iota_2 + \iota_3) - \pi.$$

We are adapting to the triangle the definitions in 8.3; thus $\iota_j + \varepsilon_j = \pi$ gives the exterior angle formula.   ∎

If the edge curves of the triangle are geodesics, then of course the geodesic curvature term vanishes. In particular, for a geodesic triangle in a surface with constant curvature K, this result reduces to

$$\iota_1 + \iota_2 + \iota_3 = \pi + KA$$

where A is the area of the triangle. Thus the well-known theorem of plane geometry that the sum of the interior angles of a triangle is π depends on

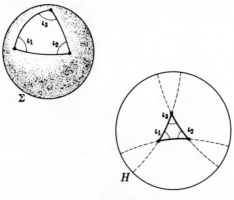

FIG. 7.49

the fact that $\mathbf{E}^2$ is flat. Examples show easily enough how a geodesic triangle can manage to have $\iota_1 + \iota_2 + \iota_3$ larger than π on a sphere $(K > 0)$ and smaller than π on a hyperbolic plane $(K < 0)$ (Fig. 7.49).

EXERCISES

1. Find the total Gaussian curvature of:
   (a) An ellipsoid.
   (b) The surface in Fig. 4.10.
   (c) $M: x^2 + y^4 + z^6 = 1$.

2. Prove that for a compact orientable geometric surface M:

   $K > 0 \Rightarrow M$ is diffeomorphic to a sphere (text)

   $K = 0 \Rightarrow M$ is diffeomorphic to a torus

   $K < 0 \Rightarrow M$ is a sphere with $h \geqq 2$ handles

3. (a) Let M be a compact orientable geometric surface with h handles. Prove that there exists a point $\mathbf{p}$ of M at which

   $$K(\mathbf{p}) > 0 \quad \text{if } h = 0,$$

   $$K(\mathbf{p}) = 0 \quad \text{if } h = 1,$$

   $$K(\mathbf{p}) < 0 \quad \text{if } h \geqq 2.$$

   (b) If M is a compact orientable surface in $\mathbf{E}^3$ which is not diffeomorphic to a sphere, show that there is a point $\mathbf{p}$ of M at which $K(\mathbf{p}) < 0$ (compare Theorem 3.5 of Chapter VI).

4. (a) For a regular curve segment $\alpha\colon [a,\, b] \to M$, show that the total geodesic curvature $\int_\alpha \kappa_g\, ds$ is

$$\int_a^b \frac{\alpha'' \cdot J(\alpha')}{\alpha' \cdot \alpha'}\, dt.$$

(*Hint:* Exercise 9 of Section 4.)

(b) Let $\mathbf{x}$ be a (positively oriented) orthogonal patch in M. Deduce the following formulas for total geodesic curvatures of parameter curves:

$$\int_{u_1}^{u_2} \frac{\mathbf{x}_{uu} \cdot \mathbf{x}_v}{\sqrt{EG}}\, (u, v_0)\, du, \qquad -\int_{v_1}^{v_2} \frac{\mathbf{x}_{vv} \cdot \mathbf{x}_u}{\sqrt{EG}}\, (u_0, v)\, dv$$

(Note that $\frac{1}{2}E_v = -\,\mathbf{x}_{uu}\cdot\mathbf{x}_v$ and $\frac{1}{2}G_u = -\,\mathbf{x}_{vv}\cdot\mathbf{x}_u$, and if M is in $\mathbf{E}^3$, either intrinsic or Euclidean derivatives give the same results.)

5. Let $\mathbf{x}\colon R \to \Sigma$ be the geographical patch (Example 2.2 of Chapter IV) restricted to the rectangle $R\colon 0 \leq u, v \leq \pi/4$. Compute separately each term in the Gauss-Bonnet formula for $\mathbf{x}$.

6. If $F\colon M \to N$ is a mapping of compact oriented surfaces, the *degree* d_F of F is the algebraic area of $F(M)$ divided by the area of N. Thus d_F represents the total algebraic number of times F wraps M around N. If M is a compact oriented surface in $\mathbf{E}^3$, prove the *Hopf theorem*: The degree d_G of the Gauss mapping is the integer $\chi(M)/2$. (It can be shown that degree is always an integer.)

An *oriented polygonal region* $\mathcal{P}$ in a surface M is an oriented region that has a rectangular decomposition $\mathbf{x}_1, \cdots, \mathbf{x}_k$ which we always arrange to be positively oriented. Then the boundary $\partial\mathcal{P}$ of $\mathcal{P}$ is the formal sum of those edge curves appearing in exactly one of the boundaries $\partial\mathbf{x}_1, \cdots, \partial\mathbf{x}_k$. We exclude the situation shown in Fig. 7.50, so $\partial\mathcal{P}$ consists of simple closed (broken) curves. These definitions are such that if α is one of the edges in $\partial\mathcal{P}$ then $J(\alpha')$ always points *into* the region $\mathcal{P}$. (This rigorizes the rough rule: "Travel around the boundary keeping the region always on the left.")

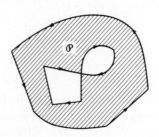

FIG. 7.50

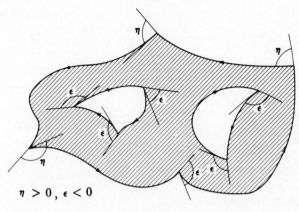

$$\eta > 0, \; \epsilon < 0$$

FIG. 7.51

7. (a) If ϕ is a 1-form on an oriented polygonal region $\mathcal{P}$, prove the generalized Stokes' theorem

$$\iint_{\mathcal{P}} d\phi = \int_{\partial \mathcal{P}} \phi.$$

   (If $\partial \mathcal{P} = \Sigma \, \alpha_i$, then $\int_{\partial \mathcal{P}} \phi$ means $\Sigma \int_{\alpha_i} \phi$.) (*Hint:* Lemma 6.6 of Chapter IV produces some cancellation in pairs, as in the proof of Theorem 8.8.)

   (b) Deduce that if ϕ is any 1-form on a compact oriented surface M, then $\iint_M d\phi = 0$.

   (c) Two different (positively oriented) rectangular decompositions of the same region $\mathcal{P}$ produce technically different boundaries of $\mathcal{P}$; however, both occupy the same set of points. Prove that for any 1-form on $\mathcal{P}$, $\int_{\partial \mathcal{P}} \phi$ is the same for both.

8. (*Generalized Gauss-Bonnet theorem*). If $\mathcal{P}$ is an oriented polygonal region in a geometric surface, prove that

$$\iint_{\mathcal{P}} K \, dM + \int_{\partial \mathcal{P}} \kappa_g \, ds + \sum \varepsilon_j = 2\pi \chi(\mathcal{P})$$

where $\Sigma \, \varepsilon_j$ is the sum of the exterior angles of $\mathcal{P}$ as defined in Definition 8.3 for the special case of a rectangular region (Fig. 7.51).

   (*Hint:* Refine the proof of Theorem 8.8, classifying edges and vertices into those in $\partial \mathcal{P}$ and those in the interior of $\mathcal{P}$. Note that on each simple closed boundary curve, the number of edges is the same as the number of vertices.)

9. Prove that the following properties of a compact orientable surface M are equivalent:

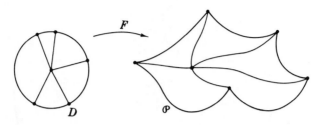

FIG. 7.52

(a) There is a nonvanishing tangent vector field on M.

(b) $\chi(M) = 0$.

(c) M is diffeomorphic to a torus.

(*Hint:* For (a) $\Rightarrow$ (b), introduce a geometric structure and use Exercise 7.) Properties (a) and (b) are, in fact, equivalent for any compact manifold.

10. (a) If a region $\mathcal{R}$ has a rectangular decomposition, derive a triangular decomposition, and show that $v - e + f$ is the same for both.

   (b) Do the same with "rectangular" and "triangular" reversed.

The notion of *simple region* (Ex. 12 of VI.7) may be extended by allowing the mapping $F: D \to M$ to be nondifferentiable (but still continuous) at n points on the circle $u^2 + v^2 = 1$. This permits n corners to appear in the boundary $\partial\mathcal{P}$ of $\mathcal{P} = F(D)$. We call $\mathcal{P}$ in this case an *n-polygon* ($n \geq 0$) The Euler-Poincaré characteristic of an n-polygon is $+1$, since for a triangular decomposition as in Fig. 7.52, we have $v - 1 = f = e/2$.

11. If $\mathcal{P}$ is an oriented geodesic n-polygon (the edges are geodesics) in a geometric surface, show that

$$\iint_{\mathcal{P}} K \, dM = 2\pi - \sum_{j=1}^{n} \varepsilon_k = (2-n)\pi + \sum_{j=1}^{n} \iota_j$$

where ε_j and ι_j are the exterior and interior angles of $\mathcal{P}$.

12. (*Continuation*). (a) If $\mathcal{P}$ is a geodesic n-polygon in the plane, prove that $n \geq 3$ and that the sum of the exterior angles of $\mathcal{P}$ is 2π.

   (b) If M is a surface with *constant* Gaussian curvature $K \neq 0$, show that the area of a geodesic polygon is determined by its exterior or interior angles.

   (c) In the sphere Σ of radius r, find a geodesic 3-polygon $\mathcal{P}$ whose interior angles are each $3\pi/2$. What is the area of $\mathcal{P}$?

13. (a) In a surface M with $K \leq 0$ prove that there exist no geodesic n-polygons with $n \leq 2$. Thus, in particular, two geodesics in M cannot meet to form the boundary of a simple polygonal region.

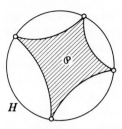

FIG. 7.53

(b) On a sphere Σ, for which values of $n \geqq 0$ do there exist geo-desic n-polygons? (Do not count "removable vertices"—those with exterior angle 0.)

14. In the hyperbolic plane, let $\mathcal{P}_n$ ($n \geqq 3$) be a "geodesic n-polygon" whose vertices are on the rim $u^2 + v^2 = 4$ of H, hence are not actually in H (Fig. 7.53). Find the area of $\mathcal{P}_n$.

15. (*Hopf Umlaufsatz*). If β is a simple closed curve in $\mathbf{E}^2$, then the total geodesic curvature of β is $\pm 2\pi$. Thus the unit tangent T of β turns through one full circle in traversing β. Prove this result assuming β is the boundary curve of a simple region S. (*Hint:* S is a 0-polygon as defined in the paragraph preceding Ex. 11.)

The assumption above is always true, but its proof requires rather deep topological methods.

9 Summary

A geometric surface—that is, a 2-dimensional Riemannian manifold—generalizes the Euclidean plane by replacing $\mathbf{E}^2$ by any surface and re-placing the dot product on tangent vectors by arbitrary inner products. In the resulting Riemannian geometry, the *length* of a curve is defined as before, and gives a notion of intrinsic distance directly generalizing the familiar Euclidean distance in the plane. The *acceleration* of a curve is also a geometric notion, but it is not so obvious how an inner product on tangent vectors can lead to a measurement of the turning of a curve. For 70 or 80 years after Riemann this was accomplished by rather complicated formulas in terms of coordinate patches (4.2 is a sample). In the Cartan approach, the inner products serve to define the notion of frame field, and the rate of rotation of a frame field is expressed by its connection form. The con-nection equation $\nabla_V E_1 = \omega_{12}(V)E_2$ then defines the covariant derivative—with acceleration as a special case.

In Riemannian geometry as in Euclidean geometry, geodesics are the curves with acceleration zero. Geodesics are not only *straightest* curves, however; they are also *shortest* curves in the sense discussed in Sections 5 and 6. The simple Euclidean rule that "a straight line is the shortest distance between two points" is no preparation for the new and subtle behavior of geodesics on an arbitrary geometric surface—or even a surface as simple as a sphere or a cylinder. Some idea of how far the analysis of geodesics can lead may be found in Milnor [7].

It is by now hardly necessary to say that the Gaussian curvature K of a geometric surface M is its most important geometric property, for we have seen that sooner or later curvature enters into almost every geometrical investigation. Indeed K could be *defined*, for example, in terms of parallel vector fields (holonomy) or radial geodesics (the Jacobi equation) or polar circles. (For a surface in $\mathbf{E}^3$ we used the shape operator, and could have used the Gauss mapping.) In the Cartan approach, however, curvature is defined by the structural equation $d\omega_{12} = -K\,\theta_1 \wedge \theta_2$, which presents K (in a sense discussed earlier) as the common "second derivative" of all frame fields on M. And it is this definition that leads most directly to the central result of two-dimensional Riemannian geometry, the Gauss-Bonnet theorem. Leaving aside trigonometric consequences such as Corollary 8.10, the content of the theorem is that curvature determines topology, at least in the compact orientable case.

Generally speaking, the results of this chapter are valid for Riemannian manifolds of arbitrary dimension n, and in most instances the definitions and proofs need scarcely any readjustment. Dimension 2 has simplified certain consistency proofs such as Theorems 2.1 and 3.2, but these can be avoided entirely by more advanced methods. As one might expect, it is the Gauss-Bonnet theorem whose generalization offers the most difficulty (see Hicks [5]), and in higher (even) dimensions the curvature of M influences but does not control the topological configuration of M.

Bibliography

1. H. Flanders, "Differential Forms: With Applications to the Physical Sciences." Academic Press, New York, 1963.
2. G. Birkhoff and S. MacLane, "A Survey of Modern Algebra." Macmillan, New York, 1953.
3. T. J. Willmore, "An Introduction to Differential Geometry." Oxford Univ. Press, London and New York, 1959.
4. R. Courant and H. Robbins, "What is Mathematics?" Oxford Univ. Press, London and New York, 1941.
5. N. J. Hicks, "Notes on Differential Geometry." Van Nostrand, Princeton, New Jersey, 1965.
6. D. J. Struik, "Lectures on Classical Differential Geometry." Addison-Wesley, Reading, Massachusetts, 1961.
7. J. W. Milnor, "Morse Theory." Princeton Univ. Press, Princeton, New Jersey, 1963.
8. S. Lefschetz, "Introduction to Topology." Princeton Univ. Press, Princeton, New Jersey, 1949.

The books of Willmore and Struik are at about the same level of difficulty as this one. Hicks' book *begins* at a level comparable with our Chapter VII and gives a very concise exposition of multidimensional Riemannian geometry; its bibliography lists a number of more detailed works on the subject.

Answers to Odd-Numbered Exercises

(These answers are not complete; and in some cases where a proof is required, we give only a hint.)

Chapter I

Section 1

1. (a) $x^2y^3 \sin^2 z$, (c) $2x^2y \cos z$
3. (b) $2xe^h \cos (e^h)$, $h = x^2 + y^2 + z^2$

Section 2

1. (a) $-6U_1(\mathbf{p}) + U_2(\mathbf{p}) - 9U_3(\mathbf{p})$
3. (a) $V = (2z^2/7)U_1 - (xy/7)U_3$
  (c) $V = xU_1 + 2yU_2 + xy^2U_3$
5. (b) Use Cramer's rule.

Section 3

1. (a) 0, (b) $7 \cdot 2^7$, (c) $2e^2$
3. (a) y^3, (c) $yz^2(y^2z - 3x^2)$, (e) $2x(y^4 - 3z^5)$
5. Use Ex. 4.

Section 4

1. $\alpha'(\pi/4) = (-2,0,\sqrt{2})_p$, where $p = (1,1,\sqrt{2})$
3. $\beta(s) = (2(1 - s^2),\ 2s\sqrt{1 - s^2},\ 2s)$
5. The lines meet at $(11,7,3)$
7. $\mathbf{v}_p = (1,0,1)_p$
9. At $\alpha(0)$: $t \rightarrow (2,2t,t)$

Section 5

1. (a) 4, (b) -4, (c) -2
5. (b) $(x\,dy - y\,dx)/(x^2 + y^2)$

7. (a) $dx - dz$, (b) not a 1-form, (c) $z\,dx + x\,dy$, (d) $2(x\,dx + y\,dy)$, (e) 0, (f) not a 1-form

9. $\pm\ (0,\ 1,\ \tfrac{1}{2})$

11. (a) Use the Taylor approximation of the function $t \to f(\mathbf{p} + t\mathbf{v})$
   (b) Exact: -0.420, approximate: $-\tfrac{1}{2}$

Section 6

1. (a) $\phi \wedge \psi = yz \cos z\,dx\,dy - \sin z\,dx\,dz - \cos z\,dy\,dz$
   (b) $d\phi = -z\,dx\,dy - y\,dx\,dz$. Note that $d(dz) = d(1 \cdot dz) = 0$ by I.6.3.

7. Apply this definition to the formula following I.6.3.

9. Assuming the formula, set $f = y$, $g = x$.

Section 7

1. (a) $(0,0)$, (b) $(-3,1)$, $(3,-1)$, (c) $(0,0)$, $(1,0)$

5. (a) $(2,0,3)$ at $(0,0,0)$, (b) $(2,2,3)$ at $(0,2,\pi)$

7. $GF = (g_1(f_1,f_2),\ g_2(f_1,f_2))$

11. (a) $F^{-1} = (v,\ ue^{-v})$, (b) $F^{-1} = (u^{1/3},\ v + u^{1/3})$, (c) $F^{-1} = ((9 - u - 2v)/2,\ 5 - u - v)$. F is a diffeomorphism for (a) and (c) only, since in (b), F^{-1} is not differentiable (when $u = 0$).

Chapter II

Section 1

1. (a) -4, (b) $(6,-2,2)$, (c) $(1,2,-1)/\sqrt{6}$, $(-1,0,3)/\sqrt{10}$, (d) $2\sqrt{11}$, (e) $-2/\sqrt{15}$

5. If $\mathbf{v} \times \mathbf{w} = 0$, then $\mathbf{u} \cdot \mathbf{v} \times \mathbf{w} = 0$ for *all* $\mathbf{u}$; use Ex. 4.

7. $\mathbf{v}_2 = \mathbf{v} - (\mathbf{v} \cdot \mathbf{u})\mathbf{u}$

Section 2

3. $\beta(s) = (\sqrt{1 + s^2/2},\ s/\sqrt{2},\ \sinh^{-1}(s/\sqrt{2}))$

5. If β_i is based at $t_i\,(i = 1,2)$, then s_0 is plus or minus the arc length of α from t_1 to t_2.

9. (b) The condition is certainly necessary; to see that it is sufficient, show that a unit speed reparametrization of α has acceleration zero.

11. (b) $L(\alpha) \geq \displaystyle\int_a^b \alpha' \cdot \mathbf{u}\,dt = \sum \int_a^b \frac{d\alpha_i}{dt}\,u_i\,dt = (\mathbf{q} - \mathbf{p}) \cdot \mathbf{u} = d(\mathbf{p},\mathbf{q})$

Section 3

1. $\kappa = 1$, $\tau = 0$, $B = (-\tfrac{3}{5},0,-\tfrac{4}{5})$, center $(0,1,0)$, radius 1.

7. (a) $1 = \|\alpha(h)'\| = \|\alpha'(h)h'\| = |h'|$, hence $h = \pm 1$
   (b) Let $\epsilon = \pm 1$, then $\bar{\alpha} = \alpha(h)$ implies $T = \alpha'(h)h' = \epsilon T(h)$; hence $\bar{\kappa}\bar{N} = \kappa(h)\,N(h)$, and so on.

9. The orthogonal projection on the $N_0 B_0$ plane (the *normal plane* of β at $\beta(0)$) is $s \to \kappa_0(s^2/2)N_0 + \kappa_0\tau_0(s^3/6)B_0$ (cusp at $s = 0$).

11. $B = \bar{B}$ implies $\tau N = \bar{\tau}\bar{N}$, hence either (1) $\bar{\tau} = \tau$ and $\bar{N} = N$, or (2) $\bar{\tau} = -\tau$ and $\bar{N} = -N$.

Section 4

1. Let $f = t^2 + 2$; then $\kappa = \tau = 2/f^2$; $B = (t^2, -2t, 2)/f$
3. (a) $N(0) = (0, -1, 0)$, $\tau(0) = \frac{3}{4}$
9. (a) $\vartheta = \pi/4$, $\mathbf{u} = (1, 0, 1)/\sqrt{2}$, $\gamma(t) = (t - (t^3/6), t^2, -t + (t^3/6))$
15. (c) The evolute is also a cycloid.
17. $\alpha'(t) = (f(t) \sin t, f(t) \cos t, f(t)g(t))$

Section 5

1. (a) $2U_1(\mathbf{p}) - U_2(\mathbf{p})$, (b) $U_1(\mathbf{p}) + 2\,U_2(\mathbf{p}) + 4\,U_3(\mathbf{p})$
5. (a) $8U_1(\mathbf{p}) - 4(U_3)(\mathbf{p})$

Section 6

1. Show that $V \cdot \tilde{W} = 0$, and use II.1.8.
3. For instance, $E_2 = -\sin z\, U_2 + \cos z\, U_3$, and $E_3 = E_1 \times E_2$.

Section 7

1. $\omega_{12} = 0$, $\omega_{13} = \omega_{23} = (df)/\sqrt{2}$
3. $\omega_{12} = -df$, $\omega_{13} = \cos f\, df$, $\omega_{23} = \sin f\, df$
5. By (3) of II.5.4, $\nabla_V(\sum f_i E_i) = \sum V[f_i]E_i + \sum f_i \nabla_V E_i$.
7. At an arbitrary point $\mathbf{p}$, $\alpha(t) = t\mathbf{p}$ is a curve with $\alpha' = \| \mathbf{p} \| F_1$. Show that

$$\| \mathbf{p} \| F_1[\rho] = \| \mathbf{p} \|.$$

Chapter III

Section 1

3. $(T_a)^{-1} = T_{-a}$, $C^{-1} = {}^{-t}C$, hence $F^{-1} = (T_a C)^{-1} = C^{-1}T_{-a} = T_{-C^{-1}(a)}C^{-1}$.
5. (b) Using Ex. 3, we find $F^{-1}(\mathbf{p}) = (5\sqrt{2}, -5, 4\sqrt{2})$

Section 3

1. If F and G have orthogonal parts A and B, then by **III.1.2**, $\mathrm{sgn}(FG) = \det(AB)$ $= \det A \cdot \det B$.
5. C is a rotation, through angle $\pi/2$, around the axis given by $\mathbf{a}$.
7. For $\mathbf{E}^1$: $F(s) = \epsilon s + s_0$; for $\mathbf{E}^2$: $F = TC$, where $C = \begin{pmatrix} \cos\vartheta & -\epsilon\sin\vartheta \\ \sin\vartheta & \epsilon\cos\vartheta \end{pmatrix}$.

Section 4

1. (b) By definition, $\tilde{\beta}(s)$ is the point canonically corresponding to $T(s)$, hence by III.2.1, $C(\tilde{\beta})$ corresponds to $F_*(T)$, the unit tangent of $F(\beta)$.
5. For a tangent vector $\mathbf{v}$ at $\mathbf{p}$, $F_*(\nabla_v W) = \tilde{W}(F(\mathbf{p}) + tC(\mathbf{v}))'(0) = \nabla_{F_*(v)}W$.

Section 5

1. $\beta = T_p(C(\alpha))$, where $C(\mathbf{u}_i) = \mathbf{e}_i$
3. Consequence of III.5.7
5. If τ is not identically zero, assume $\tau(0) \neq 0$ and examine the proof of 5.3.
7. Let $F = TC$, where T is translation by $(0,0,bs_0/c)$ and

$$
C = \begin{pmatrix} \cos\ (s_0/c) & -\epsilon\sin\ (s_0/c) & 0 \\ \sin\ (s_0/c) & \epsilon\cos\ (s_0/c) & 0 \\ 0 & 0 & \epsilon \end{pmatrix}
$$

where $\epsilon = \pm 1$. Then $F(\beta) = \beta(\epsilon s + s_0)$
9. $\alpha(s) = (\int \cos\varphi(s)\ ds, \int \sin\varphi(s)\ ds)$, where $\varphi(s) = \int \kappa(s)\ ds$
11. Use Ex. 9

Chapter IV

Section 1

1. (a) The vertex, $\mathbf{0}$, (b) all points on the circle $x^2 + y^2 = 1$, (c) all points on the z-axis
5. (b) $c \neq -1$
9. Use Ex. 7
11. $\mathbf{q}$ is in $F(M)$ if and only if $F^{-1}(\mathbf{q})$ is in M, that is, $g(F^{-1}(\mathbf{q})) = c$. Use the hint to apply IV.1.4.

Section 2

1. (c) $\mathbf{x}(u,v) = (u,v,u^2 + v^2)$ is one possibility; a parametrization derived from IV.2.5 will omit a point of the surface.
5. $\mathbf{x}_u \times \mathbf{x}_v = v\delta' \times \delta$
7. (b) Straight lines (rulings) and helices, (c) $M\colon x\sin\ (z/b) = y\cos\ (z/b)$
9. $\mathbf{x}(u,v) = (\cos u - v\sin u,\ \sin u + v\cos u,\ v)$
13. (a) If g' is never 0, reparametrize the profile curve to obtain $u \to (u,f(u),0)$, and use IV.2.5.

Section 3

1. (a) $r^2\cos^2 v$, (b) $r^2(1 - 2\cos^2 v\cos u\sin u)$
3. (a) $\bar{u}$ and $\bar{v}$ are the Euclidean coordinate functions of $\bar{\mathbf{x}}^1\mathbf{y}$
   (b) Express $\mathbf{y} = \mathbf{x}(\bar{u},\bar{v})$ in terms of Euclidean coordinates, and differentiate.
5. (a) M is given by $g = z - f(x,y) = 0$, with $\nabla g = (-f_x,-f_y,1)$, and $\mathbf{v}$ is tangent to M at $\mathbf{p}$ if and only if $\mathbf{v}\cdot\nabla g(\mathbf{p}) = 0$.
7. $\nabla g = (-y,-x,1)$ is a normal vector field; V is a tangent vector field if and only if $V\cdot\nabla g = 0$, for example, $V = (x,0,z)$.
9. (a) $\bar{T}_p(M)$ consists of all points $\mathbf{r}$ such that $(\mathbf{r} - \mathbf{p})\cdot\mathbf{z} = 0$; hence $\mathbf{v}_p$ is in $T_p(M)$ (that is, $\mathbf{v}\cdot\mathbf{z} = 0$) if and only if $\mathbf{p} + \mathbf{v}$ is in $\bar{T}_p(M)$.
11. (a) 2π

Section 4

3. $d(f\phi)(\mathbf{x}_u,\mathbf{x}_v) = \dfrac{\partial f(\mathbf{x})}{\partial u}\,\phi(\mathbf{x}_v) - \dfrac{\partial (f(\mathbf{x}))}{\partial v}\,\phi(\mathbf{x}_u) + f(\mathbf{x})\left[\dfrac{\partial}{\partial u}\,\phi(\mathbf{x}_v) - \dfrac{\partial}{\partial v}\,\phi(\mathbf{x}_u)\right]$

$$= (df \wedge \phi + f d\phi)\,(\mathbf{x}_u,\,\mathbf{x}_v).$$

5. If α is a curve with initial velocity $\mathbf{v}$ at $\mathbf{p}$, then

$$\mathbf{v}_p[g(f)] = (gf\alpha)'(0) = g'(f\alpha)(0)(f\alpha)'(0) = g'(f(\mathbf{p}))\mathbf{v}_p[f].$$

7. On the overlap of $\mathcal{U}_i$ and $\mathcal{U}_j$, $df_i - df_j = d(f_i - f_j) = 0$

9. (b) $d\tilde{u}(\mathbf{x}_u) = \mathbf{x}_u[\tilde{u}] = \dfrac{\partial(\tilde{u}(\mathbf{x}))}{\partial u} = \dfrac{\partial u}{\partial u} = 1.$

Section 5

1. If $\mathbf{x}: D \to M$ is a patch, then $F(\mathbf{x}): D \to N$ is (by 3.2) a differentiable mapping. Hence $\mathbf{y}^{-1}F\mathbf{x}$ is differentiable for any patch $\mathbf{y}$ in N.

3. If $\bar{\mathbf{x}}$ and $\bar{\mathbf{y}}$ are patches in M and N, respectively, then $\bar{\mathbf{y}}^{-1}F\bar{\mathbf{x}} = (\bar{\mathbf{y}}^{-1}\mathbf{y})(\mathbf{x}^{-1}\bar{\mathbf{x}})$ is differentiable, being a composition of differentiable functions.

7. (b) $\mathbf{x}^*(\nu) = r^3 \sin^2 v \cos v \, du \, dv$

11. Only (a) is not a diffeomorphism.

13. (b) $F_*(a\mathbf{x}_u + b\mathbf{x}_v) = a\mathbf{y}_u + b\mathbf{y}_v$ implies linearity.

Section 6

7. (a) $2\pi m$, (b) $2\pi n$

13. (b) Show $\int_\alpha \phi = \int_\mathbf{x} d\phi$ for a suitable $\mathbf{x}$.

15. Use simple connectedness to prove the formula given in the hint—see Fig. 4.46.

Section 7

1. (a) Connected, not compact, (c) connected and compact, (e) connected, not compact.

3. If ν is nonvanishing on N, show that $F^*\nu$ is nonvanishing on M.

5. (a) If Z is a nonvanishing normal, then let $\pm U = \pm Z/\|Z\|$. If V is an arbitrary unit normal, write $V = (V \cdot U)U$ and use Ex. 4(b).

(b) The image $\mathbf{x}(D)$ of a patch with D connected.

9. (c) Use Ex. 7

Section 8

1. Modify the proof given for the Möbius band in IV.7.

9. $(\mathbf{x} \times \mathbf{y})^{-1}(\bar{\mathbf{x}} \times \bar{\mathbf{y}}) = (\mathbf{x}^{-1}\bar{\mathbf{x}}) \times (\mathbf{y}^{-1}\bar{\mathbf{y}})$, a differentiable function.

Chapter V

Section 1

1. Use Method 1 in text.
3. (a) 2, (c) 1

Section 2

1. (b) If $\mathbf{e}_1$, $\mathbf{e}_2 = (\mathbf{u}_1 \pm \mathbf{u}_2)/\sqrt{2}$, then $S(\mathbf{e}_1) = \mathbf{e}_1$ and $S(\mathbf{e}_2) = -\mathbf{e}_2$

Section 3

5. (b) An ellipse on one side and no points on the other; the two branches of a hyperbola (asymptotes the two lines in (a)); two parallel lines on one side, and no points on the other.
7. (a) If α is a curve in M with initial velocity $\mathbf{v}$ at $\mathbf{p}$, then $F_*(\mathbf{v}) = F(\alpha)'(0) = (\alpha + \epsilon U_a)'(0) = \mathbf{v} - \epsilon S(\mathbf{v})$ at $F(\mathbf{p})$.

Section 4

5. $K = -36r^2/(1 + 9r^4)^2$; not minimal
7. Compute speed from $\alpha' = a_1'\mathbf{x}_u + a_2'\mathbf{x}_v$
13. $\mathbf{p} = \mathbf{x}(u,v)$ is umbilic if and only if $S(\mathbf{x}_u) = k\mathbf{x}_u$ and $S(\mathbf{x}_v) = k\mathbf{x}_v$ at (u,v). Dot with $\mathbf{x}_u$ and $\mathbf{x}_v$.
15. (a) None, since $K < 0$, (b) origin (a planar point),

$$(c) \left(0, \pm \frac{b}{2}\sqrt{a^2 - b^2}, \frac{a^2 - b^2}{4}\right) \text{ if } a \geq b$$

Section 5

3. A meridian α lies in a plane orthogonal to M along α, hence α'' is tangent to this plane, and (with constant speed parametrization) orthogonal to α'; thus α'' is orthogonal to M.
7. $S(T) = -U'$, hence by orthonormal expansion, $U' = -S(T)\cdot T\ T - S(T)\cdot V\ V$. Continue as in the proof of the Frenet formulas.
15. On the ruling through $\sigma(u)$, the formula for K in Ex. 14 shows that *either* K is identically zero, *or* $K < 0$ has minimum value $-1/p(u)^2$ at $\sigma(u)$ and rises symmetrically toward zero as $v \to \pm\infty$
17. (a) For $(u,0,0) + v(0,1,u)$: the x axis, with $p(u) = 1 + u^2$. (b) Use Ex. 15. For fixed u, $K = -(1 + u^2 + v^2)^{-2}$ is a minimum when $v = 0$.
19. (c) $\mathbf{x} = \alpha + v\delta$ is noncylindrical, and we can assume that α is a (unit speed) striction curve. But $\alpha'\cdot\delta \times \delta' = 0$ (since $K = 0$) and $\alpha'\cdot\delta' = 0$ (striction), hence $T = \alpha'$ and δ are collinear.

Section 6

1. $K = (1 - x^2)(1 + x^2 \exp(-x^2))^{-2}$
3. Use the results of V.2. Note that meridians are normal sections.
5. M has parametrization $\mathbf{x}(u,v) = (u \cos v, u \sin v, f(u))$.

7. With the usual parametrization, an argument like that for V.6.2 reduces to the extreme cases: g' always zero, g' never zero. In the former, M is part of a plane (special case of cone).

9. (c) If $y = f(x)$ has unit speed parametrization (g,h), where $h(u) = ce^{-u/c}$, show that f (not h!) satisfies the differential equation in VI.6.6.

Chapter VI

Section 1

1. (a) $\alpha'' = \omega_{12}(T)E_2 + \omega_{13}(T)E_3$, hence α'' is normal to M if and only if $\omega_{12}(T) = 0$.

5. If the cylindrical frame field is restricted to M and the indices 1 and 3 are reversed, we obtain the frame field in (1) of VI.1.3. By the computations in II.7,

$$\omega_{12} = \omega_{13} = 0, \qquad \omega_{23} = -d\vartheta$$

Section 2

3. (a) $\psi = \psi(E_1)\theta_1 + \psi(E_2)\theta_2 = -h_2\theta_1 + h_1\theta_2$.

(b) $\zeta = h_1\omega_{23} - h_2\omega_{13}$

Section 3

1. If $K = H = 0$, then $k_1k_2 = k_1 + k_2 = 0$. Thus $k_1 = k_2 = 0$, and $S = 0$.

3. Assume $k_1 \neq k_2$, and use the Hilbert Lemma (3.6) to get a contradiction.

5. In the case $k_1 \neq k_2$, use VI.2.7 to show that, say, $k_1 = 0$. By Ex. 2 the k_1 principal curves are straight lines. Show that the k_2 principal curves are circles, and that the (k_1) straight lines are parallel in $\mathbf{E}^3$.

Section 4

1. $(d) \Rightarrow (b)$: if $\mathbf{u}$ is an arbitrary tangent vector at $\mathbf{p}$, then $\mathbf{u} = a\mathbf{v} + b\mathbf{w}$, hence

$$\| F_*\mathbf{u} \|^2 = a^2 \| F_*\mathbf{v} \|^2 + 2ab\, F_*\mathbf{v}\cdot F_*\mathbf{w} + b^2 \| F_*\mathbf{w} \|^2 =$$

$$a^2 \| \mathbf{v} \|^2 + 2ab\, \mathbf{v}\cdot\mathbf{w} + b^2 \| \mathbf{w} \|^2 = \| \mathbf{u} \|^2$$

3. If α is a curve segment from $(-1,0,0)$ to $(1,0,0)$ with length 2, then by the exercise mentioned, α parametrizes a straight line segment—impossible, since α must remain in M.

5. (a) Define $F(\alpha(u) + vT_\alpha(u)) = \beta(u) + vT_\beta(u)$

(b) Choose β in $\mathbf{E}^2$ with the same curvature function.

7. (a) Criterion (a) becomes $F_*(\mathbf{v})\cdot F_*(\mathbf{w}) = \lambda^2(\mathbf{p})\mathbf{v}\cdot\mathbf{w}$; criterion (c) becomes $F_*(\mathbf{e}_i)\cdot F_*(\mathbf{e}_j) = \lambda^2(\mathbf{p})\delta_{ij}$.

11. Write $F(\mathbf{x}(u,v)) = \bar{\mathbf{x}}(a(u), b(v))$ for suitable parametrizations.

13. For $\mathbf{y}$, show that the conditions $E = G$ and $F = 0$ are equivalent to $g' = \cos g$, which has solution $g(v) = 2\tan^{-1}(e^v) - (\pi/2)$ such that $g(0) = 0$. Use Ex. 7.

15. $F(\mathbf{x}(u,v)) = (f(u)\cos v, f(u)\sin v)$, where $\mathbf{x}$ is a canonical parametrization and $f(u) = \exp\left(\int_1^u (dt/h(t))\right)$.

Section 5

1. If $\alpha' = E_1$ along α, then $F(\alpha)' = F_*(E_1) = \bar{E}_1$ along $F(\alpha)$. Use VI.5.3.
3. There is no local isometry of the saddle surface M $(-1 \leq K < 0)$ onto a catenoid with $-1 \leq \bar{K} < 0$, since K has an isolated minimum point (at $\mathbf{0}$) while $\bar{K}$ takes on each of its values on entire circles. (Many other examples are possible.)
5. (b) Follows from VI.4.5, since for $\mathbf{x}_t$ we compute $E_t = \cosh^2 u = G_t$, and $F_t = 0$ (independent of t).
    (d) For $M_t: U_t = (s, -c, S)/C$, so the Euclidean coordinates of U_t are independent of t.

Section 6

1. (b) $\theta_1 = \sqrt{1 + u^2}\, du$, $\theta_2 = u\, dv$, $\omega_{12} = dv/\sqrt{1 + u^2}$, $K = 1/(1 + u^2)^2$.
3. $\omega_{12} = -\vartheta_u\, du$

Section 7

3. (a) $A = (2\pi/3)\{(1 + c^2)^{3/2} - 1\}$, (b) ∞
5. (a) Use a canonical parametrization; then $\mathbf{x}^*(K\, dM) = (-h''/h)(h\, du\, dv) = -h''\, du\, dv$. Recall that $h' = \sin \varphi$.
    (c) for the bugle surface, $\lim\limits_{a \to 0} \varphi_a = -1$, $\lim\limits_{b \to \infty} \varphi_b = 0$.
7. Use Ex. 5. At the rims of all these surfaces, $h' = \sin \varphi \to \pm 1$. For $K = 1/c^2$ (V.6.5): in Case (2), $TC = 4\pi a/c$, $A = 4\pi ac$; in Case (3), $TC = 4\pi$, $A = 4\pi c^2$. For $K = -1/c^2$ (Ex. 9 of V.6): M_a has $TC = 2\pi(a - c)/c$, M_b has $TC = -4\pi$.
9. (a) Use Ex. 14 of V.5.

$$\int_{-\infty}^{\infty} \frac{p^2}{(p^2 + v^2)^{3/2}}\, dv = 2.$$

11. Define $\mathbf{x}(u,v) = F((1 - u) \cos v, (1 - u) \sin v)$, on $R: 0 \leq u \leq 1, 0 \leq v \leq 2\pi$.
13. With outward normal, $H = -1/r$, and $h = r$.
15. Dividing if necessary by the lengths of V and W, we can assume they are unit vector fields. Thus V, $U \times V$ is a tangent frame field on α. Orthonormal expansion shows that if $f = W \cdot V$ and $g = W \cdot U \times V$, then $f^2 + g^2 = 1$.
17. (a) Use $\| F_*\mathbf{v} \times F_*\mathbf{w} \| = |J| \| \mathbf{v} \times \mathbf{w} \|$
19. (a) F carries positively oriented pavings of M to positively oriented pavings of N.
    (b) If $F: M \to \bar{M}$ is an isometry, orient M and $\bar{M}$ so that F preserves orientation (see remark following Ex. 4). Then

$$\iint_{\bar{M}} \bar{K}\, d\bar{M} = \iint_M F^*(\bar{K}\, d\bar{M}) = \iint_M \bar{K}(F)F^*(d\bar{M}) = \iint_M K\, dM$$

Section 8

1. If N is *isometric* to a sphere Σ of radius r, show that N is compact and has $K = 1/r^2$. Thus, by Liebmann's theorem, N is also a sphere of radius r, so a translation will show that N is *congruent* to Σ.
3. Except for geodesics, all follow immediately from preservation of shape operators. Only geodesics and Gaussian curvature need be preserved by arbitrary isometries (Ex. 1 of VI.5, and *theorema egregium*).

5. (b) That given by $-U$.
9. Deduce from *theorema egregium* that a Euclidean symmetry $\mathbf{F}$ of M must leave the origin fixed, hence $\mathbf{F}$ is orthogonal. Consider its effect on the natural frame at 0.

Chapter VII

Section 1

1. (b) $\alpha'\circ\alpha' = \alpha'\cdot\alpha'/g^2(\alpha)$ is speed squared.
3. (b) Let E_1 and E_2 be the vector fields on $\mathbf{x}(D)$ determined by $\mathbf{x}_u/\sqrt{E}$ and $V/\parallel V \parallel$, where $V = \mathbf{x}_v - (F/E)\mathbf{x}_u$ (Gram-Schmidt process).
5. (a) Show that the definition $J(E_1) = E_2$, $J(E_2) = -E_1$ is independent of the choice of *positively oriented* frame field. (Any two have the same orientation in the sense defined just preceding VII.1.4.)
 (b) For $J^2 = -I$, show $J(J(E_i)) = -E_i$.
7. (a) $F_*U_1 = f_uU_1 + g_uU_2$, $F_*U_2 = f_vU_1 + g_vU_2$ (Use Ex. 6).
 (b) The complex derivative of $F = f + ig$ is $dF/dz = f_u + ig_u$. Thus the magnitude $\mid dF/dz \mid$ is $(f_u{}^2 + g_u{}^2)^{1/2}$, which may be rewritten in various ways using the Cauchy-Riemann equations.

Section 2

1. $\theta_1 = du/v$, $\theta_2 = dv/v$

3. $A = \pi r^2/\left(1 - \dfrac{r^2}{4}\right)$, $(E, F, G$ computed as in Example 4.11$)$; $A(H) = \infty$.

5. $E = G = 1$ and $F = 0$, so $A = 4\pi^2$. We can redefine the geometric structure to make E and G any positive numbers.
7. Show that the parametrization $\mathbf{x}_0$ in Ex. 5 of VI.5 is an isometry.

Section 3

1. Note that $E_i = vU_i$ restricted to α is r sint U_i, thus $\alpha' = -E_1 + \cot(t)E_2$. Since $\omega_{12} = du/v$ evaluated on α' is -1, we derive from the covariant derivative formula that $\alpha'' = \cot(t)E_1 - \cot^2(t)E_2$.
3. In the proof referred to, note that $\omega_{13}(V)E_3 = -\nabla_V E_3 \cdot E_1 E_3 = S(V) \cdot E_1 E_3$.
5. (a) The proof of VII.3.6 shows that the holonomy angle of α is

$$-\int_a^b \omega_{12}(\alpha')\,dt = -\int_\alpha \omega_{12}$$

Use Stokes' theorem, recalling that $d\omega_{12} = -K\,dM$ (first structural equation).
7. $Y' = f'E_1 + f\omega_{21}(\alpha')E_2$, hence $F_*(Y') = f'\bar{E}_1 + f\omega_{21}(\alpha')\bar{E}_2$. $F_*Y = \bar{Y} = f\bar{E}_1$, hence $\bar{Y}' = f'\bar{E}_1 + f\bar{\omega}_{21}(F_*\alpha')\bar{E}_1$. Use VI.5.3.
11. W has constant length c; on any orientable region, show that the frame field $E_1 = W/c$, $E_2 = J(E_1)$ has connection form *zero*, hence curvature zero.

Section 4

1. Since $\alpha'' = 0$, $\alpha(h)'' = \alpha'(h)h''$
3. Apply Ex. 6 of Section 3.
5. All Euclidean circles through the north pole, since these correspond under stereographic projection to straight lines in the plane.
9. Using the equations for α' and α'' given in the text, we have $J(\alpha') = J(vT) = vN$, hence $\alpha'' \cdot J(\alpha') = \kappa_g v^3$.

11. (a) For $u_0 \leq u \leq u_1 - \epsilon$, $a_1' = \dfrac{\sqrt{G - c^2}}{\sqrt{EG}} \geq A_\epsilon > 0$.

(b) *If* the meeting takes place, then α' and β' are collinear, hence by VII.4.2, α and β are equal (but for parametrization)—an impossibility.
13. (a) for the usual parametrization of a surface of revolution, $G = h^2$ (h distance to the axis of revolution), and the u-parameter curves are the meridians, hence the slant c is $h \sin \varphi$.

(b) Follows from Exs. 11 and 12, since such a parallel is a barrier curve.
15. Meridians obviously approach the rim (in one direction). Use Ex. 13 to show that any other geodesic meets one barrier curve and approaches the rim in both directions.
17. $|\sin \varphi_0| < 1/\cosh u_0$.
19. (a) Use VI.2.4, (b) $\kappa_1 = 0$, $\kappa_2 = h'/h$.

Section 5

1. $\rho(0, \mathbf{p}) = \tanh^{-1}(\| \mathbf{p} \|/2)$, Euclidean norm.
3. The geodesics are helices, and $\mathbf{y}(u, v) = \gamma_{ue_1 + ve_2}(1) = (r \cos u/r, r \sin u/r, v)$. The largest normal neighborhood is $\eta_{\pi r}$; $\mathbf{y}$ is regular for any ϵ, but the one-to-one condition fails for $\epsilon > \pi r$.
5. (a) If $\mathbf{q}$ is the η_ϵ, then by 5.5, $\rho(\mathbf{p}, \mathbf{q}) < \epsilon$. If $\mathbf{q}$ is not in η_ϵ a curve from $\mathbf{p}$ to $\mathbf{q}$ meets C_δ for every $\delta < \epsilon$.

(b) If $\mathbf{p} \neq \mathbf{q}$, it follows from the Hausdorff axiom that there is a normal neighborhood of $\mathbf{p}$ which does not contain $\mathbf{q}$, hence $\rho(\mathbf{p}, \mathbf{q}) > 0$.
7. The length L of α from 0 to $r + \epsilon$ is the same as that of the *broken* curve: β from 0 to r, then α from r to $r + \epsilon$. Hence $L > \rho(\alpha(0), \alpha(r + \epsilon))$ by the remark preceding 5.8.
9. $D: u^2 + v^2 < 1$ with Euclidean geometric structure.
11. (a) There is only *one* geodesic segment (a meridian) from $\mathbf{p}$ to any other point.

Section 6

5. (a) $\mathbf{x}_u(0, v) = X(v)$, and since $\mathbf{x}(0, v) = \gamma_{X(v)}(0) = \beta(v)$, we have $\mathbf{x}_v(0, v) = \beta'(v)$. Thus $EG - F^2$ is nonzero when $u = 0$, and by continuity for $|u|$ small.

(b) (iii) β a base curve, $X = \delta$.
7. (a) The v-parameter curves are meridians of longitude.

(b) Since $K = 0$, the Jacobi equation becomes $(\sqrt{G})_{uu} = 0$, hence $\sqrt{G}$ is linear in u, and it follows that $G(u, v) = 1 - \kappa_g(v)u$.

Section 7

1. If N is geodesically complete, fix a point $\mathbf{p}$ of M; then there is a geodesic segment γ_w from $F(\mathbf{p})$ to an arbitrary point $\mathbf{q}$ of N. But for $\mathbf{v}$ such that $F_*(\mathbf{v}) = \mathbf{w}$, we have $F(\gamma_v) = \gamma_w$, hence $\mathbf{q}$ is in $F(M)$.

3. If $\mathbf{p} \neq \mathbf{q}$ in M, then any geodesic segment σ from $\mathbf{p}$ to $\mathbf{q}$ has nonzero speed. Hence $F(\sigma)$ is a nonconstant geodesic from $F(\mathbf{p})$ to $F(\mathbf{q})$, and it follows that $F(\mathbf{p}) \neq F(\mathbf{q})$, even if "two" is interpreted as "two distinct" in the uniqueness property of N.

5. Variant of VII.7.3.

7. To show that a surface M is not frame homogeneous, it suffices (by the argument given in the text for the cylinder) to show that M has some geodesics which are one-to-one and some which are not. Note that $\mathbf{x}$ in VII.2.5 is a local isometry.

9. If L is the arc length of the ellipse, show that $F(\mathbf{x}(u, v)) = \mathbf{x}(u + (L/4), v)$ is an isometry of M which does not preserve Euclidean distance.

11. By Ex. 8 (b), F is the restriction to Σ of a linear (orthogonal) transformation, hence $F(-\mathbf{p}) = -F(\mathbf{p})$. Thus we can define $\bar{F}\{\mathbf{p}, -\mathbf{p}\} = \{F(\mathbf{p}), F(-\mathbf{p})\}$, and deduce the frame-homogeneity of $\bar{\Sigma}$ from that of Σ.

13. The function is a homomorphism, so it suffices to prove it is one-to-one. But in the proof of VI.8.3, if M is not planar (so $S \neq 0$), then $\mathbf{F}$ is unique.

Section 8

1. In (a) and (c) the surface is diffeomorphic to a sphere, so $TC = 4\pi$. In (b) there are four handles, so $TC = -12\pi$.

3. (a) For $h = 1$: if K were never zero, then by an earlier exercise, either $K > 0$ or $K < 0$; both are impossible, since by VII.8.9, $\iint_M K \, dM = 0$.

(b) By VI.3.5, K is somewhere positive. But M has at least one handle, so

$$\iint_M K \, dM \leq 0;$$

hence K is somewhere negative.

5. $\iint_{\mathbf{x}} K \, dM = - \int_{\partial \mathbf{x}} \kappa_g \, ds = \pi/4\sqrt{2}$

7. (a) If $\mathbf{x}_1, \cdots, \mathbf{x}_k$ is a positively oriented decomposition, then by Stokes theorem (IV.6.5) we have $\int_{\mathscr{P}} d\phi = \Sigma \iint_{\mathbf{x}_i} d\phi = \Sigma \int_{\partial \mathbf{x}_i} \phi$. But for edges not in $\partial\mathscr{P}$, IV.6.6 produces cancellation by pairs.

(c) If $\mathbf{x}_i$ and $\mathbf{y}_j$ are two positively oriented decompositions, then by the remark following VI.7.5, we have

$$\sum \int_{\partial \mathbf{y}_j} \phi = \sum \iint_{\mathbf{y}_j} d\phi = \sum \iint_{\mathbf{x}_i} d\phi = \sum \int_{\partial \mathbf{x}_i} \phi$$

9. (a) $\Rightarrow$ (b): Any geometric structure on M (such exist) can be modified so that the nonvanishing vector field E_1 has unit length; then $E_1, J(E_1)$ is a frame field. Its connection form is defined on the entire surface, so $2\pi\chi(M) = \iint_M K \, dM = -\iint_M d\omega_{12} = 0$.

11. Consequence of Ex. 8.

13. By the uniqueness of geodesics, interior angles of a geodesic polygon cannot be $\pm\pi$ (that is, there are no cusps), so $-\pi < \epsilon < \pi$. Use Ex. 11.

(b) all $n \neq 1$ (for $n = 0$, a great circle).

INDEX